AF181752

APPLIED MICROPALAEONTOLOGY

# APPLIED MICROPALAEONTOLOGY

*Edited by*

DAVID GRAHAM JENKINS
*Department of Geology,*
*National Museum of Wales,*
*Cardiff, United Kingdom*

SPRINGER-SCIENCE+BUSINESS MEDIA, B.V.

Library of Congress Cataloging-in-Publication Data

Applied micropalaeontology / edited by David Graham Jenkins.
p. cm.
Includes index.
ISBN 978-90-481-4268-2 ISBN 978-94-017-0763-3 (eBook)
DOI 10.1007/978-94-017-0763-3
1. Micropalaeontology. I. Jenkins, D. Graham.
QE719.A67 1993
560--dc20 93-18856

*The cover-figure has been kindly provided by BP*

*Printed on acid-free paper*

All Rights Reserved
© 1993 Springer Science+Business Media Dordrecht
Originally published by Kluwer Academic Publishers in 1993
No part of the material protected by this copyright notice may be reproduced or utilized in any form or by any means, electronic or mechanical, including photocopying, recording or by any information storage and retrieval system, without written permission from the copyright owner.

## Table of contents

**Acknowledgements** ix

**Preface** xi

D. Graham Jenkins

**Chapter 1 Recent coccolith sedimentation patterns and transport in the North Sea: implications for palaeoceanographic studies of marginal and continental shelf seas** 1

*S.D. Houghton*

Abstract 1
1.1 Introduction 1
1.1.1 Background and main objectives 1
1.1.2 Geographical area 2
1.2 Hydrography of the North Sea 2
1.3 Techniques 3
1.4 Results 4
1.5 Discussion 10
1.5.1 Comparison with previous work 10
1.5.2 Relationship of the coccolith assemblages in the sediments to the overlying hydrographic regions 11
1.5.3 Transport paths of coccoliths into and within the North Sea 15
1.5.4 Sources of reworked coccoliths 16
1.5.5 Differences between North Sea and Celtic Sea coccolith assemblages 16
1.5.6 Ecological controls on the distribution of coccolithophores and other phytoplankton in the northwest European shelf regions 18
1.5.7 Characteristic features of Recent coccolith assemblages from marginal seas 19
1.5.8 Mesozoic and Cenozoic calcareous nannofossil assemblages as indicators of marginal sea/open oceanic sediments 29
1.5.8.1 Cenozoic nannofossil assemblages 29
1.5.8.2 Mesozoic nannofossil assemblages 30
1.6 Conclusions 33

**Chapter 2 Ostracods as palaeoenvironmental indicators: examples from the Tertiary and Early Cretaceous** 41

*Michael C. Keen*

2.1 Introduction 41
2.2 The late Eocene of the Hampshire basin, southern England 41
2.2.1 Ostracod assemblages 44
2.2.2 Environmental conditions 48
2.2.3 Salinity profiles, local and international correlation 49
2.2.4 Other examples 56
2.3 Late Jurassic and Early Cretaceous cypridean ostracods 59
2.3.1 Cypridean faunas of southern England and Europe 61
2.3.2 Cypridean faunas of China 63
2.3.3 Cypridean faunas of Brazil 64

**Chapter 3 Preliminary observations on benthonic foraminifera associated with biogenic gas seep in the North Sea** 69

*R.W. Jones*

Abstract 69
3.1 Introduction 69
3.1.1 Previous studies on the recent and sub-recent benthonic foraminifera of the North Sea 69
3.1.2 Previous studies on benthonic foraminfera associated with seeps 71
3.1.3 Previous studies on other micro-organisms associated with seeps 71
3.2 Present Study 71
3.2.1 Background 71
3.2.2 Material 73
3.2.3 Methods 73
3.3 Discussion of results (surface samples) 74
3.3.1 Similarity index matrix 74
3.3.2 Abundance, diversity and dominance 75
3.3.3 Taxonomic composition (ordinal level) 76
3.3.4 Taxonomic composition (specific level) 76
3.3.5 Proportions of epifaunal and infaunal morphotypes 77
3.4 Discussion of results (subsurface samples) 79
3.5 Conclusions 80
3.5.1 Surface samples 80
3.5.2 Subsurface samples 80

**Chapter 4 Application of micropalaeontology to hydrocarbon exploration in the North Sea basin** **93**

*P. Copestake*

Abstract 93
4.1 Introduction 93
4.2 Review of techniques 95
4.2.1 Correlation 95
4.2.2 Age determination 96
4.2.3 Unconformity identification 97
4.2.4 Application to sequence stratigraphy analysis 98
4.2.4.1 Sequence stratigraphy models 98
4.2.4.2 Sequence analysis of well 12/21-2, Inner Moray Firth 104
4.2.5 Characterisation of formations ("fingerprinting") 109
4.2.6 Palaeoenvironmental interpretation 114
4.3 Role of micropalaeontology in hydrocarbon exploration and development programmes 116
4.3.1 Introduction 116
4.3.2 Requirements for hydrocarbon accumulation 116
4.3.3 Exploration for prospects 117
4.3.3.1 Well evaluation 118
4.3.3.2 Drilling problems 118
4.3.4 Appraisal of discoveries 121
4.3.4.1 Well correlation 121
4.3.4.2 Reservoir distribution and reserves estimates 121
4.3.4.3 Trap evaluation 122
4.3.5 Field development 125

4.4 North Sea hydrocarbon plays and micropalaeontology 135
4.4.1 Northern North Sea 135
4.4.2 Central North Sea 138
4.4.3 Micropalaeontology 138
4.5 Pitfalls in biostratigraphic interpretations 140
4.5.1 Factors affecting data quality 140
4.5.2 Reworking and caving 140
4.5.3 Age interpretations 141
4.5.4 Taxonomic nomenclature 141
4.5.5 Preparation techniques 142
4.5.6 Microfossil and zonal identification 142
4.5.7 Palaeoenvironmental controls 142
4.5.8 Practical correlation and biostratigraphy 142
4.6 Development of micropalaeontology in North Sea exploration 143
4.7 Concluding remarks - how can micropalaeontology help to find oil 144

**Chapter 5 Palynofacies analysis** **153**

*R.C. Tyson*

5.1 Introduction 153
5.2 The uses of palynofacies 155
5.3 Kerogen - particulate organic matter in sedimentary rocks 155
5.4 Simple classifications for rapid assessment of hydrocarbon potential 158
5.5 More detailed classifications for palaeoenvironmental analysis 159
5.6 Palynofacies parameters for palaeoenvironmental analysis 161
5.6.1 Bulk kerogen parameters 161
5.6.2 Palynomorph parameters 165
5.7 Comments on data interpretation 173
5.8 Diagrammatic representation of palynofacies data 174
5.8.1 Total kerogen plots 177
5.8.2 Total palynomorph plots 177
5.8.3 Phytoplankton palynomorph plots 183
5.9 Conclusions 183

**Chapter 6 Sequence stratigraphy of Barrow Group (Berriasian-Valanginian) siliciclastics, North-West Shelf, Australia, with emphasis on the sedimentological and palaeontological characterization of systems tracts** **193**

*R.W. Jones et al.*

Abstract 193
6.1 Introduction 193
6.1.1 Present study 193
6.1.2 Sequence stratigraphy and systems tracts: definitions and discussion 197
6.1.3 Previous work on the Early Cretaceous stratigraphy of the North-West Shelf 198
6.1.4 Previous work on the sedimentological characteristics of systems tracts 200
6.1.5 Previous work on the palaeontological characteristics of systems tracts 200

6.2 Sequence stratigraphy of the Barrow Group 203
6.2.1 Introduction 203
6.2.2 Use of palaeontology in palaeoenvironmental interpretation 205
6.2.3 Sequence BD1 207
6.2.4 Sequence BD2 208
6.2.5 Sequence BD3 210
6.2.6 Sequence BD4 212
6.2.7 Sequence BD5 213
6.3 Sedimentological characteristics of systems tracts 214
6.3.1 Low-stand systems tracts (LSTs) 214
6.3.2 Transgressive systems tracts (TSTs) 214
6.3.3 Maximum flooding surfaces (MFSs) 215
6.3.4 High-stand systems tracts (HSTs) 215
6.4 Palaeontological characteristics of systems tracts 215
6.4.1 Low-stand systems tracts (LSTs) 215
6.4.2 Transgresssive systems tracts (TSTs) 216
6.4.3 Maximum flooding surfaces (MFSs) 216
6.4.4 High-stand systems tracts (HSTs) 217
6.4.5 Significance of barren or impoverished samples 217
6.5 Conclusions 218
6.5.1 Sequence stratigraphy 218
6.5.2 Sedimentological characteristics of systems tracts 218
6.5.3 Palaeontological characteristics of systems tracts 219

**Chapter 7 Engineering and economic geology** **225**

Malcolm B. Hart

7.1 Introduction 225
7.2 The Channel Tunnel site investigation 228
7.3 The Thames Barrier site investigation 234
7.4 Cambridge-bypass site investigation 244
7.5 Isle of Wight landslips 247
7.6 Geological resources 247
7.7 Conclusions 261

## Acknowledgements

The editor wishes to thank all the authors, especially for their patience over the years. Dr Sally Rudford read all the typescripts and offered valuable advice to the authors. Mrs Judith Jenkins gave invaluable help in the editing.

# PREFACE

D. Graham Jenkins

According to Daniel J. Jones in his Introduction to Microfossils in 1956, micropalaeontology became an applied science in 1877 "when the age of the strata in a water well near Vienna, Austria, was determined as Middle Miocene by the use of Foraminifera". The real impetus into research and applied micropalaeontology came during the 1920s in the United States of America when microfossils were used to determine the age of drill cuttings in the oil industry. Major oil companies established their own micropalaeontological laboratories but from early on, independent consultant firms were also set up to examine micropalaeontological samples. Micropalaeontology has not only established biostratigraphic frameworks for oil producing basins all over the world but it has also made major advances in our understanding of palaeoenvironments associated with both source and reservoir rocks. Micropalaeontology has also become a useful aid to engineering and economic geology.

As a micropalaeontologist who has worked both for a major oil company and as a consultant, I have selected seven case studies to illustrate the modern use of microfossils. The basis of palaeoecology is the study of modern taxa in known environments. The first chapter by Simon Houghton is a study of Recent coccolith sedimentation patterns and transport in the North Sea: these data could be used in evaluating Cenozoic marginal and continental shelf areas. Ostracods are also very good palaeoenvironmental indicators and Michael Keen provides examples from the Tertiary and Early Cretaceous, in Chapter 2. Offshore biogenic gas seeps provide clues about petroleum reservoirs, and Robert Jones describes a study of benthonic foraminifera associated with such a seep in the North Sea in Chapter 3. Philip Copestake then describes the application of micropalaeontology in the search for hydrocarbons in the North Sea Basin in Chapter 4. This is followed by Richard Tyson's description of palynofacies analysis of total kerogen and palynomorph assemblages in order to determine depositional environments and hydrocarbon source rock potential. In Chapter 6 Robert Jones and his ten colleagues interpret the sequence stratigraphy of the Early Cretaceous Barrow Group of the North-West Shelf, Australia, using palaeontological and sedimentological data. Finally, Malcolm Hart provides examples of the application of micropalaeontology to both engineering and economic geology, including the Channel Tunnel site investigation, in Chapter 7.

D. Graham Jenkins
Dept. of Geology
National Museum of Wales
Cardiff CF1 3NP
Wales
UK

# 1 RECENT COCCOLITH SEDIMENTATION PATTERNS AND TRANSPORT IN THE NORTH SEA: IMPLICATIONS FOR PALAEOCEANOGRAPHIC STUDIES OF MARGINAL AND CONTINENTAL SHELF SEAS

Simon D. Houghton

**Abstract**

Ten species of Recent coccolith occur in the sediments of the North Sea. The assemblages are dominated by *Emiliania huxleyi,* which frequently forms >90% of the total flora. Coccolith abundance and diversity in the sediments decreases from north to south, reflecting the decreasing Atlantic influence and increasing tidal current velocities in the overlying waters. South of latitude 55°N, Recent coccoliths are either rare or absent in the sediments. The coccolith assemblages in the sediments are supplied from Atlantic water which flows into the North Sea between Shetland and Norway.

Recent marginal sea coccolith assemblages are distinguished from their open-oceanic counterparts mainly by an increase in dominance of one or two species. The Asian marginal seas of the western Pacific and Indian oceans are characterized by a dominance of *Gephyrocapsa oceanica,* whereas *E. huxleyi* is usually rare. In contrast, Atlantic Ocean marginal seas from the tropics to the subarctic, are distinguished by a dominance of *E. huxleyi*, whereas *Gephyrocapsa* species are either rare or absent. These variations in dominance of *E. huxleyi* and *G. oceanica* are attributed to changes in water fertility and stratification. *G. oceanica* favors low-latitude, nutrient-enriched marginal seas which are influenced by monsoons. *E. huxleyi* prefers low to medium fertility, highly stratified seas and marginal basins with lagoonal circulation patterns (e.g. northern Red Sea and the Mediterranean).

Ancient marginal sea assemblages are often difficult to discern from oceanic assemblages because they may contain similar dominant coccolith species (e.g. *Watznaueria* and *Ellipsagelosphaera*). However some marginal nannofossil assemblages may be distinguished by an increase in density and diversity of non-coccolith nannofossils (e.g. nannoconids).

## 1.1 Introduction

### 1.1.1 Background and main objectives

The most extensive record of coccolithophore distribution in the North Sea was published by Braarud et al., (1953), who described the plankton in the waters of the North Sea in May 1948. Previous work describing sediment assemblages is restricted to Murray (1985) and Houghton (1991b). Most other records of coccolithophore distribution in the region are based on plankton samples from the coastal waters and fjords of Norway and from the Norwegian Sea (Birkens and Braarud, 1952; Berge, 1962; Braarud et al., 1974, and Schei, 1975).

The main objectives of this chapter are to define the species composition and abundance of coccolith assemblages in the recent sediments of the North Sea. The relationship between the coccolith assemblages and the overlying hydrological regimes and the directions of coccolith transport into and within the North Sea are discussed. Finally, the characteristic features of the marginal sea coccolith assemblages, (including Cenozoic and Mesozoic examples) the hydrologic controls on their species composition, and the palaeoecological significance of these controls are presented.



*D. G. Jenkins (ed.), Applied Micropaleontology,* 1–39.
© 1993 *Kluwer Academic Publishers.*

### 1.1.2 Geographical area

The North Sea is an epicontinental sea, bordered on three sides by land and open to the Atlantic waters in the north, where it continues into the Norwegian Sea (Figure 1.1). In the south, the North Sea has a narrow connection with the English Channel through the Strait of Dover. To the east, it is open to the Baltic through the Skagerrak. Lying between latitudes 51°N and 61°N, the area undergoes pronounced seasonal changes. A general trend of decreasing depth from north to south occurs throughout the central North Sea. The greater part of the northern North Sea away from the coasts has a depth of between 100m and 140m. Depths in excess of 140m occur to the north and west of the Shetland Isles and in the Norwegian Channel at the eastern margin of the North Sea. The area of the North Sea sampled in this study (Figure 1.2) lies to the west of 4°E longitude and excludes the regions off the continental coast from Holland to Norway. Two samples from the Sea of the Minches, off the northwestern coast of Scotland, are also included here.

## 1.2 Hydrography of the North Sea

The mean distribution of surface salinity in the North Sea (August Chart) is shown in Figure 1.1. There are three major inflows of water to the North Sea (Lee, 1980);

1) high salinity (ca. 35.3‰) North Atlantic water entering from the north between Shetland and Norway;
2) high salinity (ca. 35‰) English Channel water entering from the south through the Strait of Dover; and
3) the low salinity Baltic outflow from the Skagerrak.

Maximum temperatures in the surface waters occur in August when mean values range from 17°C to 19°C along the German and Dutch coasts, and from 12°C to 13°C along the Scottish coast (Figure 1.3). Minimum winter temperatures in the waters of the North Sea occur in February, and range from 7°C east of Shetland, to 2°C to 3°C along the coast of Denmark. The waters of the southern North Sea, south of ca. 54°N, and along the English and continental coast are homohaline and homothermal throughout the year. An exception to this occurs in a small region off the Dutch coast, where low salinity coastal water may ride over high salinity Atlantic-derived waters (Postma, 1954; Fonds and Eisma, 1967). Waters in the central and northern North Sea are thermally stratified in the summer with the lower limit of the thermocline generally lying at a depth of between 20m and 50m (Tomczack and Goedecke, 1964).

Figure 1.1: Map of the North Sea showing water-depth contours (isobaths) in metres, and mean distribution of sea-surface salinity ($^o/_{oo}$) for August (modified after Lee, 1980). FIC = Faire Isle Channel, SM = Sea of the Minches.

Figure 1.2: Location of sample sites in the North Sea. Solid circles represent open marine samples, triangles represent estuarine samples, circles with dots represent filter samples. Solid circles represent open marine samples, triangles represent estuarine samples, open circles respresent filter samples.

Figure 1.1

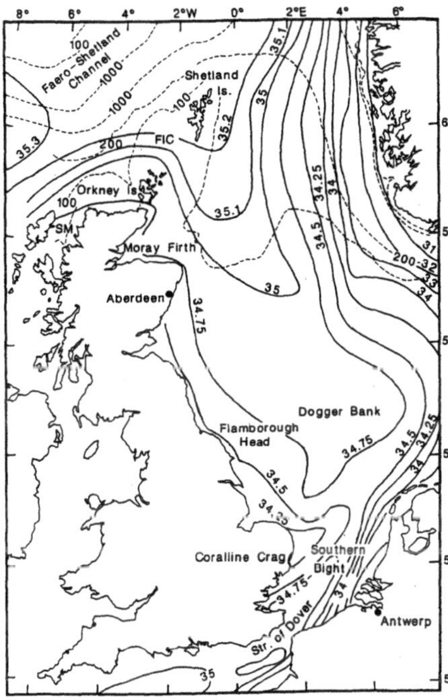


Figure 1.2

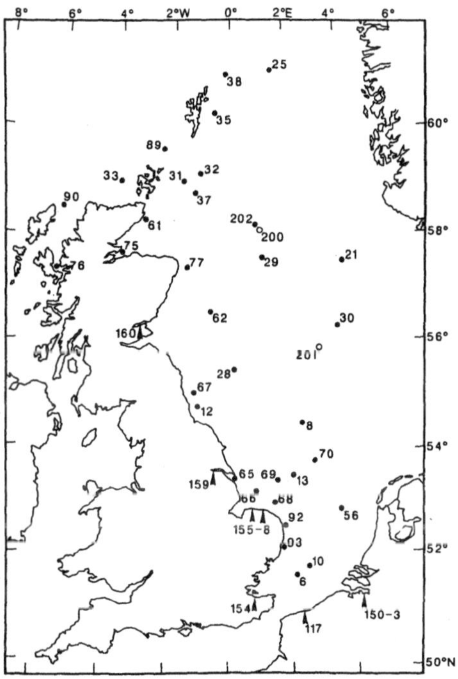


## 1.3 Techniques

Forty-five Recent marine and estuarine sediments and four filter samples were studied. The nannofossil-size fraction (2-20μm) was separated using a short-centrifuge technique (Houghton, 1986). Quantitative coccolith abundances were expressed as numbers per gram of fine silt (2-20μm). The percentage coccolith component of the fine fractions and the percentage abundance of coccolith species in the total assemblage was also determined. Coccolith abundances of $10^5$/g are the lowest abundance level detectable by the counting technique used (i.e. one coccolith counted per slide).

Figure 1.3: Mean distribution of sea surface temperatures (°C) for August (modified after Lee, 1980).

Figure 1.4: Distribution of maximum tidal streams during mean spring tides (knots) (modified after Lee and Ramster, 1976).

Figure 1.3

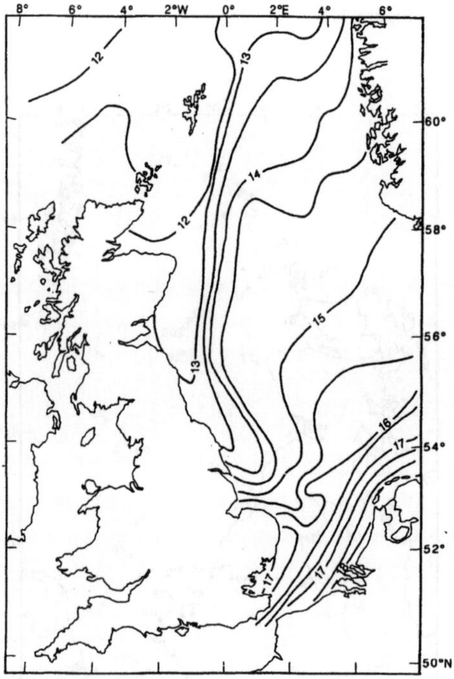


Figure 1.4

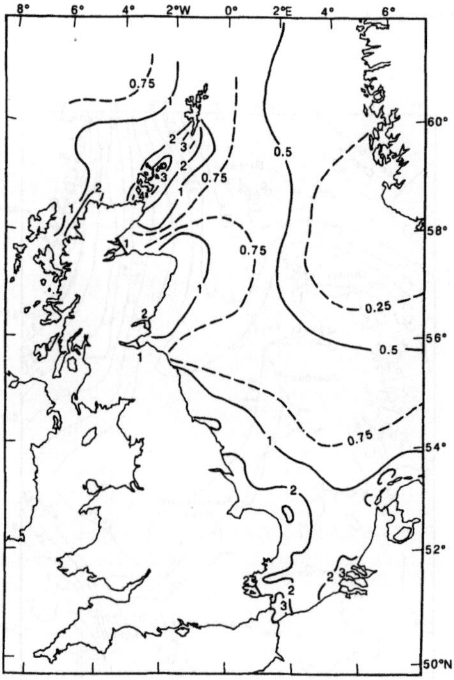


## 1.4 Results

Ten species of Recent coccolith are recorded from the North Sea sediments (Table 1.1). The species diversity in the sediments decreases from north to south (Figure 1.5), with an area of sediment containing five or more species being traced as far south as latitude 55°N. In the central North Sea, assemblages typically contain 1 or 2 species, but sediments from the greater part of the southern North Sea do not contain Recent coccoliths. Exceptions to this lie along the French and Belgian coasts.

Highest numbers of Recent coccoliths ($10^9$/g; >15% of the fine fractions) occur in sediments west of the Orkney Isles, and in the central North Sea as far south as latitude 57°30'N (Figures 1.5 and 1.7). Further south, and also in the coastal waters of the Moray Firth and around Orkney and Shetland, abundances decline to $10^8$/g. Coastal sediments from eastern England and sediments from the central North Sea contain between $10^6$ and $10^7$ coccoliths/g (<1% of the fine fractions).

Table 1.1: List of Recent coccolith species present in the surface sediments of the northern North Sea.

| SPECIES | Mean % |
|---|---|
| *Emiliani huxleyi* | 93.5 |
| *Coccolithus pelagicus* | 2.1 |
| *Syracosphaera pulchra* | 2.3 |
| *Gephyrocapsa caribbeanica* | 1.1 |
| *Calcidiscus leptoporus* | 0.7 |
| *Helicosphaera carteri* | P |
| *Braarudosphaera bigelowii* | P |
| *Rhabdosphaera clavigera* | P |
| *Coronosphaera mediterranea* | P |
| *Discosphaera tubifera* | P |

Numbers refer to mean % of each species of the total assemblage, P = <0.3%

The coccolith assemblages are dominated by *Emiliania huxleyi,* the highest dominance of this species (>95%) occurring in Atlantic-influenced assemblages deposited west of Orkney. Slightly lower values (mean 93%) are typical of assemblages deposited in the central northern North Sea and along the east coast of Scotland. South of latitude 55°N, *E. huxleyi* is usually absent from the assemblages, except for an isolated occurrence in the Scheldt estuary (Antwerp Harbour). The quantitative abundance of *E. huxleyi* in the sediments parallels the variations in total coccolith numbers. Preservation of the species usually occurs as discrete coccoliths, although coccospheres were also noted at the northern margin of the North Sea.

*Coccolithus pelagicus* (Figure 1.6C) is usually the second most abundant species in the assemblages. Except for two sites in the southern North Sea, the species forms less than 5% of the assemblages. Lowest values of the species occur in coastal sediments of the Moray Firth (0.5%) around the Orkney Isles and in the Sea of the Minches (1%). Higher relative abundances (mean 2.6%) occur in the central North Sea region. The greatest quantitative abundance of the species ($10^8$ coccoliths/g) occurs at the northern margin of the North Sea (Figure 1.8). Abundance values decline to $10^7$/g in sediments from Orkney and Shetland and in the central North Sea. A gradient of decreasing coccolith abundance continues southward; the species is absent over large areas of the southern North Sea, except for sporadic occurrences along the French and Belgian coastlines. Here $10^5$ coccoliths/g occur, which represents the lowest abundance level detectable by the present technique (i.e. 1 coccolith/slide). *Syracosphaera pulchra* (range 1.0-8.0%, mean 2.3%) is ubiquitous in sediments from the Minches and throughout the North Sea as far south as 55°N. Highest numbers of the species ($10^7$/g) occur in the North Minch, west of Orkney and throughout the Northern North Sea (Figure 1.9). Lower values occur in sediments between the Isles of Orkney and Shetland, along the Scottish coast and south of latitude 57°N in the central North Sea ($10^6$ to $10^7$/g). The species was not recorded in sediments from the southern North Sea.

Figure 1.6: Scanning electron micrographs of coccolith and dinoflagellate floras from the North Sea. Figures 1.6A and 1.6B represent filter samples from lat. 58°N, long. 00°20'W; Figure 1.6C shows sediment sample from the Forties region, northern North Sea and Figure 1.6D represents filter sample from lat. 56°N 20' N, long 02°50' E.
1.6A View (x 3,500) of fine fraction showing abundant coccoliths of *Emiliania huxleyi*. Neg. No. 489-2.
1.6B View (x 10,000) of fine fraction showing coccoliths of *E. huxleyi*. Note that fragments of *E. huxleyi* still retain their T-shaped ends suggesting that simple breakage rather than dissolution is responsible for their disaggregation. Neg.No. 490-7.
1.6C View (x 3,500) of fine fraction showing *Coccolithus pelagicus* (large coccolith, left), *E. huxleyi* (center top and bottom) and *Coronosphaera mediterranea* (top right). Neg.No. 447-5.
1.6D View (x 750) showing aggregate of dinoflagellate motile thecae dominated by *Ceratium* spines, also showing theca of *Dinophysis acuta*. Neg.No. 496-4.

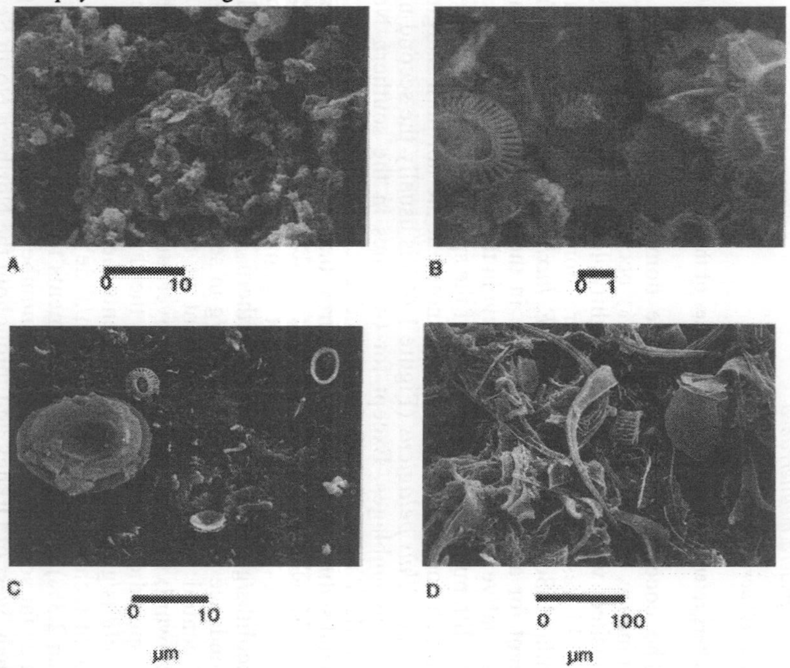

Figure 1.5: Species diversity and quantitative abundance (x $10^n$) of Recent coccoliths in the sediments of the North Sea. Vertical shading represents 5 or more coccolith species in the Recent assemblages, horizontal shading 1-2 species, diagonal shading no species. (Key in figure 1.2)

Figure 1.7: Percentage abundance of Recent coccoliths in the fine silt fraction (2-20 $\mu$m) from the surface sediments. (Key in figure 1.2)

Figure 1.5

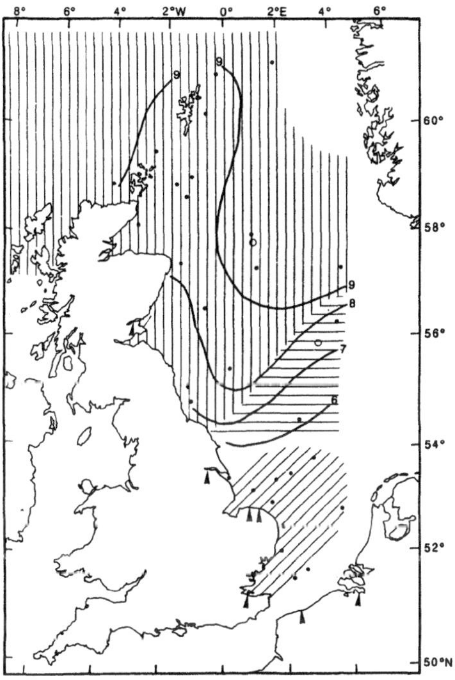


Figure 1.7

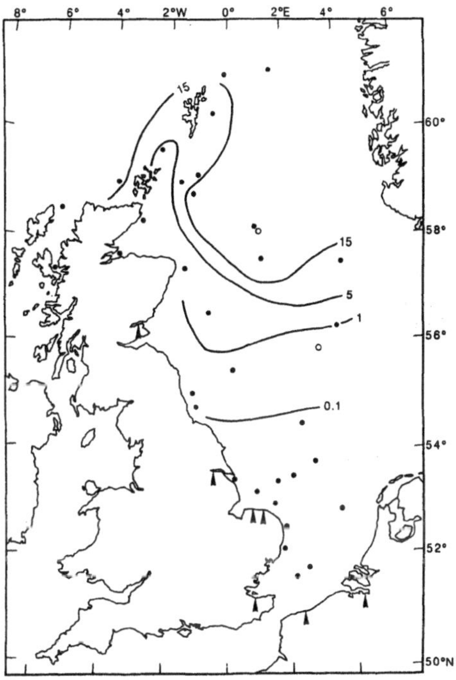


Minor species in the coccolith assemblages (0.3 to 4%) include *Braarudosphaera bigelowii, Calcidiscus leptoporus, Gephyrocapsa caribbeanica, Helicosphaera carteri* and *Rhabdosphaera clavigera.*

Figure 1.8: Map showing the abundance of coccoliths of *Coccolithus pelagicus* (x $10^n$ per gram of fine silt in the Recent sediment). (See figure 1.5 for key details.)

Figure 1.9: Map showing the abundance of coccoliths of *Syracosphaera pulchra* (x $10^n$ per gram of fine silt (2-20 $\mu$m) in Recent sediment). (See figure 1.5 for key details.)

Figure 1.8

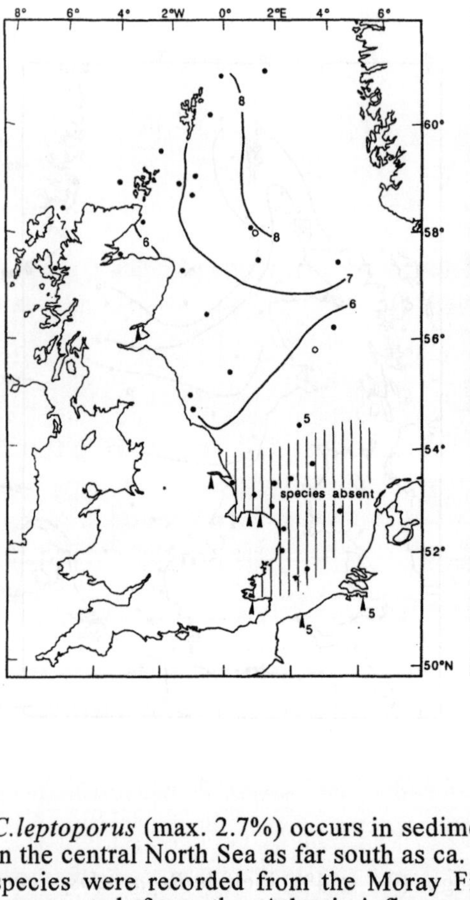


Figure 1.9

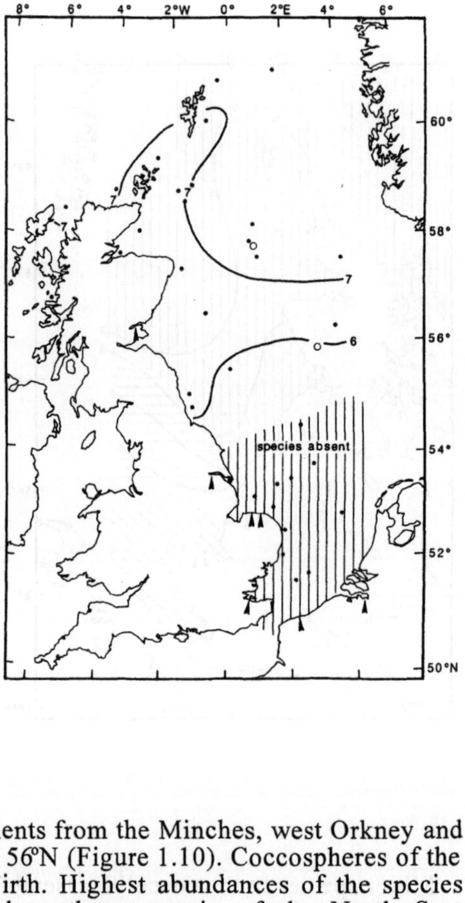


*C.leptoporus* (max. 2.7%) occurs in sediments from the Minches, west Orkney and in the central North Sea as far south as ca. 56ºN (Figure 1.10). Coccospheres of the species were recorded from the Moray Firth. Highest abundances of the species were noted from the Atlantic-influenced northern margin of the North Sea. *Gephyrocapsa caribbeanica* (mean 1.8%) occurs in sediments from the Minches and in the central North Sea as far south as ca. latitude 56ºN. *Helicosphaera carteri* has a very limited distribution in the sediments and is recorded around Shetland and Orkney and in the central North Sea.

Figure 1.10: Distribution of *Calcidiscus leptoporus* in the Recent sediments. Numbers refer to the species % abundance in the total assemblage. (See figure 1.5 for key details.)

Figure 1.11: Map showing hydrographic regions of the North Sea (after Lee, 1980). Arrows indicate transport direction of coccoliths in the North Sea. Numbers indicate volume ($km^3/yr^{-1}$) of Atlantic inflow to the North Sea. (See figure 1.5 for key details.)

Figure 1.10

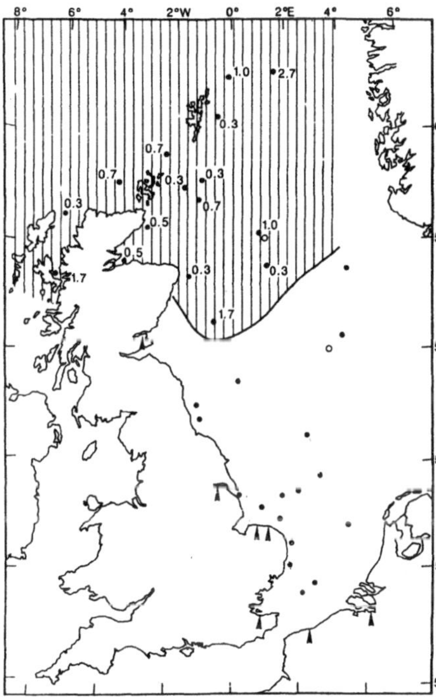


Figure 1.11

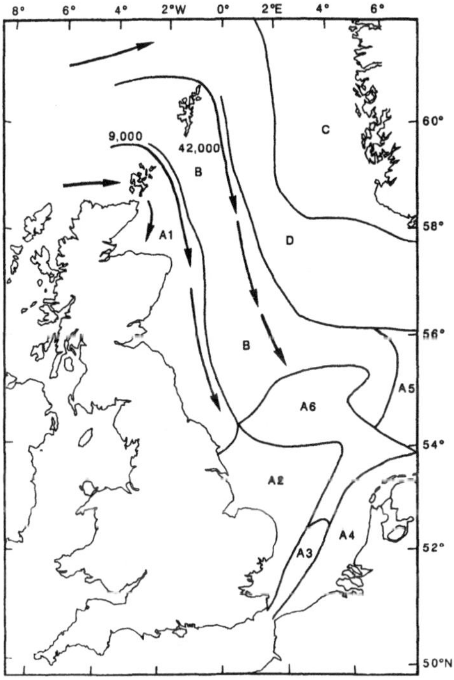


*Rhabdosphaera clavigera* has a sporadic distribution in the North Sea, being recorded from coastal sediments around Orkney and Shetland, in the Moray Firth and off the northeastern coast of England. *Braarudosphaera bigelowii* occurs in coastal sediments off eastern Britain and in the Minches. Isolated pentaliths are also found at the northern margin of the North Sea and near the entrance to the Dover Strait. *Discosphaera tubifera* and *Coronosphaera mediterranea* (Figure 1.6C) are scarce members (>0.3%) of the coccolith assemblages, being found occasionally from the northern North Sea sediments.

Reworked coccoliths are scarce (usually <1%) in assemblages from the Sea of the Minches and in the northern North Sea. Abundances increase southward, particularly along the eastern coast of England. Reworked coccoliths form 100% of the nannofloral assemblages throughout much of the southern North Sea as Recent coccoliths are frequently absent. The reworked flora is dominated by Cretaceous forms similar to those found in the eastern English Channel and in the Solent (Houghton, 1986). Reworked Palaeogene coccoliths are also found in coastal sediments of eastern England and northern France; these include *Chiasmolithus* spp., *Reticulofenestra umbilica* and *Neochiastozygus dubius*.

Table 1.2: Percentage abundance and species-composition of coccolith assemblages in filter samples from the two sites in the North Sea.

| | | |
|---|---|---|
| Latitude | 58°30'N | 56°20'N |
| Longitude | 00°20'W | 02°50'E |
| *Emiliana huxleyi* | 97.3 | 100 |
| *Coccolithus pelagicus* | 1.5 | - |
| *Calcidiscus leptoporus* | 0.9 | - |
| *Rhabdosphaera clavigera* | 0.2 | - |
| *Scyphosphaera apsteinii* | p* | - |
| *Pontosphaera discopora* | p* | - |
| Number of species | 6 | - |
| % coccoliths (<40$\mu$m) | 20 | <0.1 |

* is <0.1% of total assemblage.

Coccolith assemblages (Table 1.2) were also noted in filter samples (sampled spring/summer 1985, sample depth 55m) from the northern North Sea (latitude 58°30'N, longitude 00°20'W) and in the central North Sea (lat. 56°20'N, long. 02°50'E). The northern North Sea filter contained abundant coccoliths (10-25% of the fine fractions) and a flora dominated by *Emiliania huxleyi* (Figure 1.6A,B). Minor species included *C. pelagicus, C. leptoporus, R. clavigera, Scyphosphaera apsteinii* and *Pontosphaera discopora.* Occasional coccospheres of *E. huxleyi* were noted in the samples. Preservation of the coccoliths of *E. huxleyi* ranges from moderate to poor, with fragments of the spoke-like elements of the cold-water form (McIntyre and Bé, 1967) and also of the partially-fused transitional forms observed. The occurrence of both coccoliths and sub micron skeletal elements of these coccoliths leads to a bimodal distribution of carbonate in the samples, with peak occurrences in the 2 to 4 micron range. The central North Sea filter contained a sparse (<0.1% of the fine fractions) monospecific flora of *E. huxleyi.*

## 1.5 Discussion

### 1.5.1 Comparison with previous work

Murray (1985) noted the species diversity of coccolith assemblages in selected sediment samples from the region; greatest species diversity (3 to 5 species) occurred in sediments from the west coast of Scotland and in the Forties region (lat. 57°44'N, long. 00°54'E) whereas lower diversities (1 or 2 species) occurred in the Fair Isle Channel and at Ekofisk (lat. 56°N, long. 03°13'E). *E. huxleyi* dominated all assemblages. The main difference recorded here is the increased species diversity of assemblages in the Fair Isle Channel (5 or more species). Eight species of Recent coccolith were noted by Murray from the North Sea, but two of these, *Caneosphaera mollichii* and *Gephyrocapsa oceanica,* were not recorded by this author. Additional species reported from the North Sea sediments include *R. clavigera, S. pulchra, B. bigelowii* and *D. tubifera* (Houghton, 1991b).

Braarud et al. (1953) reported 13 coccolithophore species in the plankton of the North Sea. These communities were overwhelmingly dominated by *E. huxleyi*. The species was ubiquitous in the plankton throughout the northern North Sea and extended as far south as latitude 56°N. An isolated occurrence of the species was noted in waters off the coast of Denmark. Most other reported occurrences of *E. huxleyi* in the North Sea are from coastal waters of Norway, where the species is undoubtedly the most dominant coccolithophore (Berge, 1962; Schei, 1975). Abundant *E. huxleyi* (max 14.5 million cells/$m^3$) was also reported by Cadee (1985) in sediment traps attached to a drifting buoy launched during May (1983) at 58°N, 0°28.5'E.

*Anthosphaera robusta* was generally the second most dominant coccolith species recorded in the plankton survey of Braarud et al. (1953). Peak abundances (5-25,000 cells/l) occurred at the northern boundary of the North Sea. The species is also a common component of the plankton from the coast of Norway (Braarud et al., 1974) but has not been recorded from the southern North Sea. *Coccolithus pelagicus* was recorded by Braarud et al. (1953) in the waters of the northern North Sea, particularly between the Faeroe Isles and the Shetland Isles. The other species recorded in the plankton survey were rare and all had a very sporadic distribution in the waters of the northern North Sea and in the Norwegian Sea. Seven of these species: *Acanthoica quatrospinosa, Anthosphaera robusta, Calyptrosphaera hyalina, Lohmannosphaera michaelsari, Oolithotus fragilis, Caneosphaera mollischi* and *Syracosphaera caudata* were not recorded in the Recent sediments. With the exception of *O. fragilis*, all these species are either caneoliths or holococcoliths, which are the most fragile of all types of coccolith structure (McIntyre and McIntyre, 1971). These taxa were reported as rare components in the plankton in the northern North Sea and are probably destroyed during descent through the water column. Disintegration would be enhanced by the generally turbulent and high-energy conditions which are present in the shelf waters of the North Sea. *G. caribbeanica, S. pulchra, H. carteri* and *D. tubifera* are present in the Recent sediments of the North Sea, but were not recorded in the plankton survey of Braarud et al. (1953). *D. tubifera* was, however, found at sites in the central and northeastern North Sea in April, 1920 by Wulff (1925).

1.5.2 Relationship of the coccolith assemblages in the sediments to the overlying hydrographic regions.

The North Sea was divided into a set of hydrographic regions by Laevastu (1963), and the most recent concept of these regions (Lee, 1980) is shown in Figure 1.11. Each region represents a diagnostic water type which can be distinguished by its properties of salinity, temperature and nutrient concentration. Recent sediments were sampled from regions A1, A2, A3, A4, and A6 in this study (Table 1.3).

Table 1.3: Hydrographic regions of the North Sea (after Lee, 1980), their water mass characteristics and coccolith assemblages.

| Hydrographic Region | | Salinity | Seas. Therm. | Nutrients | Turbidity | Coccolith Diversity | Coccolith abundance |
|---|---|---|---|---|---|---|---|
| B | North Atlantic | > 35 | yes | low | low | > 5 | $10^9$-$10^8$ |
| A1 | Scottish coastal | 34-35 | yes | med-low | med-low | > 5 | $10^9$-$10^7$ |
| A2 | English coastal | 34-34.5 | no | high | high | - | - |
| A3 | Channel water | > 34.75 | no | low | low | - | - |
| A4 | Continental coastal | 34 | no | high | high | 1-2 | $10^5$ |
| A6 | Mixed comp. | 34.5-34.9 | no | med. | med. | 1 | $10^5$ |
| | Sea of Minches | < 34.75 | yes | med. | med. | > 5 | $10^9$ |
| C | Skagerrak | < 34 | yes | low | med-low | * | nd |

Seas. Therm. = Seasonal thermocline, Abund. = Abundance, nd = not determined, * = dominated by *E. huxleyi*, comp. = composition.

Region A includes the coastal zones (except off Norway) and the shallow areas of the southern North Sea from Dogger Bank southward. The region is subdivided into four sub-regions. Region A1 extends along the Scottish and English coast from Orkney to Flamborough Head, and is characterized by the southward flow of Scottish coastal water (34‰ to 35‰) with mixed components containing both oceanic and coastal components further offshore. Highest coccolith abundances ($10^{9-8}$/g) are deposited off the northeastern coast of Scotland. These waters are adjacent to the main inflow of Atlantic water into the northern North Sea and are thermally stratified in the summer months. Coccospheres of *C. leptoporus* were recorded in sediments from the Moray Firth. The decrease in Atlantic influence is reflected southward as coccolith values drop to $10^7$/g, indicating that the main oceanic component is present further offshore. This area of decreased coccolith numbers also occurs in a region of increased tidal-mixing (tidal current velocities >1 knot), where the water column is transitional in character between a stratified region and a tidally mixed region (Figures 1.4 and 1.12). The plankton of this transitional region is characterized by a rich diversity diatom flora, including the species *Asterionella japonica, Chaetoceros* spp. and *Nitzchia delicatissima* (Braarud et al., 1953).

Region A2 extends along the English coast south of Flamborough Head and is primarily occupied by English coastal water with mixed coastal water and Channel components further offshore. The waters are homohaline and are tidally mixed throughout the year. No Recent coccoliths were recorded from the sediments of this region. The neritic character of this region is confirmed by the occurrence of an estuarine-type diatom flora, where *Chaetoceros danicus* and benthic diatoms mixed with *Asterionella* spp. dominate the plankton (Braarud et al., 1953).

Figure 1.12: Map showing the summer distribution of stratified waters (vertical shading), transitional waters (horizontal shading) and tidally mixed waters (stippled shading) in the North Sea. The continuous line represents the frontal boundary (modified after Pingree and others, 1978).

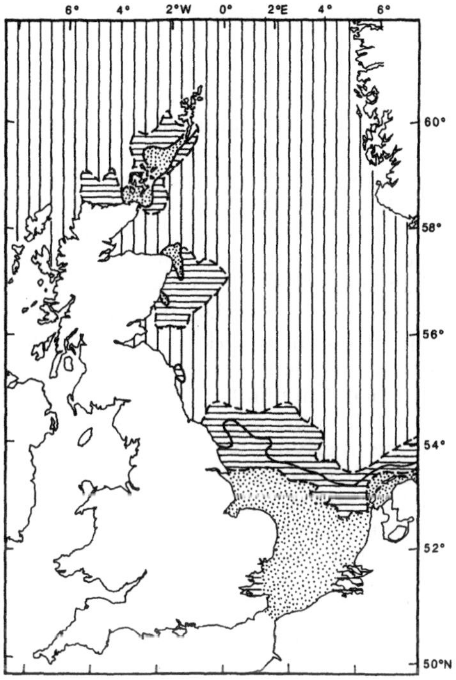


Region A3 extends from the Strait of Dover through the middle of the Southern Bight. The waters are homohaline and represent the northward flowing body of high-salinity channel water. No Recent coccoliths were noted from the sediments of this region. Coccolithophores were absent from these waters in the survey of Braarud et al. (1953), this region being characterized by a restricted flora containing mainly benthic diatom species.

Region A4 represents the Continental coastal water mass of Laevastu (1963) and is occupied by a northward flow of low salinity coastal water. This region is rich in nutrients but has high trace metal contents and high turbidity (Lee, 1980). Very scarce coccoliths ($10^5$/g) of *E. huxleyi*, *C. pelagicus* and *B. bigelowii* are found in sediments from the French coast and in the Scheldt Estuary. These species are probably transported as suspended sediment introduced from the English Channel. Braarud et al. (1953) did not record coccolithophores from the plankton samples from this region except for a small area off the Dutch coast (*E. huxleyi*; <5000 cell/l). This occurrence corresponds to the only region of the southern North Sea where the water column may become stratified and occurs as low salinity coastal water rides over high salinity Atlantic Water (Postma 1954; Fonds and Eisma, 1967). The plankton of Region A4 includes mainly bottom-dwelling forms such as *Melosira sulcata, Biddulphia* species and *Navicula* species (Braarud et al., 1953).

Region A6 contains the Dogger Bank and the region immediately to the south of it. Its waters are mainly homohaline and are of mixed origin, having components of Scottish coastal, English coastal, channel, and North Atlantic water. Very rare coccoliths of *Coccolithus pelagicus* ($10^5$/g) are recorded in sediments from the northern extremity of this area and are probably supplied from water admixed with a North Atlantic component. Braarud et al. (1953) noted that the plankton of this region is characterized by a mixed community of dinoflagellates (*Exuviaella baltica, Ceratium* species) and diatoms (*Rhizosolenia* species). Diatoms were particularly common in turbulent, upwelling waters on the Dogger Bank.

Region B extends from the northern margin to Dogger Bank, and represents the southward flow of high salinity (>35%), seasonally stratified North Atlantic water. Sediments underlying the northern part of region B contain high diversity (5 or more species) and abundant coccolith ($10^9$/g) assemblages. All ten coccolith species reported from the North Sea were recorded in sediments from Region B. Lower coccolith numbers ($10^8$/g) occur near Orkney and Shetland, where tidal current velocities (2-4 knots) cause turbulence with the result that the water column is practically homogeneous. Coccolith numbers and species diversity steadily decrease southward in region B, reflecting the decreasing Atlantic influence and increasing tidal current velocities in the overlying water column. The southern part of Region B is largely occupied by North Atlantic water in the winter, but in the summer the surface layer is dominated by mixed-water containing Scottish coastal, Continental coastal, Skagerrak and a minor North Atlantic component. Coccolith assemblages deposited here are limited to the species *E. huxleyi* and *C. pelagicus*. The southern margin of this region at 55°N correlates with the maximum southerly extension of recent coccolith assemblages in the North Sea from an area of tidally mixed waters in the southern North Sea (Figure 1.12).

Similar changes in the plankton present in the overlying waters of this region were recorded by Braarud et al. (1953). In the northern North Sea the inflowing Atlantic water was characterized by a flora poor in the number of species, but rich in populations of *E. huxleyi* (>125,000 cells/l) and the dinoflagellate *Exuviaella baltica*; diatoms however were extremely scarce. The plankton recorded in this region suggested that the oceanic Atlantic water had mixed only very slightly with waters containing the characteristic neritic flora of diatoms. Further south, diatoms (*Chaetoceros* species, *Thalassiosira bioculata* and *Nitzchia delicatissima*) replaced *E. huxleyi* as the dominant component of the plankton. South of latitude 58°S, *E. huxleyi* occurred only in numbers less than 5000 cells/l. Near the southern margin of Region B the plankton was limited to a poor flora of dinoflagellates (mainly *E. baltica* and *Ceratium* species). A similar coccolith-poor, but dinoflagellate-dominated flora was recorded in this study from filter samples located at the southern margin of Region B (Figure 1.6D). Abundant dinoflagellates including *Ceratium furca, C. tripos, Dinophysis acuta* and *Protoperidinium* species occurred in the filters, but coccoliths were extremely scarce (<0.1%) and limited to occasional specimens of *E. huxleyi*.

The Sea of Minches contains seasonally-stratified water of medium salinity (<34.75%.) with medium nutrient and turbidity levels and low summer temperatures. High diversity (5 or more species) coccolith assemblages dominated by *Emiliania huxleyi* (>19%) occur in the region. The occurrence of abundant Recent coccoliths in the sediment suggests a strong Atlantic component is present in the Minches region.

Region C, forming the coastal waters of west Norway and the Skagerrak, is characterized by haline stratification throughout the year. The region contains the Norwegian Rinne, the deepest parts of the North Sea. The surface layers are occupied by Skagerrak water with a large Baltic component. It has low salinity, low winter temperature, medium to high summer temperatures, low to medium turbidity and is poor in nutrients (Lee, 1980). The surface flow is westward and then along the Norwegian coast. *E. huxleyi* is undoubtedly the most important coccolithophorid in the coastal waters of Norway. Large populations of the species have been reported by Birkens and Braarud (1952) and Berge (1962). Other common coccolithophore recorded from this region include *Anthosphaera robusta, Calciopappus caudatus* and *Ophiaster hydroideus* (Schei, 1975). *E. huxleyi* is also likely to be the most abundant coccolith species in the sediments of this region; the species was found to be very dominant in latest Quaternary sediments cored from the southern slope of the Norwegian Channel in the Skagerrak (Mikkelsen, 1985).

### 1.5.3 Transport paths of coccoliths into and within the North Sea

Atlantic water having a salinity of >35% enters the North Sea through the Strait of Dover in the south and between Shetland and Norway in the North. These form the two major pathways for the transport of coccolithophore species into the North Sea basin. Using the transport path from the north the coccolith species would be transported in the southward-flowing mass of Atlantic water around the eastern coast of Shetland and enter the central area of the northern North Sea. Strong evidence for this transport direction is provided by the decreasing gradient from north to south of both coccolith diversity and abundance in the sediments. The transport of coccolith species through the Faero-Shetland Channel and into the North Sea was also proposed by Braarud et al. (1953). Large numbers of *E. huxleyi* were only found in waters which had a significant Atlantic component. The inflow of water between the coast of Norway and Shetland forms ca. 42,000 $km^3$/y of the total inflow (51,000 $km^3$/y) into the North Sea across its northern margin (Figure 1.11). The $10^9$ coccolith abundance contour traces out the path taken by the inflowing Atlantic water east of Shetland, confirming that this watermass is the major supplier of coccoliths to the region. Lower coccolith numbers ($10^8$/g) are deposited from Atlantic water flowing between Orkney and Shetland through the Fair Isle Channel, and between Orkney and Scotland.

The other main inflow of Atlantic water into the North Sea through the Strait of Dover is not thought to be an important supplier of coccolithophores. At the eastern margin of the English Channel high salinity (>35%) water deposits a very sparse nannoflora (ca. $10^5$ coccoliths/g) limited to the species *C. pelagicus*, *E. huxleyi* and *B. bigelowii* (Houghton, 1986, 1988), whereas throughout most of the southern North Sea Recent coccolith are either absent or extremely rare in the sediments. Braarud et al. (1953) found coccolithophores to be absent in the plankton at the eastern entrance to the English Channel. Very scarce coccoliths found in the Schelde Estuary at Antwerp, and along the French coastline in the southern North Sea, are thought to have been transported within the suspended load of the inflowing coastal water from the Channel. Concentrations of suspended matter in coastal waters of the Dover-Calais Strait may reach up to 11 mg/l (Eisma, 1981).

Populations of *E. huxleyi* are unlikely to survive the harsh winter conditions of the North Sea, where surface waters are characterized by low temperatures (<7°C) and are subjected to severely reduced light intensities. In Norwegian coastal waters (Balsfjord) populations of *E. huxleyi* were delimited by the 7°C isotherm (Gaarder, 1938). In the Oslofjord where *E. huxleyi* occurs in large numbers, the species was not thought to survive the winter (Birkens and Braarud, 1952). New populations of the species were introduced from the Skagerrak in spring, and these formed the basis of the summer populations. New populations of *E. huxleyi* are likely to be introduced to the North Sea each spring by Atlantic water, and these would then form the basis of the summer blooms (Houghton, 1991b).

### 1.5.4 Sources of reworked coccoliths

The reworked coccolith assemblages in the southern North sea are dominated by Cretaceous forms, with Palaeogene forms subordinate. Reworked forms are particularly abundant along the eastern coast of England (up to 100%) where there is close proximity to chalk outcrops. These Cretaceous forms are thought to have been derived from marine erosion of chalk cliffs and subsequently transported by the prevailing current system, rather than by fluviatile input into the basin. Detailed studies on the distribution of Cretaceous coccoliths in marginal and estuarine sediments of the Solent in the central English Channel (Houghton, 1986) have shown that modern-day chalk rivers supply little coccolith debris. Although chalk rivers do contain high concentrations of carbonate, it occurs mainly in dissolved form (as the bicarbonate ion) rather than particulate suspended detritus. The extensive Plio-Pleistocene deposits which flank the southern North Sea may be a potential source of the diverse coccolith assemblages found in the central and northern North Sea. However, analysis of the Pliocene-Pleistocene crags of eastern England (Jenkins and Houghton, 1987) and Pleistocene from southern England (Preece et al., 1990) indicate that they contain different nannofossil assemblages dominated by *Reticulofenestra haqii, R. minutula, Gephyrocapsa* species and *C. pelagicus*. Apart from fragments of *C. pelagicus*, these forms are usually absent from coastal sediments of the southern North Sea. Cenozoic taxa found in the southern North Sea assemblages are mainly restricted to Palaeogene forms. The absence of Recent coccolith throughout much of the southern North Sea basin cannot be attributed to estuarine dissolution effects because of the high numbers of reworked coccoliths in both marginal and estuarine sediments.

### 1.5.5 Differences between North Sea and Celtic Sea coccolith assemblages

The major differences between the coccolith assemblages of the two shelf regions is the species diversity; eighteen Recent coccolith species occur in the sediments from the Celtic Sea and western English Channel (Houghton, 1986; 1988) whereas only ten species occur in the northern North Sea. The reduced diversity in the North Sea cannot be attributed to the occurrence of a more neritic watermass, as in fact a large area of the northern North Sea comes under a strong Atlantic influence, similar to that present in the Celtic Sea and western English Channel. The drop in diversity is thought to result from the more northern latitude and cooler summer temperature of the water entering the North Sea across its northern margin. Surface water enters the northern North Sea from a latitude of 61°N, with spring/summer temperatures reaching only 12°C. However, Atlantic water enters the Celtic Sea from the southwest (lat. 48°N) and surface water temperature easily reaches 16°C during the summer. Table 1.4 lists those species present in the Celtic Sea assemblages but not recorded from the North Sea. These species prefer warm water and it is likely that the Celtic Sea is at or near their northern biogeographic range in the North Atlantic (Geitzenauer et al., 1977). Species such as *Umbilicosphaera sibogae, U. hulburtiana* and *Umbellosphaera tenuis,* all of which have a reported lower temperature limit of 13°C, but prefer much warmer temperatures, are unlikely to be consistent plankton components of the major inflow of Atlantic water entering the northern North Sea.

Table 1.4: List of coccolith species present in the Celtic Sea sediments but absent from the North Sea sediments.

| Species | Maximum temperature range (°C) |
|---|---|
| *Ceratolithus cristatus* ** | * |
| *Gephyrocapsa ericsoni* | 12-27 |
| *Gephyrocapsa oceanica* | 12-30 |
| *Syracosphaera nodosa* | 2-27 |
| *Syracosphaera histrica* | 10-27 |
| *Scyphosphaera apsteinii* | * |
| *Umbellosphaera tenuis* ** | 13-28 |
| *Umbilicosphaera sibogae* | 13-20 |
| *Umbilicosphaera hulburtiana* | 13-30 |
| *Anoplosolenia brasiliensis* ** | * |

* not precisely known but prefers tropical and subtropical water; ** species found only at southwestern extremity of Celtic Sea. Temperature data compiled mainly from Okada and McIntyre (1979).

The coccolith assemblages in the surface sediments of the North Sea are similar to those in the Celtic Sea sediments in the following respects:

1) the dominance and quantitative abundance of *E. huxleyi*
2) the relative and quantitative abundance of the subordinate species *C. pelagicus* and *S. pulchra*.
3) allowing for differences in species-diversity, both regions have a similar minor species component (0.3-2%).

Figure 1.13: Schematic representation of the spatial changes in coccolithophore abundance and diversity in the plankton and bottom sediments along a north-south traverse in the central North Sea. Water column conditions represent the summer period, with thermocline development in the northern North Sea. Blooms of coccolithophores, dominated by *Emiliania huxleyi*, occur in the nutrient-deficient, but well illuminated surface waters above the thermocline. Coccoliths are generally absent in sediments underlying the tidally mixed water mass of the southern North Sea. Dinoflagellates are more abundant within the thermocline and at the frontal boundary, where nutrient renewal across the thermocline stimulates their growth. Diatom abundance increases southward in the plankton, with peak abundances in the upwelling waters above the Dogger Bank. For simplicity, changes in water depth along the traverse have not been shown.

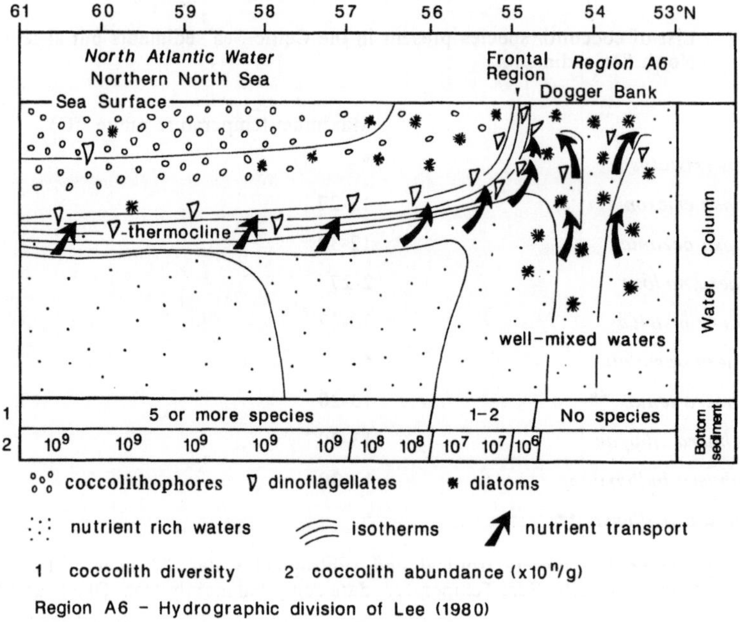


## 1.5.6 Ecological controls on the distribution of coccolithophores and other phytoplankton in the northwest European shelf regions

The patterns of coccolith accumulation in the Recent shelf sediments of northwest Europe have been explained by reference to the spatial and temporal changes in the phytoplankton populations in the overlying waters (Houghton, 1988, 1989). Seasonal coccolithophore blooms (dominated by *E. huxleyi*) on the European shelf are restricted to stratified waters with a well-developed seasonal thermocline (Holligan, 1978; Holligan et al., 1983; Murray et al., 1983). This ecological restriction is explained by the nutrient and light preference of the group. Coccolithophore growth above a thermocline is stimulated in the stable, well-illuminated but nutrient-poor environment. Oceanic nannoplankton have a considerable advantage over larger-celled algae in competition for nutrients because smaller-sized cells frequently have lower $K_s$ values (Friebele et al. 1978). The constant $K_s$ defines the minimum concentration at which a species can grow (Parsons et al., 1984). *E. huxleyi* (diameter<10μM) has been shown by Eppley et al., (1969) to have a Ks value of 0.1

μM nitrate/l, whereas *Gonyaulax polyhedra,* a large (>45μm) dinoflagellate has a Ks value of >5μM nitrate/l. These differences in ecological preference explain why the growth rates of *E. huxleyi* and other nannoplankton species are highest in regions of water stability, whereas large algal cells grow faster in upwelling and dynamic hydrographic regimes. Coccolithophore blooms are absent from tidally mixed waters on the European shelf as they would be removed from the surface waters before any significant growth could occur.

The ecological preference of coccolithophores for low nutrient levels associated with waters above a shallow thermocline explains the spatial variations in coccolith abundance and diversity observed southward in the North Sea. These spatial changes are shown schematically in Figure 1.13. High coccolith abundances ($10^9$/g) and high diversities (5 or more species) are found in assemblages deposited from the seasonally-stratified Atlantic water of the northern North Sea. Coccoliths are generally absent from the tidally mixed water mass of the southern North Sea.

The occurrences of such tidally mixed hydrographic conditions in the southern North Sea basin is not always thought to have existed in the Late Neogene. The occurrence of diverse coccolith assemblages and planktonic foraminiferal assemblages in Facies A of the Late Pliocene Coralline Crag (Jenkins and Houghton, 1987), which outcrops along the eastern coast of southern England (Figure 1.1), suggests deposition from waters with weak tidal influences (<1.5 knots) and with a well-developed thermocline. Atlantic influences must have been pronounced in the southern North Sea at this time, and may have been supplied to the Coralline Crag basin either from a southerly extension of the northern Atlantic water, or could have come from the southwest through the English Channel. The coccolith flora in the Coralline Crag however does not provide conclusive evidence that the English Channel was open during the Late Pliocene. Today, the high salinity Atlantic water flowing through the Strait of Dover does not supply any coccolithophores or planktonic foraminifera; the tidally mixed water column of the central and eastern English Channel acts as an ecological filter for these planktonic forms (Jenkins and Houghton, 1987, Houghton, 1991b).

### 1.5.7 Characteristic features of Recent coccolith assemblages from marginal seas

Okada and Honjo (1975) and Honjo (1977) studied the coccolithophorid distribution in plankton samples from the Asian marginal seas such as the China Sea and the Sunda Shelf area of the Java Sea. These marginal sea floras differed greatly from the pelagic floras at the equivalent latitude in the Pacific Ocean (Okada and Honjo, 1973). From these studies coccolithophorid floras in marginal seas are suggested to be characterized by the following features:
(1) a lower species diversity of coccolithophorids occurs than is observed in the open ocean;
(2) placolith-bearing coccolithophorids are dominant;
(3) most coccolithophorids bear malformed coccoliths
(4) the standing crop is considerably more aggregated in marginal seas than in the open ocean;
(5) *E. huxleyi* is generally scarce;
(6) *G. oceanica* is dominant.

The extent to which these points are characteristic for coccolith assemblages deposited in marginal seas, and their palaeo-ecological significance will now be discussed.

Marginal seas are generally thought to support a lower diversity coccolithophore flora than that found in the open ocean. These differences are not always apparent when the number of species recorded in the sediments is used as an index of species diversity. Twenty-two coccolith species are commonly found in the surface sediments of the North Atlantic and Pacific (Geitzenauer et al., 1977). Numbers of coccolith species recorded in various marginal sea and shelf environments are listed in Table 1.5. Marginal sea coccoliths in the western Pacific usually contain more than 28 species with a maximum of 59 being recorded from the Great Barrier Reef-Coral Sea region (Conley, 1979). All these assemblages containing high numbers of coccoliths were deposited at low-latitudes (<30ºS). Marginal sea assemblages deposited at higher latitudes contain fewer species, ranging from eight to 18. A similar decrease in numbers of coccolith species also occurs with decreasing latitude in the oceanic environment (McIntyre and Bé, 1967; Burns, 1972).

Table 1.5: Numbers of coccolithophore species recorded from marginal sea areas.

| | No. of species | Dominance * (%) | |
|---|---|---|---|
| Celtic Sea | 18 | > 90 | S |
| Irish Sea | 8 | > 90 | S |
| North Sea | 10 | > 90 | S |
| East China Sea | 35 | > 80 | S |
| South China Sea | 38 | > 80 | S |
| Taiwan Shelf | 32 | + | S |
| Great Barrier Reef | 59 | > 95 | S |
| Tasman Sea | 12 | > 80 | S |
| Western Mediterranean | 18 | + | S |
| Gulf of Mexico | | > 80 | S |
| Red Sea | 41 | > 80 | S/P |
| NW Sargasso Sea | 25 | > 80 | P |

* Dominance refers to combined abundance values of the two most common species.
+ no quantitative data presented, P- plankton survey, S- sediment survey.

Despite the relatively high species number in some marginal seas, such environments may be distinguished from open oceanic assemblages by a predominance of one or two species (usually *E. huxleyi* and *G. oceanica*). Dominance values in bottom sediments of the two most abundant coccolith species are <80% in the North Atlantic (McIntyre and Bé, 1967) and 40% to 80% in the western Pacific (Geitzenauer et al., 1977). Higher dominance values, however, may occur in subpolar and polar seas (Geitzenauer and Huddlestun, 1973). Dominance values of the two most abundant species are always high throughout marginal environments (>80%) and may be regarded as one of the most reliable indicators of such an environment.

Honjo (1977) considered that the dominance of placolith-bearing coccolithophorids in living floras was characteristic of marginal seas. The mean placolith components of shelf assemblages from northwest Europe are 97.4% for the Celtic Sea and western English Channel, 96.8% for the Irish Sea and 92.5% for the northern North Sea (Houghton, 1986). Wang and Samtleben (1983) concluded that since placoliths are always dominant in both oceanic and marginal sea assemblages, their dominance is not a reliable palaeoecological indicator of marginal environments. A greater than 98% placolith component occurs in nannofossil assemblages deposited at latitudes north of 40°N in the North Atlantic (McIntyre and Bé, 1967). However as placoliths are the most resistant of all types of coccolith structure, their relative abundances are always found to be greater in bottom sediments than in the overlying waters. Lower mean placolith values are found in planktonic communities sampled at the weather stations Charlie (84%) and Delta (79.5%) in the subarctic and transitional zones respectively of the North Atlantic (Okada and McIntyre, 1977) (Figure 1.14). The dominance of placoliths in coccolith assemblages decreases with latitude; assemblages deposited between 40°N and 2°S in the North Atlantic contain between 80% and 98% placolith forms. This decrease in dominance is caused by an increasing abundance of the forms *Rhabdosphaera clavigera, Umbellosphaera tenui,* and *Umbellosphaera irregularis* in tropical and subtropical regions. A similar dominance of placoliths occurs in tropical and subtropical seas (>75%), therefore dominance of placoliths must be considered to be characteristic of Recent coccolith assemblages in general.

Okada and Honjo (1975) regard the extensive occurrence of malformed coccoliths as one of the more significant features of assemblages in marginal seas. Malformation of coccolith species was a consistent feature of the plankton in the Asian marginal seas and took several different forms. Highest numbers of cells bearing malformed coccoliths (90 to 100%) occurred in plankton from the South and East China seas. Coccoliths of *Gephyrocapsa oceanica,* which formed up to 95% of the flora (sampled during October, 1968), had an irregular bar, or lacked the central structure completely. Additionally, these coccoliths lacked a floor in the central opening and the elements of the shield were barely flush rather than imbricated as in normal coccoliths of *G. oceanica*. Malformed coccoliths of *E. huxleyi* were characterized by twisted elements of the distal shield and grill. Wang and Samtleben (1983) however, have subsequently shown that the nannoflora described by Okada and McIntyre is not typical for the East China Sea. Studies by Nishida (1977) from another plankton sample (taken during winter) recorded *E. huxleyi* to be dominant (64%), whereas *G. oceanica* formed only 21%; there were no reported malformed coccoliths in the assemblages.

Malformed coccoliths are either very scarce or absent in sediments from the China Sea (Wang and Samtleben, 1983; Chen and Sheih, 1982). The absence of such coccoliths in bottom sediments underlying waters where reported malformation is extensive indicates that malformed coccoliths do not survive settling through the water column and are susceptible to rapid dissolution and disintegration. No malformed coccoliths have been observed in any of the species preserved in the shelf sediments from northwest Europe. Malformed coccoliths have rarely been reported in surface sediments from other shelf areas which suggests that they have a very limited fossil record (Wang and Samtleben, 1983).

Okada and Honjo (1975) suggested that malformation of coccoliths is triggered by rapid algal growth in a nutrient-deficient environment caused by overcrowding of coccolithophorid species during blooming. Although no significant correlation of malformation was recorded with concentrations of dissolved oxygen, ammonia and phosphate, some relationship was inferred with nitrate levels less than 0.1 μM/l. Similar low nitrate levels occur in the Celtic Sea following spring diatom blooms (Pingree, 1980); however no malformation of coccoliths has been reported to occur in the plankton at this time. Coccolithoporids require an exogenous supply of one, two or more vitamins (Tappan, 1980), particularly B1 (thiamin) and B2 (cobalimin). Perhaps a deficiency in one or more of these vitamins causes coccolith malformation in marginal seas. Malformed coccoliths of *E. huxleyi* have also been reported from the Subarctic Pacific (Okada and Honjo, 1973), indicating that malformation is not restricted to marginal environments but may be triggered by temperature extremes.

Laboratory experiments by Watabe and Wilbur (1966) on the effects of temperature on calcification and coccolith form in *E. huxleyi* also suggest a temperature control on coccolith malformation. Abnormal coccoliths were produced at all temperatures, but were minimal in number (<20%) at 18°C. The normal form of coccoliths of *E. huxleyi* depends on a delicately balanced growth of crystals which originate from a circle of nucleation centres in the basal region of the central cylinder (Wilbur and Watabe, 1963). Abnormal coccoliths are likely to be produced if crystal growth does not begin simultaneously in all centres in the basal region, or if the crystal of the upper and lower disc grow at different rates (Watabe and Wilbur, 1966). Most abnormalities are probably the result of an acceleration or an inhibition of growth of individual crystals resulting in asymmetry of coccoliths. Possible conditions which could cause growth rates to vary include local differences in carbonate concentration and the presence of increased lithogenic particle levels (Houghton and Guptha, 1991).

Figure 1.14: The location of the five weather stations and the distribution of the coccolithophore flora zones in the North Atlantic (after McIntyre and Bé, 1967). Note that Station Bravo (ca. lat. 56°30' N, long. 51°00' W) represents Northern Subpolar water; Station Delta (ca. lat. 44°00' N, long. 43°00'W) represents North Atlantic drift; Station Charlie (lat. ca. 52°45'N, long. 35°30'W) is located near the boundary of the Subpolar water and N. Atlantic drift; Station Echo (ca. lat. 35°00' N, long. 48°00'W) represents Northern Subtropical water; and Station Hotel (ca. lat. 35°00' N, long. 71°00'W) represents the tropical Gulf Stream.

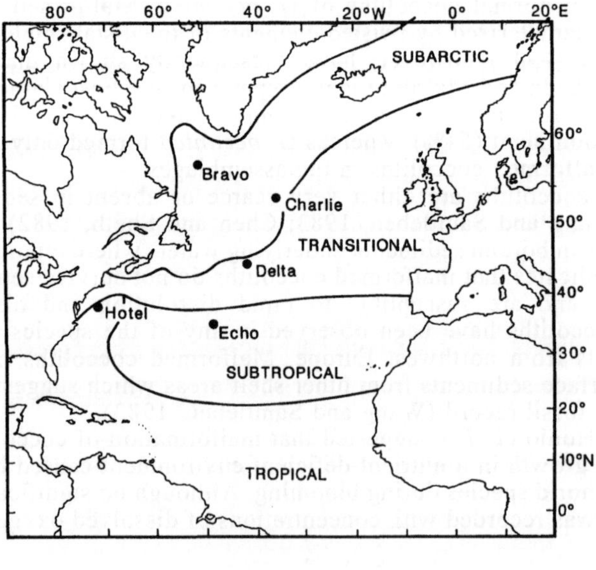

Honjo (1977) considered that the standing crop of coccolithophorids is significantly more aggregated (i.e. more localized in distribution) in marginal sea facies than in open oceanic facies. Similar results were reported by Okada and Honjo (1975) who concluded that the coccolithophores in the marginal seas of the western Pacific were highly concentrated when compared with those found in the open ocean. In the South China Sea, the distribution varied with the distance from the coast: largest crops of coccolithophores were found close to the western and northern shores. Wang and Samtleben (1983) suggested that any initial aggregation must be averaged with time, as no such results were recorded from bottom sediments. Additionally, large surface water crops of coccolithophores close to the shore were not reflected in coccolith abundance in the bottom sediments because of dilution from terrigenous sediment. Little evidence can be found of aggregated coccolith distribution in the sediments from the shelf of northwest Europe. Conversely, a characteristic feature is their homogeneity, particularly in the outer shelf environments of the Celtic Sea and the northern North Sea. In these outer shelf regions coccolithophore blooms have been reported over a large area lying between 45°N and 60°N latitude (Holligan et al., 1983).

Apart from the results obtained from the plankton survey of Okada and Honjo (1975) coccolith abundance decreases in both water and sediment samples across shelf regions towards land. Murray et al. (1983) reported a decrease in coccolith diversity and abundance northward and eastward away from oceanic conditions present at the southwestern margin of the Celtic Sea. A similar decrease occurs in the North Sea southward away from the Atlantic-influenced northern margin. Other records of coccolith abundance increasing with water depth in shelf seas include those from the Tasman Sea (Burns, 1975), the Capricorn Basin (Heckel, 1973), Jason Bay, South China Sea (Varol, 1985) and the East China Sea (Wang and Samtleben, 1983).

Honjo (1977) concluded that a sharp biological boundary exists between oceanic and marginal sea coccolithophore assemblages. This boundary is marked in the Asian marginal seas by a sudden increase in the dominance of *G. oceanica* in the assemblages. In the northeast Atlantic, oceanic coccolith assemblages are deposited on the continental slope (depth 1200 m) at the western margin of the Celtic Sea (Houghton, 1986). These assemblages contain high numbers of coccoliths with three species dominant: *E. huxleyi* (67%), *G. caribbeanica* (20%) and *Calcidiscus leptoporus* (12%). A sharp biological boundary must occur near the break of slope, at the 200 m contour of water depth, as shelf assemblages throughout the Celtic Sea are characterized by an abrupt increase in the dominance of *E. huxleyi* (>90%). Satellite studies on coccolithophore production along the continental shelf have also picked out this sharp biological boundary (Holligan et al., 1983). Large surface water blooms of *E. huxleyi* were recorded on the Celtic and Amorican shelf regions during April and May, close to and inshore of the 200 m contour. The outer limit of these *E. huxleyi* blooms was the shelf-slope boundary.

Okada and Honjo (1975) concluded that the extremely rare occurrence of *E. huxleyi* and the dominance of *G. oceanica* was characteristic of coccolithophore populations in the Asian marginal seas. With minor exceptions, *G. oceanica* was found to be the dominant in the plankton of the East China Sea, South China Sea, Java Sea, off the northern coast of Australia and in the Yellow Sea. *G. oceanica* formed almost monospecific communities in plankton samples from the South China Sea. A trend of increasing dominance towards land was also noted. Wang and Samtleben (1983), although noting *E. huxleyi* as one of the more dominant species in bottom sediments from the South China Sea, also found a trend of increasing relative abundance of *G. oceanica* towards the coast; *G. oceanica* became dominant over *E. huxleyi* in the middle shelf region. A similar reversal in dominance was also noted in sediments from the Great Barrier Reef (Heckel, 1973; Conley 1979), the west Tasman Sea (Burns, 1975) and the East China Sea (Wang and Samtleben, 1983). *G. oceanica* is also the most abundant species in assemblages deposited from the Bay of Bengal, in the Arabian Sea and the Andaman Sea in the northern Indian Ocean (Guptha, 1976, 1981, 1983, 1985).

The coccolithophores in the Australian coastal waters show a latitudinal dominance change from *G. oceanica*-dominated (North West Shelf, Gulf of Carpentaria, Coral Sea, northern Great Barrier Reef region) to *E. huxleyi*-dominated in the Tasman Sea and Great Australian Bight (Hallegraeff, 1984).

The abundance of *G. oceanica* in the marginal seas of the Indo-Pacific is certainly not characteristic of shelf waters and sediments from the Atlantic Ocean. Throughout the shelf assemblages from northwest Europe *E. huxleyi* is by far the most dominant species (>90%), whereas *G. oceanica* is only a rare component. Numerous plankton studies on coccolithophore distribution in shelf and marginal sea regions of the North Atlantic have been made, for example in the Norwegian coastal waters (Braarud et al., 1958), the coast of southern Europe and North Africa (Gaarder, 1954), the eastern coast of USA (Hulburt, 1967; Woods, 1968; Marshall, 1976), the Gulf of Mexico (Gaarder and Hasle 1971), Honduras lagoon (Kling, 1975) and in the Caribbean Sea (Thronsen, 1972). All these studies reported that *E. huxleyi* was the dominant species (sometimes forming a monospecific flora), whereas *G. oceanica* was usually either rare or absent.

*G. oceanica* is a warm-water species, with a maximum reported temperature range in the Atlantic of between 12°C and 30°C (Okada and McIntyre, 1979). The species is therefore not expected to be a major component of shelf sediment at temperate latitude. *G. oceanica*, however, is replaced at higher latitude by *G. caribbeanica*, a colder-water preferring *Gephyrocapsa* species which forms a major component (max. 50%) of North Atlantic assemblages (Geitzenauer et al., 1977). In shelf regions of the North Atlantic *G. caribbeanica* is a rare component of coccolith assemblages and is certainly not as adaptive as *G. oceanica* to the near-shore environment. The high dominance of *E. huxleyi*, coupled with the rarity of *Gephyrocapsa* species in sediment assemblages may be considered as one of the most characteristic features of Recent coccolith assemblages deposited in Atlantic shelf regions.

The rarity and sometimes absence of *E. huxleyi* in the Asian marginal seas occurs in waters which are well within the temperature and salinity tolerance of the species. The abundance distribution of *E. huxleyi* and *G. oceanica* is perhaps better explained by an ecological control related to water mass fertility and water quality. Winter (1982) correlated the percentage abundance of *G. oceanica* and *E. huxleyi* in bottom sediments with changes in the nutrient characteristics of the overlying watermass, along a traverse from the Gulf of Aden-Red Sea to the Gulf of Aqaba. Winter suggested *E. huxleyi* had a preference for low fertility waters and that *G. oceanica* favours high fertility waters. High values of *E. huxleyi* occurred in the Gulf of Aqaba (phosphate values 0.1 μM/l), whereas *G. oceanica* was dominant in the Gulf of Aden (1 μM/l phosphate). Studies on sediment cores from the Gulf of Elat also indicated that *G. oceanica* is the dominant species in glacial periods. During glacial periods, higher fertility was attributed to changes in hydrography, sea-level, and increase in nutrient supply from the land to the Gulf. It is significant that Winter (1982) found greatest percentages of *G. oceanica* in a core located near a major Wadi (Wadi Dahab), which transports nutrient-rich waters into the Gulf. A good correlation was found by Winter between $\delta^{18}O$ curve of benthic and planktonic foraminifera and the distribution of *E. huxleyi*. Large $\delta^{18}O$ values, which represent periods of increased stratification, are characterized by an increase in the abundance of *E. huxleyi* and a corresponding decrease in *G. oceanica*. Plankton studies (Winter et al., 1979) tend to confirm the association, as living *G. oceanica* is absent during the summer period of albeit slight water stratification. A localized *G. oceanica* assemblage was noted by Friedinger and Winter (1987) in plankton samples taken in the southwest Indian Ocean off South Africa. The stable coastal Augulhas Current was characterized by dominant *E. huxleyi*; however off the Great Fish River mouth an isolated *G. oceanica* dominant flora was present. Additionally, the highest abundance of *G. oceanica* in the plankton of the Caribbean Sea was noted by Thronsen (1972) in a land-influenced locality off Panama. An adaption for higher nutrient levels or high levels of lithic particles, or both of these, may explain the dramatic increase in landward dominance of *G. oceanica* (over *E. huxleyi*) recorded from several marginal seas from Indo-Pacific and Atlantic regions (Houghton and Guptha, 1991).

The optimum nutrient requirements of *E. huxleyi* in oceanic waters have not been precisely defined, probably because the species has at least two different variants and has a cosmopolitan distribution ranging from tropical to subpolar latitudes. Berger (1976), however, regards *E. huxleyi* to have a positive response to nutrient-enrichment (cf., fig. 21.11, p. 291 of Berger) and suggests that the species may behave like a diatom in some cases, rather than like a typical oceanic coccolithophore. Smayda (1963) also noted that *E. huxleyi* only occurred in the Gulf of Panama (8°45'N, 79°23'W) during upwelling conditions (December-April); the species was absent during the rainy season. *G. oceanica* was the dominant species in the Gulf and occurred all year round. These results suggest that there are ecological variants of *E. huxleyi* which are adjusted to various nutrient conditions. Alternatively, *E. huxleyi* may have a very low tolerance for warm coastal waters with a strong continental influence. At high water temperatures the species may become very susceptible to changes in the light quality in the surface waters. *E. huxleyi* is absent from the plankton of the Gulf of Panama during the rainy season when there is a vastly increased runoff, resulting in an increase in surface water turbidity.

Data compiled from the North Atlantic (Houghton, 1991a) do not uphold the suggestion of Berger (1976) that *E. huxleyi* has a nutrient preference similar to that of a diatom. The coccolith is particularly dominant in the warm, highly-stratified oligotrophic waters of the Sargasso Sea and in the western Caribbean. *E. huxleyi* dominated the plankton during the late autumn, in the winter and throughout the spring, with the minor exception of April, when a moderate bloom of diatoms occurred (Hulburt et al., 1960).

Laboratory studies (Watabe and Wilbur, 1966) have indicated that cultures of *E. huxleyi* have low percentages of calcified cells at upper (24°-27°C) and lower temperature (7-12°C) extremes indicating a marked suppression of total calcification. The suppression could result from two controls: 1) absence of calcification in some cells which at intermediate temperatures would calcify, and 2) a preferential increase in naked cells lacking the capability to calcify. If the temperature control identified in cultures also occurs in the oceans, then we can expect low percentages of coccolith-producing cells of *E. huxleyi* in the tropical plankton. The reduced abundance of *E. huxleyi* in the tropical seas of the Indo-Pacific could be related to the occurrence of mixed clonal populations of naked and coccolith-forming cells.

In the Indo-Pacific marginal seas, naked cells of *E. huxleyi* may outgrow the coccolith formers, thus reducing the contribution of coccoliths to the bottom sediments.

The dominance of *E. huxleyi* and *G. oceanica* in Atlantic marginal seas and the Indo-Pacific marginal seas respectively, tends to reflect the species' abundance patterns shown in oceanic sediments. *E. huxleyi* is more abundant in the Atlantic than in the Pacific, this difference being particularly discernable when comparing low-latitude assemblages. In high latitudes, *E. huxleyi* tends to dominate assemblages from both the Atlantic and the Pacific, while *G. oceanica* is a warm-water species, and is absent from such assemblages. Schniederman (1977) found *E. huxleyi* to be ubiquitous in Atlantic Ocean assemblages; the species constitutes over 50% of the Atlantic flora in tropical regions and increases to near 100% in subarctic and subantarctic latitudes. Geitzenauer et al. (1977) also recorded *E. huxleyi* to be the dominant coccolith in Atlantic Ocean assemblages. Okada and McIntyre (1979) recorded the seasonal distribution of coccolithophores at five weather stations from the western North Atlantic (Figure 1.14). A mean species composition was compiled for surface water and 100m levels from the seasonal data at the five weather stations (Table 1.6). *E. huxleyi* dominated the flora at all stations, ranging from tropical to subarctic environments. *G. oceanica* was a rare component of the coccolithophore communities and had a maximum mean abundance of 6.2 % at Station Hotel in the tropics.

Table 1.6: **Mean (%) abundance of *E. huxley*, *G. oceanica* and *G. caribbeanica* of the nannoflora at the five weather stations in the western North Atlantic (Figure 1.14). Values represent data combined from surface water and 100m levels (compiled from Okada and McIntyre, 1979).**

| | *E. huxleyi* | *G. oceanica* | *G. caribbeanica* |
|---|---|---|---|
| Hotel (Tropical) | 75.3 | 6.1 | R* |
| Echo (Subtropical) | 79.5 | 1.3 | - |
| Delta (Transitional) | 69.7 | 1.1 | 2.5 |
| Charlie (Subarctic) | 67.9 | R* | R* |
| Bravo (Subarctic) | 74.0 | - | - |

R* is < 0.2% of total assemblage.

*G. oceanica* usually forms 10-25 % of coccolith assemblages deposited from tropical to warm transitional waters of the Atlantic (Schneiderman, 1977). Highest abundance values for *G. oceanica* (43.2 %) were found in sediments underlying the upwelling region off northwest Africa (Geizenauer et al., 1977). In the Pacific Ocean, abundant *E. huxleyi* at low to mid-latitudes is generally restricted to assemblages underlying waters of low fertility. Highest relative abundance of the species (>40 %) occurs in sediments from the East Pacific Rise, near the Galapagos Islands and below the central Pacific gyre at 20ºN to 30ºN latitude (Roth and Coulbourne, 1982). The mean relative abundance of *G. oceanica*, *G. caribbeanica* and *E. huxleyi* for each of the major water masses in the North Pacific (Figure 1.15) is shown in Table 1.7. *G. oceanica* is most abundant in assemblages deposited below the equatorial water mass.

Table 1.7: Mean (%) of *G. oceanica*, *G. carbibbeanica* and *E. huxleyi* for each of the major water masses of the North Pacific Ocean (after Roth and Coulbourne, 1982).

| | Equatorial | Central | Transitional | Subarctic |
|---|---|---|---|---|
| *G. caribbeanica* | 3.9 | 7.1 | 37.4 | 21.3 |
| *G. oceanica* | 57.0 | 32.4 | 4.5 | 3.0 |
| *E. huxleyi* | 12.4 | 30.5 | 23.7 | 10.8 |

Figure 1.15: Generalized diagram of the water masses of the North Pacific (modified after Sverdrup et al., 1942 and Roth and Coulborn, 1982). Vertical shading delineates water masses. Note that EQ = Pacific Equatorial water, WNPC = Western North Pacific Central water, ENPC = Eastern North Pacific Central water, TRANS = Transitional Region, SAP = Subarctic Pacific water. Compiled from sources listed in the text.

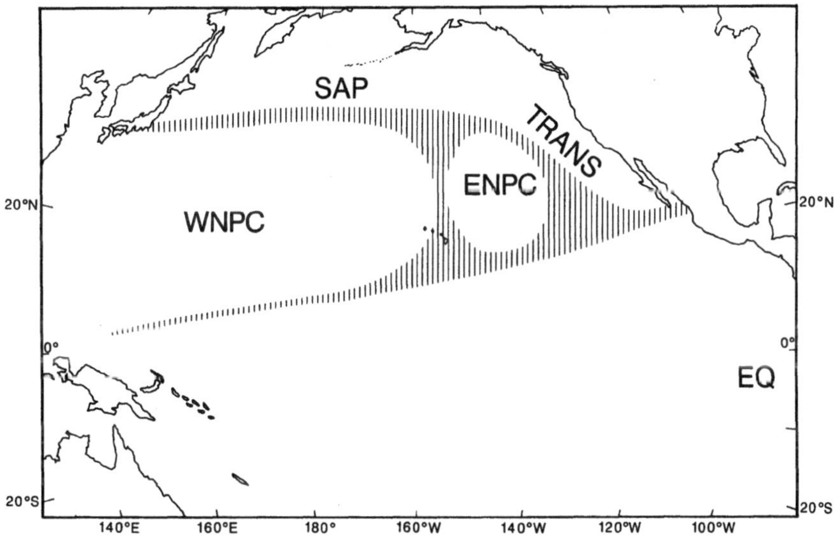


The basin-wide variations in the abundance of *G. oceanica* and *E. huxleyi* in the Atlantic and the Indo-Pacific may be related to major sedimentation facies patterns and oceanic circulation. Sverdrup et al., (1942) noted that carbonate sediments predominate in the Atlantic, while siliceous oozes are more common in the Pacific and Indian Oceans. These differences in biogenic sedimentation are attributed to fractionation of both carbonate and silica between the ocean basins as a result of deepwater circulation. The North Atlantic is dominated by a lagoonal (anti-estuarine) circulation, whereas the North Pacific has an estuarine circulation (Redfield et al., 1963). These differences result in a transfer of nutrient elements in the direction of the major circulation of deep water from the North Atlantic thorough the South Atlantic, Indian Ocean and South Pacific to the North Pacific. The dominance of *E. huxleyi* at all latitudes in the North Atlantic suggests that the nutrient-deficient lagoonal circulation represents the optimum oceanic environment for the species (Houghton, 1991a).

The dominance of *G. oceanica* in low-latitude marginal seas of the Indo-Pacific occurs in a region of strong seasonal variation in phytoplankton crop, related to the occurrence of monsoons. These changes are particularly pronounced in the northwest Indian Ocean during the winter (southwest monsoon) period, roughly May to August,when there is strong upwelling and nutrient-enrichment in the surface water (McGill, 1973). The richness in nutrient elements in the open monsoonal upwelling regions contrasts strongly with the general paucity of surface nutrient elements over large areas of the open tropical and subtropical ocean. Monsoonal upwelling occurs off Somalia, in the northern Arabian Sea, in the Bay of Bengal, off the Andaman Islands, off northwest Australia and in the seas of southeast Asia; all these regions have coccolith assemblages dominated by *G. oceanica*. The high nutrient-preference of *G. oceanica* could explain its dominance in monsoonal-influenced regions and also explain the change to *E. huxleyi*-dominated assemblages observed southward in the subtropical regions of the southern Great Barrier Reef, Tasman Sea and southern coasts of Australia, where there is no monsoonal influence (Houghton, 1991a; Houghton and Guptha, 1991).

The suggestion that *E. huxleyi* thrives in low-nutrient, well-stratified waters is also supported by empirical data. Parsons and Takahashi (1973) computed growth rates for a relatively large-celled species (*Ditylum brightwellii*) and *E. huxleyi*, applying mean values for incident radiation, mean light intensity in the water column, mixed layer depth, nitrate concentration and upwelling intensity. The growth rate of *E. huxleyi* was estimated to be higher in regions of water stability, whereas the larger celled algae appeared to grow faster in upwelling and coastal water. The low abundances of *E. huxleyi* in the Indo-Pacific region occur in marginal seas which are subjected to extreme climatic variability due to the occurrence of monsoons. *E. huxleyi* is known to have a very low $K_s$ value of ca. 0.1 $\mu$M nitrate/l (Eppley et al., 1969) which may explain why the species is dominant in oligotrophic stratified waters.

The sporadic distribution of *E. huxleyi* in assemblages deposited throughout the marginal seas of the Indo-Pacific may have important consequences for Late Pleistocene biostratigraphy. The persistent local dominance of *G. oceanica* in latest Pleistocene times may prevent recognition of the *E. huxleyi* Acme Zone of Gartner (1977) or even the *E. huxleyi* NN21 Zone of Martini (1971). In such instances where *E. huxleyi* is either rare or absent because of an ecological control, an overestimation of the age of the sediments may occur. Heckel (1973) also indicated that the absence of *E. huxleyi* from near-shore sediments from the Capricorn Basin was probably a facies control and should not necessarily be used as evidence for exclusion of an assemblage from the *E. huxleyi* Zone. Other nannofossil species may have to be erected as biostratigraphic markers in Indo-Pacific Pleistocene sediments from marginal seas, e.g. *Geminilithella subtilis,* which was used by Hughes et al., (1986) to date uplifted, latest Pleistocene sediments from the Solomon Islands.

Okada (1983) suggested that the abundance of *Florisphaera profunda* may be used as an indicator of palaeodepth of sediments. In marginal seas from the western Pacific the species showed a positive correlation with depth, increasing to over 50% of the assemblages at depths of 2000m. The species was absent in shallow water sediments. Separate counts of *F. profunda* var. *elongata* did not show any correlation between abundance and water depth. This may indicate that the variety *elongata* is less sensitive to changes in water depth than the nominate type of *F. profunda*. The species *F. profunda* ranges from the Upper Miocene to the Recent, therefore it is a potential tool for the recognition of palaeodepth during the Late Neogene. However, the abundance of *F. profunda* in Neogene sediments has probably been underestimated by nannopalaeontologists because of their failure to distinguish the shape from other micron-sized calcitic detritus.

1.5.8 Mesozoic and Cenozoic calcareous nannofossil assemblages as indicators of marginal sea/ open oceanic sediments

1.5.8.1 Cenozoic nannofossil assemblages

Recent calcareous nannofossil assemblages are distinguished from their open oceanic counterparts mainly by an increase in dominance of the species *E. huxleyi* and *G. oceanica,* this dominance change being particularly discernable at low-latitude sites. Both these species evolved during the Pleistocene, so other nannofossil species will have to be used as indicators of marginal seas and continental shelf in older Cenozoic sediments. *E. huxleyi* has a first occurrence datum late in oxygen isotope stage 8, dated at 0.268 Ma (Theirstein et al., 1977). The species occurs during the first two-thirds of its range in relatively low abundances and is subordinate to *Gephyrocapsa* species. However, subsequently *E. huxleyi* became the dominant form. The *Gephyrocapsa/E. huxleyi* dominance reversal is time-transgressive; in tropical and subtropical sediments it occurs between oxygen isotope stages 5a and 5b (0.085 Ma) whereas in transitional sediments it correlates with oxygen isotope stage 4c (ca. 0.073 Ma). The widespread dominance of *G. oceanica* in the Asian region during the latest Pleistocene suggests that *E. huxleyi* has failed to dislodge *G. oceanica* from the tropical and subtropical marginal sea niche of the western Pacific and northern Indian Ocean, a niche which *E. huxleyi* occupies with overwhelming success in the North Atlantic. These abundance patterns indicate that ecological preference and oceanic fertility patterns may have a more important influence on the abundance of coccoliths in fossil assemblages than evolutionary trends.

Between 0.085 Ma and 0.275 Ma *E. huxleyi* usually occurs in low numbers in coccolith assemblages and its dominance cannot be used as an indicator of marginal environments. However in cores from the Gulf of Elat (Aqaba) *E. huxleyi* is abundant (up to 60 %) in isotope stage 6 dated at 0.14 Ma, some 25 K years earlier than the first peak in abundance from open oceanic sediments (Winter, 1982). The much earlier abundance peak of *E. huxleyi* in the Gulf of Elat suggests that the Gulf may have acted as the initial evolutionary centre for the species. In Pleistocene European sediments prior to the evolution of dominant *E. huxleyi*, *Gephyrocapsa oceanica* and other *Gephyrocapsa* species usually dominate shelf and marginal sea assemblages (Preece et al., 1990). Other species common in Pleistocene and Neogene shelf assemblages include *Dityococcities productus, Coccolithus pelagicus* and *Reticulofenestra* species. All these forms are also abundant in Atlantic oceanic sediments, particularly at high latitudes. However, differences in the percentage abundance of these species in marginal and oceanic assemblages is not precisely known because of the lack of detailed quantitative studies on coccolith dominance in these environments. Further complications also arise from differences in the taxonomic identification used by nannofossil workers for the species of *Gephyrocapsa* and *Reticulofenestra*.

In the Cenozoic, genera which are more abundant in shelf areas than in oceanic conditions include the *Braarudosphaeracea (Braarudosphaera, Micranolithus, Pemma, Pentaster), Pontosphaera, Scyphosphaera, Traversopontis* and some *Rhabdospheara* (Bybell and Gartner, 1972; Tappan, 1980). The *Braarudosphaeracea* is probably the most widely reported nannofossil group from near shore environments. Representatives of this group have been reported from hemipelagic Eocene deposits in the USA, Mexico, Britain, northern and central Europe, southwest areas of the former Soviet Union, India and the Blake Plateau (Bybell and Gartner, 1972). These forms are conspicuously absent from oceanic deposits of a similar age. The only notable exception of this distribution is the widespread presence of *Braarudosphaera rosa* in an Oligocene deep-sea chalk in the south Atlantic (Maxwell et al., 1970). The Oligocene *Braarudosphaera* blooms are thought to have been triggered by environmental stress, possibly because of a salinity control.

Modern *Braarudosphaera biglowii* occurs in coastal embayments such as the Bay of Fundy (Gran and Braarud, 1935), and is widely distributed in Atlantic coastal waters of southwest Ireland, off the Azores and on the border of the Sargasso Sea (Gaarder, 1954). In sediments from the shelf of northwest Europe the species is a consistent but rare component (mean 0.5 %). Other records of the species occurrence in shelf assemblages from the Indo-Pacific are the Great Barrier Reef (Conley, 1979), east China Sea (Wang and Samtleben, 1983) and Sendai Bay, Japan (Takayama, 1972). Living *B. bigelowii* can tolerate relatively low salinities; the species is present in the Black Sea at 17-18 $^{o}/_{oo}$ salinity (Bukry, 1974). However the species is probably not so tolerant of elevated salinities; *B. bigelowii* is not found in the Red Sea (salinity 37 to 41 $^{o}/_{oo}$).

It is also generally accepted that holococcoliths (coccoliths constructed of calcitic rhombs of similar size) are usually only found in assemblages deposited at depths of up to a few hundred metres (Perch-Nielsen, 1985b). However, some care must be taken as holococcoliths are also known to have survived transport into oceanic basins by turbidity currents.

1.5.8.2 Mesozoic nannofossil assemblages

Although there are a few reports of isolated coccolith-like fossils from Palaeozoic sediments, calcareous nannofossils are generally thought to have originated in epicontinental seas during the Late Triassic. The earliest nannofossil to appear in the Triassic was *Prinsiosphaera triassica*, a form comprising "spherical to hemispherical solid nannofossils often containing a depression at one end and consisting of a parallely stacked group of calcitic plates orientated in a random fashion" (Jafar, 1983). These forms were reported from the Carnian to the Rhaetian of the Alps. Jafar assumed these forms to be of organic origin because of their abundance, stratigraphical occurrence, structural regularity and their distribution in the Alpine region. The biological affinity of the *Prinsiosphaera* is unclear, but it is likely to be the product of some planktonic algal form, possibly related to a primitive calcareous dinoflagellate (Bown, 1987). *Schizosphaerella puncticulata* is the next nannofossil form to appear and it occurs in Upper Triassic/Lower Jurassic boundary sediments. *Schizosphaerella* replaced *Prinsiosphaera* as the dominant spherical nannofossil form in the earliest Jurassic and both forms probably occupied the same ecological niche in the warm epicontinental seas. The first *Thoracosphaera (T. geometrica)*, another spherical-shaped form, also evolved in the Late Triassic (Jafar, 1983; Bown, 1987). The first true coccoliths *Archaeozygodiscus keossensis* and *Crucicirhadus minutis* occur in the Norian *Rhadocera suessi* ammonite Zone. The comparatively sudden appearance of coccoliths and other calcareous nannofossils in the Late Triassic followed a period of heavy salt precipitation throughout the Permian and Triassic periods; perhaps the salinity of the Late

Triassic epicontinental seas was suitable for the secretion of calcitic plates (Perch-Nielsen, 1986). The initial location of calcitic-secreting nannoplankton appears from the fossil record to have been in the Tethys ocean. The tropical and subtropical waters of much of the Tethys were probably the optimum water-mass best suited for the secretion of calcareous plates.

The first new coccolith lineage to appear in the Early Jurassic was the tiered placolith group represented by the genus *Mazaganalla*, although it has only been reported from the Sinemurian and Pliensbachian of the southern Tethys region. Placoliths with radiating crystal sutures appeared in the Late Sinemurian and are represented by *Biscutum novum* (Bown, 1987). The final coccolith lineage to appear in the early Jurassic (Pliensbachian) was placoliths with imbricated crystal sutures (e.g. *Lotharingius imprimus*). Calcareous nannofossil productivity during the Early and Middle Jurassic continued to be centered in the warm shallow epicontinental seas of the Tethys region. However in latest Jurassic the sink for carbonate deposition shifted from shallow epicontinental seas to the deep-sea (Keunen, 1941). The migration of calcareous nannoplankton from shelf and epicontinental seas into the oceanic regime in the latest Kimmeridgian to Tithonian is reflected in abrupt change in facies; from impure limestone, ribbon cherts to pure nannofossil carbonates with nannoconids. Nannofossil productivity was sufficiently high in the pelagic regime to depress the carbonate compensation depth in the Tethyan Ocean (Roth, 1986).

Late Jurassic-Early Cretaceous marginal assemblages are usually enriched with non-coccolith forms, especially the genus *Nannoconus*. Cooper (1989) studied nannofossil provincialism in the Late Jurassic-Early Cretaceous (Kimmeridgian to Valanginian) period. The Kimmeridgian and Tithonian high latitude nannofloras, with the exception of Boulogne, were dominated by *Ellipsagelosphaera*, whereas in low-latitude sites *Watznaueria* was dominant. At Boulogne *Cyclagelosphaera margerelii* was the dominant species. Keupp (1976) regards a dominance of *C. margerelii* as characteristic of a restricted environment, while the dominance of *Ellipsagelosphaera* species indicates a direct connection to the sea. *Watznaueria* and *Ellipsagelosphaera* may also be recognised in shelf limestones during the Late Jurassic. The Lower Tithonian, Solnhofen lithographic limestones of Bavaria, Germany have long been known to contain abundant coccoliths (Gumbel, 1873). The fine fractions of these limestones contain ca. 500,000 coccoliths per $mm^3$ (Flugel and Franz, 1967). The coccolith laminae were dominated by *W. barnesae* and *Ellipsagelosphaera lucasi* with smaller contributions from other species. The coccolith species are thought to have been transported into the stratified, stagnant lagoonal environment of the Solnhofen Basin (Barthel, 1970) from offshore waters, as the coccoliths are mainly preserved as fragments, rather than as complete coccoliths. Abundant coccoliths also occur in chalk bands in the Upper Jurassic, Kimmeridge Clay of southern Britain and consist predominantly of *Ellipsagelosphaera britannica* (Gallois, 1976). *E. britannica* formed 99 % or more of most of the assemblages; up to 5 % of other forms occurred including the forms *Actinozygus crux*, *A. geometricus, Podorhabdus cylindratus, Staurorhabdus, Stradnerlithsi* and *Zeugrhabdotus*. The coccolith limestones accumulated under highly anaerobic bottom water conditions (Tyson et al., 1979); very quiet water conditions are indicated by the abundant occurrence of well-preserved coccospheres. Tidal currents must have been remarkably weak in the Kimmeridge shelf seas; comparisons with the modern shelf of the Celtic Sea show that even in the deepest waters of the shelf where maximum tidal streams are ca. 0.3 m/s, coccospheres form $< 0.1$ % of the preserved nannoflora.

Lower Cretaceous nannofloras at both high and low latitude sites are dominated by *Ellipsagelosphaera*. The only exception to this distribution is on the Norwegian shelf where *Watznaueria* is more abundant (Cooper, 1989). Tithonian to Lower Cretaceous shallow marginal marine sites are usually characterized by an increase in density and diversity of nannoconids.

Theirstein (1976) recorded the differences between Cretaceous nannofossil assemblages in the Tethys and those assemblages found in Pacific sites recovered by the Deep Sea Drilling Project. These differences were attributed to the paleogeography of the two regions at the time of deposition; the Tethys was considered to be a marginal environment, while the Pacific sites were considered to be oceanic. Nannofossil species which were common in the Tethys but absent in the Pacific sites included *Nannoconus* species, *Conusphaera mexicana, Micranolithus obtusus, Lithraphidites bollii, Calcicalathina oblongata* and *Lithastrinus floralis*. Other species which have a similar preference for marginal conditions include *Lucianorhabdus cayeuxii, Tetralithus obscurus* and *B. biglowii*. All these forms which are more abundant in Cretaceous marginal environments are not true coccoliths, but various-shaped nannofossil forms, ranging from spherical-shaped forms to calcitic rods. Coccolith species which are usually more common in marginal assemblages than in oceanic environments include *Arkhangelskiella cymbiformis* and *Kamptneriums magnificus* (Perch-Nielsen, 1985a).

Detailed studies were carried out by Roth and Bowdler (1981) and Roth (1986) on the biogeography and palaeoceanography of nannofossil assemblages in the North Atlantic and Tethys Oceans during the Middle Cretaceous (Aptian to lower Cenomanian). They found that the palaeogeographic distribution of nannofossil species showed weak latitudinal gradients and more pronounced neritic-oceanic gradients related to oceanic fertility patterns. A high-latitude assemblage with dominant *Seribiscutum primitivum* and *Lithasstrinus floralis* was found on the Falkland Plateau and in the Indian Ocean at palaeolatitudes of more than 50°S. *Seribiscutum primitivum* was also found in lower numbers in the contemporaneous boreal Albian of the North Sea region. Middle Cretaceous assemblages enriched in *Zygodiscus erectus* and *Biscutum constans* occur along the continental margin of the eastern Atlantic, of the Iberian Peninsula and North Africa. These two species are also dominant in the equatorial Pacific (Roth, 1981). Assemblages with dominant *Z. erectus* and *B. constans* are thought to have underlain nutrient-rich upwelling waters. Atlantic Ocean assemblages characterizing areas far removed from continents, contain dominant *Watznaueria barnesae*, with *Rhagodiscus splendens* and *Rhagodiscus asper* also abundant. In shallow marginal and epicontinental sea environments *Broinsonia* and *Nanoconnus* species are most common (Roth and Bowdler, 1981). These two genera are considered good indicators of neritic conditions.

During the Late Cretaceous there was a general rise in sea-level which led to the spread of the pelagic carbonate sedimentation (Chalk facies) to the epicontinental seas of the European platform, the Middle East and North America. At this time, northwest Europe lay at the margin of the European continent which was submerged beneath a widespread epicontinental sea. The Chalk sea in Europe was of normal marine salinity (ca. 35 ‰) as shown by the widespread occurrence of echinoderms and brachiopods which cannot thrive in reduced salinities (Hancock, 1975). Most estimates of palaeodepth suggest that the Chalk was deposited at depths between 50m and 200m. The marine transgression of the Upper Cretaceous submerged large areas of land, thus reducing supply of detrital sedimentation. The lack of continental influences and nutrient supply allowed the blooming of calcareous nannoplankton in the warm, highly stratified epicontinental Chalk seas. Current or tidal action in at least part of the Chalk sea is strongly indicated by the mode of preservation of the nannofossils. Hancock (1975) describes the Chalk as being predominantly composed of two particle-size ranges, 0.5-4 microns and 10-100 microns. The finer grain-size is composed of individual coccoliths and their skeletal fragments which account for the sub-micron particles. Whole coccolith detritus suggests that most coccospheres disintegrated during sedimentation. Such processes occur in the modern-day shelf seas of northwest Europe, even at current speeds of ca. 0.3 m/s. The reported absence of coccospheres from the Upper Cretaceous Chalk distinguishes these sediments from similar coccolith-rich thin limestone bands in the Upper Jurassic Kimmeridge Clay. These differences probably reflect variations in the bottom-water conditions. The bottom waters of the Chalk seas were oxygenated and influenced at least periodically by currents; the Jurassic limestones were deposited under quiet, stagnant, bottom waters.

The nannofossil assemblages of British Chalk sequences are usually dominated by *Retacapsa, W. barnesae, Prediscosphaera, Zygodiscus, Eiffellithus, Ellipsagelosphaera*, and *Litastinus floralis*. The Upper Cretaceous is also the first period in which holococcoliths represent an important component of the preserved assemblages, represented by the genera *Lucianorhabdus, Orastrum, Phanolithus* and *Russellia* (Crux, 1982). *Lucianorhabdus cayeuxii* is particularly abundant in Santonian/Campanian boundary assemblages where it dominates nannofloral assemblages of the Yorkshire coast. In offshore chalks of the Central Graben and in the Dutch sectors of the North Sea, this form is a relatively minor component (J. Jeremiah, pers. comm., 1992).

## 1.6 Conclusions

Coccolith abundance and species diversity in the North Sea sediments are controlled by the seasonal supply of coccolithophorids from Atlantic water to the basin, and by the degree of tidal influence in the overlying water column. High diversity assemblages are deposited by seasonally-stratified Atlantic water. Recent coccoliths are either very rare or absent in sediments underlying the tidally mixed water column of the southern North Sea.

The abundance of *E. huxleyi* (>90 %) in the Recent sediments of the North Sea again emphasizes the dominance of the species in European marginal sea coccolith assemblages and in the Atlantic Ocean in general. However, marginal sea coccolith assemblages from the western Pacific and northern Indian Ocean typically contain *G. oceanica* as the dominant form. These abundance patterns are related to the species' nutrient preference. The lagoonal circulation of the Atlantic is thought to be the optimum oceanic environment for *E. huxleyi*. *G. oceanica* prefers low-latitude, high to moderately fertile waters, with greatest dominance in assemblages underlying the Pacific high-productivity belt, and in the monsoon-influenced marginal seas of the Indo-Pacific. Changes in the ratio of *E. huxleyi*/*G. oceanica* at low and mid-latitudes may therefore be used as an index of watermass fertility in both marginal sea and oceanic environments.

Calcareous nannofossils first evolved in the Tethyian marginal seas during the Late Triassic. The earliest nannofossils which flourished in the epicontinental seas are represented by unusual-shaped (mainly spherical) non-coccolith forms. Consistent and accurate recognition of distinct marginal and oceanic nannofossil assemblages during the Mesozoic and Cenozoic is often hampered by poor preservation of the sediments and also by a lack of detailed quantitative studies on nannofossil abundance. Additionally, environmental information may be lost because some oceanic species thrived in marginal environments, e.g. *Watznaueria* and *Ellipsagelosphaera* during the Mesozoic, and *Reticulofenestra* and *Gephyrocapsa* during the late Cenozoic. However, the association of non-coccolith forms with marginal environments, first recognized from the Late-Triassic, seems to have lasted through to the modern day. Thus nannofossil assemblages enriched with non-coccolith forms or holococcoliths, frequently represent deposition from epicontinental seas.

## Acknowledgements

The research on northwest European coccolith assemblages was carried out while the author was on a NERC studentship at the Department of Geology, Southampton University. Samples were made available by Dr. J.B. Wilson (I.O.S.), H. Buckley (British Museum, Nat. History), Prof. J. Murray (Southampton), Dr. J. McManus (Dundee), Dr. J. Senten (Antwerp), Dr. R. Garden (Badley, Ashton) and Prof. I. McCave (Cambridge). The following are thanked for their comments on an earlier draft of this chapter: Dr. D.G. Jenkins, Dr. N. Scheiderman, J. Jenkins, and G.R. Ward. The figures were drafted by John Taylor and Andrew Lloyd (Open University).

### References

Barthel, K.W. 1970. On the deposition of the Solnhofen lithographic limestone (Lower Tithonian, Bavaria, Germany). Neues Jb. Geol. Palaont. Abh., 135: 1-18.

Berge, G. 1962. Discoloration of the sea due to a *Coccolithus huxleyi* bloom. Sarsia, 6, 27-40.

Berger, W.H., 1976. Biogenous deep sea sediments: production, preservation and interpretation. In Riley, J.P. and Chester, R. (eds). Chemical Oceanography, vol. 5, 2nd edition, Academic Press, London, 265-388.

Birkens, E. and Braarud, T. 1952. Phytoplankton in the Oslo Fjord during a "Coccolithus huxleyi" - summer. Avh. Norske Videnske Akad. Oslo, I Naturk. Kl., 2, 3-23.

Bown, P.R. 1987. Taxonomy, evolution and biostratigraphy of Late Triassic-Early Jurrasic calcareous nannofossils. Spec. Pap. Paleontology 38, 1-118.

Braarud, T., Gaarder, K.R. and Grontved, J. 1953. The phytoplankton of the North Sea and adjacent waters in May 1948. Journal du Conseil Permanent International pour l'Exploration de la Mer 113, 87 pp.

Braarud, T., Gaarder, K.R. and Nordli, O. 1958. Seasonal changes in the phytoplankton at various points off the Norwegian west coast (observations at the permanent oceanographic stations 1945-46). Rept. Norwg. Fish. mar. Invest. 12, 1-77.

Braarud, T., Hofsvang, B.F., Helmfoss, P. and Overland, A.A.K. 1974. The natural history of the Hardangerfjord 10 Phytoplankton in 1955-56. The quantitative phytoplankton cycle in the fjord waters and in offshore coastal water. Sarsia, 55, 63-98.

Bukry, D. 1974. Coccoliths as paleosalinity indicators - evidence from the Black Sea. Mem. Am. Ass. Petrol. Geol., 20, 353-363.

Burns, D.A. 1972. The latitudinal distribution and significance of calcareous nannofossils in bottom sediments of the southwest Pacific Ocean (lat. 15-55°S) around New Zealand. In Frazer, R. (ed.). Oceanography of the South Pacific. N.Z. Nat. Com. for UNESCO.

Burns, D.A. 1975. The abundance and species-composition of nannofossil assemblages from continental shelf to offshore basin, western Tasman Sea. Deep Sea Res., 22, 425-431.

Bybell, L. and Gartner, S. 1972. Provincialism among the mid-Eocene calcareous nannofossils. Micropaleontol., 18, 319-336.

Cadee, G.C. 1985. Macroaggregates of *Emiliania huxleyi* in sediment traps. Mar. Ecol. Prog. Ser. 24, 193-196.

Chen, M.-P. and Sheih, K.S. 1982. Recent nannofossil assemblages in sediments from the Sunda Shelf to the abyssal plain, South China Sea. Proc. Nat.Sci. Council Republic China, part A, Appl. Sci., 6, 251-285.

Conley, S.M. 1979. Recent coccolithophores from the Great Barrier Reef-Coral Sea Region, Micropaleontol., 25, 20-43.
Cooper, K.E. 1989. Nannofossil provincialism in the Late Jurassic (Kimmeridgian to Valanginian) Period. In Crux, J.A., van Heck, S.A., (eds.). Nannofossils and their applications. Ellis Horwood, Chichester, 143-211.
Crux, J.A. 1982. Upper Cretaceous (Cenomanian to Campanian) Calcareous Nannofossils. In Lord, A.R. (ed.). A Stratigraphical Index of Calcareous Nannofossils. Ellis Horwood, Chichester, 81-135.
Eisma, D. 1981. Supply and deposition of suspended sediment matter in the North Sea. Special Publication of the International Association of Sedimentologists, 5, 415-428.
Eppley, R.W., Rogers, J.N. and McCarthy, J.J. 1969. Half-saturation constants for the uptake of nitrate and ammonia by marine phytoplankton. Limnol. Oceanogr., 14, 912-920.
Flugel, E. and Franz, H.E. 1967. Elecktronmikroskopischer Nachweis von coccolithen in Solnhofen Plattenkalk (Ober-Jura). Neues Jb. Geol. Paleont. Abh. 127, 145-263.
Fonds, M. and Eisma, D. 1967. Upwelling water as a possible cause of plankton blooms along the Dutch coast. Neth. J. Sea Res., 3, 458-463.
Friebele, E.S., Correl, D.L. and Faust, M.A. 1978. Relationship between phytoplankton cell size and rate of orthophosphate uptake: in-situ observations of an estuarine population. Mar. Biol., 45, 39-52.
Friedinger, P.J.J. and Winter, A. 1987. Distribution of modern coccolithophore assemblages in the southwest Indian Ocean off southern Africa. J. Micropaleontol., 6, 49-56.
Gaarder, K.R. 1938. Phytoplankton studies from the Tromso district 1930. Tromso Mus. Arsh. (Nat. Hist. Avd. II), 55, 1-159.
Gaarder, K.R. 1954. Coccolithineae, Silicoflagellatae, Pterospermatacaea and other forms from the Michael Sars North Atlantic Deep-Sea Expedition, 1910, 1-20.
Gaarder, K.R. and Hasle, G.R. 1971. Coccolithophores of the Gulf of Mexico. Bull. Mar..Sci., 21, 519-544.
Gallois, R.N. 1976. Coccolith blooms in the Kimmeridge Clay and origin of the North Sea Oil. Nature, 259, 473-475.
Gartner, S. Jr. 1977. Calcareous nannofossil biostratigraphy and revised zonation of the Pleistocene. Mar. Micropaleontol., 2, 1-25.
Geitzenauer, K.R. and Huddlestun, P. 1973. An Upper Pliocene-Pleistocene calcareous nannoplankton flora from a subantarctic Pacific deep-sea core. Micropaleontol., 2, 973-1008.
Geitzenauer, K.R., Roache, M.B. and McIntyre, A. 1977. Coccolith biogeography from North Atlantic and Pacific surface sediments. In Ramsay, A.T.S. (ed) Oceanic Micropaleontology, 2, Academic Press, London, 973-1008.
Gran, H.H. and Braarud, T. 1935. Qualitative study of the phytoplankton in the Bay of Fundy and the Gulf of Maine (including observations on hydrography, chemistry and turbididty). J. Biol. Bd.Can., 1, 279-467.
Gumbel, C.W. 1873. Coccolithen im Eocanmergel. Neues Jb. Miner. Geol. Paleont., 299-304.
Guptha, M.V.S. 1976. Nannofossils from sediments of the continental slope off Bombay (India). Riv. Ital. Paleontol., 82, 417-430.
Guptha, M.V.S. 1981. Nannoplankton from Recent sediments off the Anadaman Islands. Ind. J. Mar. Sci., 10, 293-295.
Guptah, M.V.S., 1983. Nannoplankton from Rc9-156 in the southwest Arabian Sea. J. Geol. Soc. India, 24, 468-475.
Guptha, M.V.S. 1985. Distribution of calcareous nannoplankton from the sediments of the northwestern continental shelf of India. J. Geol. Soc. India, 26, 267-274.
Hallegraeff, G.M. 1984. Coccolithophorids (Calcareous nannoplankton) from Australian waters. Bot. Mar., 33, 229-247.
Hancock, J.M. 1975. Petrology of the Chalk. Proc. Geol. Assoc. 86, 499-535.
Heckel, H. 1973. Late Oligocene to Recent nannoplankton from the Capricorn Basin, Great Barrier Reef area. Geol. Surv. of Queensland Publ. 359, Pal. Pap. 33, 1-23.
Holligan, P.M. 1978. Patchiness in subsurface phytoplankton populations of the northwest European continental shelf. In Steel, J.H. (ed). Spatial patterns in plankton communities, NATO, 221-231.
Holligan, P.M., Viollier, M., Harbour, D.S., Camus, P. and Champagne-Phillip, M. 1983. Satellite and ship studies of coccolithophore production along a continental shelf edge. Nature, 304, 339-342.

Honjo, S. 1977. Biogeography and provincialism of living coccolithophorids in the Pacific Ocean. In Ramsay, A.T.S. (ed.). Oceanic Micropaleontology 2, Academic Press, London, 951-972.
Houghton, S.D. 1986. Coccolith assemblages in Recent marine and estuarine sediments from the continental shelf of Northwest Europe. Ph.D. thesis, University of Southampton, 465 pp.
Houghton, S.D. 1988. Thermocline control on coccolith abundance and diversity in Recent sediments from the Celtic Sea and English Channel, Mar. Geol. 83, 313-319.
Houghton, S.D. 1989. Coccolith sedimentation and transport in the Irish Sea. Mar. Geol., 86, 67-74.
Houghton, S.D. 1991a. Calcareous nannofossils. In Riding, R. (ed). Calcareous Algae and Stromatolites, Springer-Verlag, Berlin, 217-266.
Houghton, S.D. 1991b. Coccolith sedimentation and transport in the North Sea, Mar. Geol. 99, 267-274.
Houghton, S.D. and Guptha, M.V.S., 1991. Monsoonal and fertility controls on Recent marginal sea and continental shelf coccolith assemblages from the western Pacific and northern Indian oceans. Mar. Geol., 251-259.
Hughes, G.W., Varol, O. and Dunkley, P.N., 1986. Pleistocene uplift of Tetepare Island, Solomon Group as implified by foraminifera and calcareous nannofossil evidence. Roy. Soc. NZ Bull., 24, 397-408.
Hulburt, E.M. 1967. Some notes on the phytoplankton off the southeastern coast of the US. Bull. Mar. Sci., 77, 330-337.
Hulburt, E.M., Ryther, J.H. and Guillard, R.R.L., 1960. The Phytoplankton of the Sargasso Sea off Bermuda. J. Cons. Perm. Int. Explor. Mer., 25, 115-128.
Jafar, S.A. 1983. Significance of the Late Triassic calcareous nannoplankton component from Austria and southern Germany. Neues Jahrb. Geol. Paleontol. Abh. 166, 218-259.
Jenkins, D.G. and Houghton, S.D. 1987. Age, correlation and paleoecology of the St. Erth Beds and the Coralline Crag of England. Mededelingen Werkgroep voor Tertiaire en Kwartaire Geologie, 24, 147-256.
Keupp, H. 1976. Kalikiges Nannoplankton aus den Solnhofer Schichten (Unter-Tithon, Sudliche Frankenalb). Neues jahrb. Geol. Palaeontol. Montshefte, 6, 361-381.
Kling, S.A., 1975. A lagoonal coccolithophore flora from Belize (British Honduras). Micropaleontol., 21, 1-13.
Kuenen, P.H. 1941. Geochemical calculations concerning the total mass of sediments in the earth. Am. J. Sci., 239, 161-190.
Laevastu, T., 1963. Surface water types of the North Sea and their characteristics. Serial Atlas of the marine Environment Folio 4, American Geographical Society, New York.
Lee, A.J. 1980. North Sea: Physical Oceanography. In Banner, F.T., Collins, B.M. and Massie, K.S. (eds). The North-west European Shelf Sea: The Seabed in motion II/ Physical and Chemical Resources, Elsevier, Amsterdam, 467-515.
Lee, A.J. and Ramster, J.W., 1976. Atlas of the seas around the British Isles. Fish. Res. Tech. Rept., MAAF Direct. Fish. Res., Lowestoft, 20, 1-4.
Marshall, H.G. 1976. Phytoplankton distribution along the eastern coast of the USA. I. Phytoplankton Composition. Mar. Biol., 38, 81-89.
Martini, E. 1971. Standard Tertiary and Quaternary calcareous nannoplankton zonation. In Farinacci, A. (ed). Proc. Planktonic Conf., 2nd (Rome), Tecnoscienze, Rome, Vol. 12, 739-785.
Maxwell, A.E., Herzen, R.P., Von, R.P., Andrews, J.E., Boyce, R.E., Milow, E.D., Hsu, K.J., Percival, S.F. and Saito, T. 1970. Summary and conclusions. DSDP Init. Repts. Leg 3, Washington D.C. U.S Govt. Printing Office, 441-471.
McGill, D.A. 1973. In B. Zeitschel (ed) The Biology of the Indian Ocean. Chapman and Hall, London, Berlin, 52-102.
McIntyre, A. and Bé, A.W.H. 1967. Modern coccolithophorida of the Atlantic Ocean-1 Placoliths and Crytoliths. Deep-Sea Research, 14, 561-597.
McIntyre, A. and McIntyre, R., 1971. Coccolith concentrations and differential solution in oceanic sediments. In Funnel, B.M. and Riedel, W.R. (eds). The micropalaeontology of the Oceans, Cambridge University Press, 253-261.
Mikkelsen, N. 1985. Late Quaternary evolution of the Skagerrack area as mirrored by calcareous nannoplankton. Norske Geologisk Tidsskrift, 65, 87-90.
Murray, J.W., 1985. Recent foraminifera from the North Sea (Forties and Ekofisk areas) and the continental shelf to the west of Scotland. J. Micropalaeontol., 4, 117-125.
Murray, J.W., Weston, J. and Sturrock, S. 1983. Sedimentary indicators of water movement in the western approaches to the English Channel. Continental Shelf Res., 1, 330-352.

Nishida, S.H. 1977. Preliminary report on the biocoenosis and thantocoenosis of Coccolithophorids. News Osaka Micropaleontol., 6, 5-17 (in Japanese, with English title).
Nishida, S.H. 1979. Atlas of Pacific nannoplankton. News Osaka Micropaleontol. Spec. Pap. 3, 1-33.
Okada, H. 1983. Modern nannofossil assemblages of coastal and marginal seas along the western Pacific Ocean. Utrecht Micropal. Bull. 30, 171-187.
Okada. H. and Honjo, S. 1973. Distribution of oceanic coccolithophorids in the Pacific. Deep Sea Res., 20, 355-374.
Okada, H. and Honjo, S. 1975. Distribution of coccolithophores in the marginal seas of the Pacific Ocean. Mar. Biol., 31, 272-285.
Okada, H. and McIntyre, A. 1977. Modern coccolithophores of the Pacific and Atlantic Oceans. Micropaleontol., 23, 1-55.
Okada, H. and McIntyre, A., 1979. Seasonal distribution of modern coccolithophores in the western North Atlantic. Mar. Biol. 54, 319-328.
Parsons, T.R. and Takahashi. M. 1973. Environmental control on phytoplankton cell size. Limnol. Oceanogr., 18, 511-515.
Parsons, T.R., Takahashi, M. and Hargrave, B. 1984. Biological Oceanographic Processes (3rd Edition), Pergammon Press, Oxford, 320 pp.
Perch-Nielsen, K. 1985a. Mesozoic calcareous nannofossils. In Bolli, H.M., Saunders, J.B. and Perch-Nielsen, K. (eds). Plankton Stratigraphy, Cambridge University Press, 329-426.
Perch-Nielsen, K. 1985b. Cenozoic calcareous nannofossils. In Bolli, H.M., Saunders, J.B. and Perch-Nielsen, K. (eds). Plankton Stratigraphy, Cambridge University Press, 427-554.
Perch-Nielsen, K. 1986. Geological events and the distribution of calcareous nannofossils - some speculations. Bull, Centres Rech. Explor. Prod. Elf-Aquitaine, 10, 421-432.
Pingree, R.D. 1980. Physical oceanography of the Celtic Sea and the English Channel. In Banner, F.T., Collins, M.B. and Massie, K.S. (eds) The North-west European Shelf Sea: Seabed and the seas in motion II. Physical and Chemical Resources, Elsevier, Amsterdam, 415-465.
Pingree, R.D., Holligan, P.M. and Mardell, G.T. 1978. The effects of vertical stability on phytoplankton distribution in the summer on the northwest Euopean Shelf. Deep-sea Res., 25, 425-477.
Postma, H. 1954. Hydrography of the Dutch Waddensea. A study of the relation between water movement and the transport of suspended materials and the production of organic matter. Archives Neerlandaises de Zoologie, 10, 405-511.
Preece, R.C., Scource, J.D., Houghton, S.D., Knudsen, K.L. and Penny, D.N. 1990. The Pleistocene sea-level and neotectonic history of the eastern Solent, southern England. Phil. Trans. Roy. Soc. London, B. 238, 425-477.
Raymont, J.E.G. 1963. Plankton and productivity in the Oceans (2nd Edition) Vol. 1 Phytoplankton, Pergammon Press, Oxford, 489 pp.
Redfield, A.C., Ketchum, B.H. and Richards, F.A. 1963. In Hill, M.N. (ed). "The Sea" Vol. 2, Interscience, New York, vol. 2, 26-27.
Roth, P.H. 1981. Mid-Cretaceous calcareous nannoplankton from the central Pacific: Implications for paleoceanography. Init. Repts. DSDP, vol. 62, 471-489.
Roth, P.H. 1986. Mesozoic palaeoceanography of the North Atlantic and Tethys Ocean. In Summerhays, C.P. and Shackleton, N.J. (eds). Geol. Soc. London, Spec. Publ., 21, 299-320.
Roth, P.H. and Bowdler, J.R. 1981. Middle Cretaceous calcareous nannofossil biogeography and oceanography of the Atlantic Ocean. Spec. Publ. Soc. Econ. Paleontol. Mineral., 32, 517-546.
Roth, P.H. and Colbourne, W.T. 1982. Floral and solution patterns of coccoliths in surface sediments of the North Pacific. Mar. Micropaleontol., 7, 1-52.
Schei, B. 1975. Coccolithophorid distribution and ecology in the coastal waters of North Norway. Norweg. J. Bot., 22, 217-225.
Schniederman, N. 1977. Selective dissolution of Recent coccoliths in the Atlantic Ocean. In Ramsay, A.T.S. (ed). Oceanic Micropaleontology, Academic Press, London, 1009-1053.
Scholle, P.A. and Kling, S.A. 1972. Southern British Honduras: Lagoonal coccolith ooze. J. Sed. Petrol., 42, 195-204.
Smayda, T.J. 1963. A quantitative analysis of the phytoplankton of the Gulf of Panama I. Results of the regional phytoplankton survey during July and November 1957 and March 1958. Bull. Int. Trop. Tuna comm., 7, 191-253.
Sverdrup, H.U., Johnson, M.W. and Flemming, R.H. 1942. The Oceans - their Physics, Chemistry and General Biology, Prentice-Hall, New Jersey, 1087 pp.

Takayama, T. 1972. A note on the distribution of *Braarudosphaera bigelowii* (Gran and Braarud) Deflandre in the Bottom Sediments of Sendai Bay, Japan. Trans. Proc. Palaeontol., Japan, N.S., 87, 429-435.

Tappan, H. 1980. The paleobiology of Plant Protists. Freeman and Company, San Francisco, 1028 pp.

Theirstein, H.R. 1976. Mesozoic calcareous nannoplankton biostratigraphy of marine sediments. Mar. Micropaleontol., 1, 352-362.

Theirstein, H.R., Geitzenauer, K.R., Molifino, B. and Shackleton, N.J. 1977. Global synchroneity of coccolith datum levels: validation by oxygen isotopes. Geology 5, 400-404.

Thronsen, J. 1972. Coccolithophorids from the Caribbean Sea. Norw. J. Bot., 19, 51-60.

Tomczack, G. and Goedecke, E. 1964. Die thermische Schichtung der Nordsee auf Grund des mitteren Jahrganges unter Temperatur in 1/2 aus 1 Felden. Ergan. Deutsche Hydrographische Zeitschrift, 8, 249-259.

Tyson, R.V., Wilson, R.V.L. and Downie, C. 1974. A stratified water column model for the Kimmeridge Clay. Nature, 277, 377-380.

Varol, O. 1985. Distribution of calcareous nannoplankton in surface sediments from intertidal and shallow marine regimes of a marginal sea: Jason Bay, South China Sea. Mar. Micropaleontol., 9, 369-374.

Wang, F. and Samtleben, C., 1983. Calcareous nannoplankton in surface sediments of the East China Sea. Mar. Micropaleontol., 8, 249-259.

Watabe, N. and Wilbur, K.M. 1966. Effects of temperature on growth, calcification and coccolith formation in *Coccolithys huxleyi*. Limnol. Oceanogr., 11, 567-577.

Wilbur, K.M. and Watabe, N. 1963. Experimental studies on calcification in molluscs and the algae *Coccolithus huxleyi*. Ann. N.Y. Acad. Sci., 109, 82-112.

Winter, A. 1982. Paleoenvironmental interpretation of Quaternary coccolith assemblages from the Gulf of Aqaba (Elat), Red Sea. Rev. Esp. Micropaleontol., 14, 291-314.

Winter, A., Reiss, Z. and Lutz, B. 1979. Distribution of living coccolithophore assemblages from the Gulf of Aqaba (Elat), Red Sea. Mar. Micropaleontol., 4, 197-223.

Woods, E.J.F. 1968. Studies of the phytoplankton ecology in tropical and subtropical environments in the Atlantic Ocean part 3: phytoplankton communities in the providence channels and the tongue of the ocean. Bull. Mar. Sci. Mar. Sci., 18, 481-543.

Wulff, A., 1925. Nannoplankton-Untersuchungen in der Nord-see. Wiss Meeresunters., Abt. Helgoland, w.f. 1-15.

Simon D. Houghton
OHS Ltd
Institute of Research and Development
University of Birmingham
Research Park, Vincent Drive
Birmingham, B15 2SQ

## Appendix 1

Percentage abundance and species-composition of coccolith assemblages in the surface sediments of the North Sea.

| Sample | E.hx | C.pl | Sy.p | H.ct | C.lp | G.cr | B.bg | Rh.c | Total Number counted |
|---|---|---|---|---|---|---|---|---|---|
| 6 | - | - | - | - | - | - | - | - | |
| 8 | - | 100 | - | - | - | - | - | - | 50 |
| 10 | - | - | - | - | - | - | - | - | - |
| 12 | 87.0 | 2.7 | 8.3 | - | - | - | 0.7 | 1.3 | 300 |
| 13 | - | - | - | - | - | - | - | - | 400 |
| 21 | 85.3 | 1.7 | 8.3 | 0.3 | - | 4.0 | - | 0.3 | 300 |
| 25 | 92.0 | 2.0 | 1.5 | - | 2.7 | 1.5 | 0.3 | - | 300 |
| 28 | 97.0 | 2.0 | 0.3 | - | - | 0.3 | - | - | - |
| 29 | 95.3 | 2.7 | 1.0 | 0.3 | 0.3 | 0.3 | - | - | 300 |
| 30 | P* | P* | - | - | - | - | - | - | 300 |
| 31 | 95.3 | 2.8 | 1.6 | - | 0.7 | 0.3 | - | 0.3 | 300 |
| 32 | 95.0 | 2.7 | 1.0 | - | 1.0 | 0.3 | - | 0.3 | 300 |
| 33 | 95.0 | 1.0 | 1.3 | 0.3 | 0.3 | 1.7 | - | 0.3 | 300 |
| 35 | 92.4 | 3.7 | 1.0 | 0.3 | 0.3 | 2.0 | - | 0.3 | 300 |
| 37 | 93.0 | 3.0 | 2.0 | 0.3 | 0.7 | 1.0 | - | - | 300 |
| 38 | 89.0 | 5.0 | 2.0 | 0.3 | 1.0 | 2.7 | - | - | 300 |
| 56 | - | - | - | - | - | - | - | - | - |
| 61 | 97.5 | 0.5 | - | - | 0.5 | 1.5 | - | - | 200 |
| 62 | 92.0 | 2.0 | 2.0 | - | 1.7 | 2.0 | 0.3 | - | 300 |
| 65 | - | - | - | - | - | - | - | - | - |
| 66 | - | - | - | - | - | - | - | - | - |
| 67 | 95.0 | 2.0 | 2.3 | - | - | - | 0.3 | 0.3 | 300 |
| 68 | - | - | - | - | - | - | - | - | - |
| 69 | - | - | - | - | - | - | - | - | - |
| 70 | - | - | - | - | - | - | - | - | - |
| 75 | 96.5 | 0.5 | 2.0 | - | 0.5 | - | - | 0.5 | 250 |
| 76 | 91.7 | 1.3 | 2.3 | - | 1.7 | 2.3 | 0.7 | - | 300 |
| 77 | 94.7 | 1.3 | 2.0 | - | 0.3 | 1.3 | 0.3 | - | 350 |
| 89 | 95.7 | 1.7 | 1.7 | 0.3 | 0.3 | 0.3 | - | - | 300 |
| 90 | 96.7 | 1.0 | 1.7 | - | 0.3 | 0.3 | - | - | 350 |
| 92 | - | - | - | - | - | - | - | - | - |
| 93 | - | - | - | - | - | - | - | - | - |
| 117 | - | 98.0 | - | - | - | - | 2.0 | - | 50 |
| 150 | P* | P* | - | - | - | - | - | - | - |
| 151 | P* | P* | - | - | - | - | - | - | - |
| 152 | P* | P* | - | - | - | - | - | - | - |
| 153 | P* | P* | - | - | - | - | - | - | - |
| 154 | - | - | - | - | - | - | - | - | - |
| 155 | - | - | - | - | - | - | - | - | - |
| 156 | - | - | - | - | - | - | - | - | - |
| 157 | - | | - | - | - | - | - | - | - |
| 159 | - | - | - | - | - | - | - | - | - |
| 160 | - | - | - | - | - | - | - | - | - |
| 200 | 91.7 | 6.2 | 0.1 | 0.1 | 1.0 | 0.4 | 0.1 | 0.1 | 730 |

E.hx = *E. huxleyi*, C.pl = *C. pelagicus*, Sy.p = *Syracosphaera pulchra*, *H.ct* = *Helicosphaera carteri*, C.lp = *Calcidiscus leptoporus*, *G.cr* = *G. caribbeanica*, B.bg = *B. bigelowii*, Rh.c = *Rhabdosphaera clavigera*. P* = species present but accurate percentage not determined

# 2 OSTRACODS AS PALAEOENVIRONMENTAL INDICATORS: EXAMPLES FROM THE TERTIARY AND EARLY CRETACEOUS

Michael C. Keen

## 2.1 Introduction

Ostracods have a long history of service to geology. Charles Lyell recognised their value for palaeoecological reconstructions as early as 1824 with his observations on a Scottish Quaternary lake deposit, while Edward Forbes in 1851 used them to subdivide the Upper Jurassic Purbeck Beds of Dorset in southern England with possibly the oldest zonal scheme based upon microfossils. The first of these two themes will be developed here, namely the application of Ostracod studies to the fields of environmental analysis. Ostracods are probably best known for this, where their distribution through a wide variety of aqueous habitats allows a unique contribution from micropalaeontology to the understanding of ancient environments.

## 2.2 The late Eocene of the Hampshire Basin, southern England

The late Eocene sediments of the Hampshire Basin (Figure 2.1), were deposited in a variety of coastal environments, ranging from flood plain lakes, rivers and swamps, through coastal lagoons and barrier islands, to a shallow embayment of the sea (Figure 2.2).

Figure 2.1: Map showing Tertiary outcrops and localities of the Hampshire Basin.

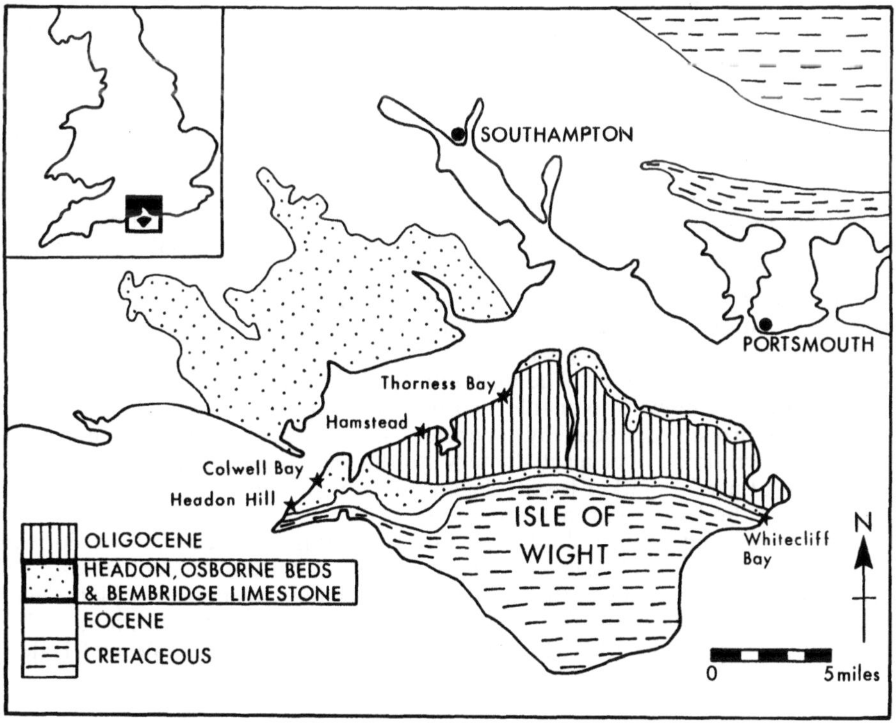




*D. G. Jenkins (ed.), Applied Micropaleontology*, 41–67.
© 1993 *Kluwer Academic Publishers.*

Figure 2.2: The palaeogeography of the Hampshire Basin during the time of maximum transgression of the Brockenhurst Beds (from Keen, 1977).

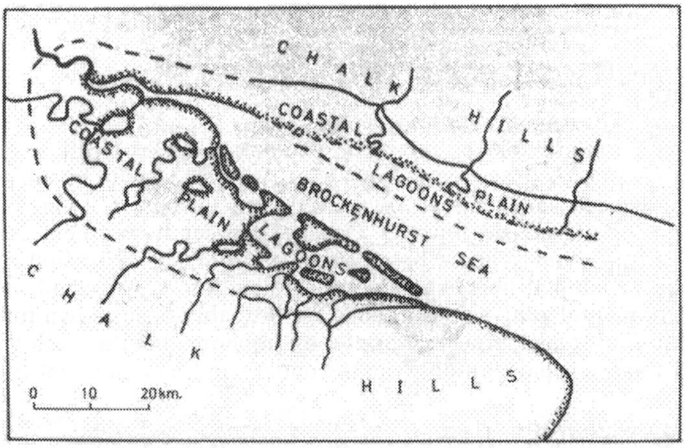


The succession is very complex, with rapid vertical and horizontal facies changes, and few horizons can be traced over more than a few miles. The problems then are firstly an understanding of the relationships between the different facies and the conditions in which they were deposited, secondly those of correlation within the sequence, and finally correlation with international zonal schemes. Complex sequences such as this require a thorough understanding of their lithostratigraphy and biostratigraphy before such concepts as chronostratigraphy or facies relationships can be applied. Conflicts related to the correlation of these sediments have been around since they were first studied by geologists in the early part of the nineteenth century. Webster published the earliest work in 1814, basing his classification and correlations almost entirely on environmental criteria, so all horizons with polyhaline-marine faunas were assumed to be of the same age, and similarly all the thick freshwater limestones were regarded as occurring at the same horizon. The lithostratigraphy, and to a large extent the biostratigraphy, were clarified by Edward Forbes in 1853 and 1856, essentially using the principle of superposition. Forbes' detailed work in measuring and correlating sections exposed around the coasts of the Isle of Wight is a classic, and remains virtually unchanged at the present day. However the conflict between lithostratigraphical correlation and ecostratigraphical correlation surfaced in the latter part of the nineteenth century with the publications of Judd in 1880 and 1882. The discussions printed in the Quarterly Journal of the Geological Society of London following Judd's papers make fascinating reading today. The problem then, as now, was the lack of a good biostratigraphical framework to enable a chronostratigraphy to be developed. Oddly enough, the approach adopted by Webster, i.e. ecostratigraphy, assumes an importance in correlation once a satisfactory stratigraphy has been developed. Thus, in the absence of useful marker fossils, basin-wide events such as transgressions or regressions give the key to chronostratigraphy.

The lithostratigraphy devised by Forbes and slightly modified by the Geological Survey (Bristow et al., 1889) survived more or less intact until recent times. The modern tendency for revising stratigraphical nomenclature has created a plethora of new lithostratigraphical units for these late Eocene sediments (Stinton, 1975; Cooper,1976; Curry et al., 1978; Melville and Freshney, 1982; Insole and Daley,

1985; Edwards and Freshney 1987a,b.); the scheme adopted here is a mixture of the Survey's nomenclature, incorporating some of the names proposed by Insole and Daley where this clarifies the situation. The strata dealt with are the lower part of the Solent Group, namely the Headon and Osborne Beds, the Bembridge Limestone, and the lower part of the Bembridge Marls (Figure 2.3).

Figure 2.3: Stratigraphical units of the late Eocene sediments of the Isle of Wight.

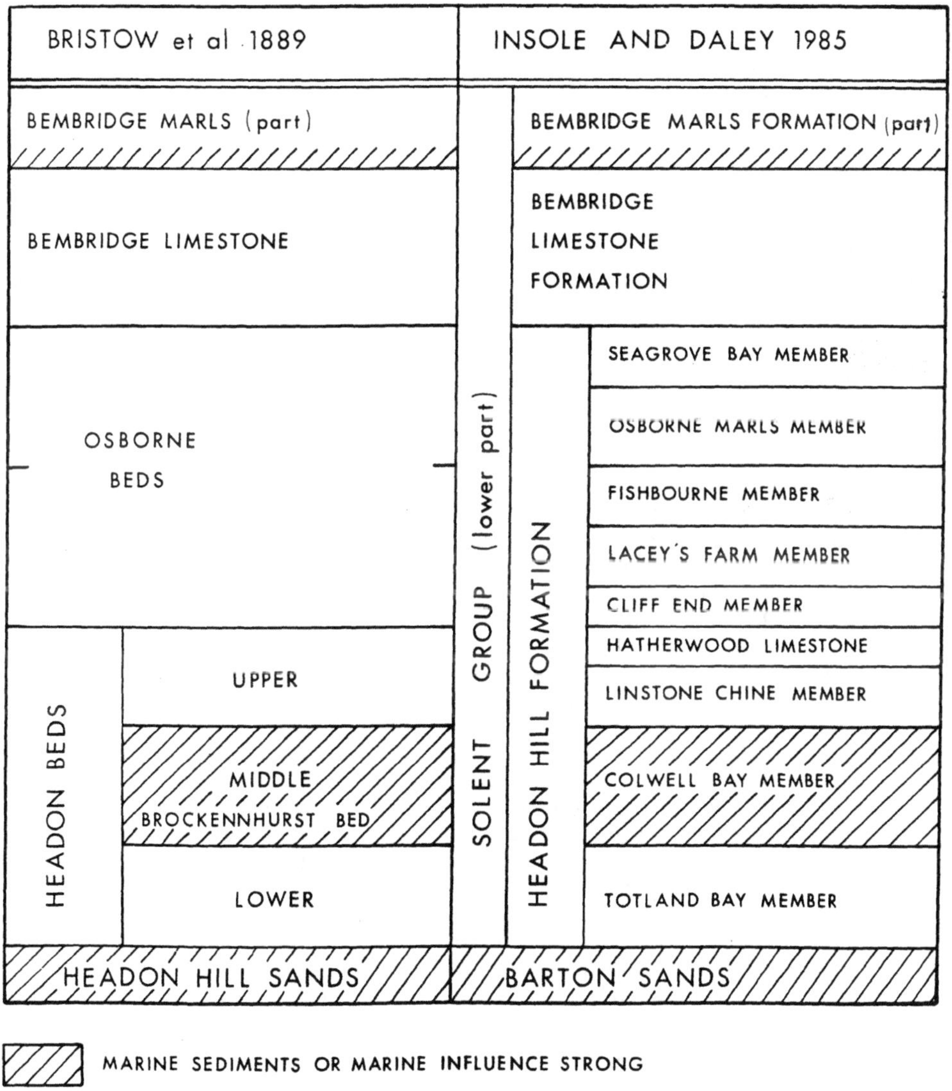

2.2.1 Ostracod assemblages

Ostracods occur abundantly at many horizons throughout the succession. Many belong to extant genera whose ecology and mode of life are reasonably well understood, so allowing direct palaeoecological comparisons to be made. However, as with all palaeoecological studies, it is important to understand whether, and how much, movement of shells has occurred. Ostracods are bivalved crustaceans, which, like all crustaceans, grow by moulting and often exhibit sexual dimorphism. This allows an examination of the population structure of the fossils to determine whether or not they represent autochthonous or allochthonous elements (Figure 2.9). An autochthonous species in a sample will tend to have carapaces as well as single valves, and the latter should be represented by equal numbers of each valve; several different moult stages as well as adults should be present, and if the species shows sexual dimorphism both sexes should occur. Autochthonous assemblages are also likely to be represented if the species or genera present are known to occur together in Recent environments. As with all fossils, the state of preservation may also yield useful information. A second problem could arise due to environmental condensation, when very low rates of sedimentation accompanied by rapid changes in the environment would lead to the mixing of different biotopes. Marginal coastal environments are areas of rapid facies changes so this could be a problem. The principal method of recognising this is ecological incompatibility of taxa, but population studies and careful closely spaced sampling also help.

Cluster analysis of the ostracods (Keen, 1975, 1977) has allowed the recognition of six distinct assemblages (Figures 2.4, 2.5) which can be compared with modern analogues, especially with Recent faunas of the Gulf Coast of the U.S.A. This has led to an understanding of the environments in which the ostracods are likely to have lived, and in turn to a better understanding of the conditions of deposition of the succession (Figure 2.6).

Two distinct freshwater assemblages are present (Figure 2.7). The first (Assemblage I) is the more diverse and is believed to have characterised lake margins or shallow lakes with depths of no more than one metre or so. This can be interpreted as the upper infralittoral zone, or zone of emergent waterplants. The second assemblage (Assemblage II) has a much lower diversity, frequently being monospecific with only *Moenocypris* species, although ostracods can be very abundant and be seen to cover whole bedding planes. This second assemblage is taken to represent the lower infralittoral zone, with water depths of 2-15m. The first of these assemblages is found in grey-green clays or buff coloured limestones, with *Chara* and the freshwater gastropods *Galba* and *Planorbina*; the second is from grey clays with the freshwater gastropod *Viviparus* and the freshwater bivalve *Unio*.

Three brackish water Assemblages can be recognised, ranging from a low salinity assemblage, through mesohaline, to polyhaline. Assemblage III is characterised by the ostracod *Cytheromorpha*; it has the lowest diversity of the brackish water assemblages, and contains common freshwater elements which are believed to be allocthonous (Keen 1977). These facts, together with the known range of living *Cytheromorpha,* suggest salinities of less than 9‰.

Assemblage IV is characterised by *Neocyprideis,* which is an extinct genus, but is perhaps ancestral to the living *Cyprideis.* The latter is the most typical brackish water ostracod living today, and its ecology has been intensively studied (see Keen 1982 for a reference list). It is found in freshwater, is very abundant in mesohaline and hypersaline salinities, but rarer in polyhaline and euhaline waters. It is an inhabitant of quiet waters such as lagoons, creeks bordering estuaries, and brackish ponds close to the sea. Assemblage IV is regarded as a mesohaline assemblage (9-16.5‰).

Figure 2.4: Ostracod assemblages from the Headon-Bembridge Beds, with suggested salinities (From Keen, 1977). Note that where the species is labelled "sp. nov", this was published in Keen, 1977.

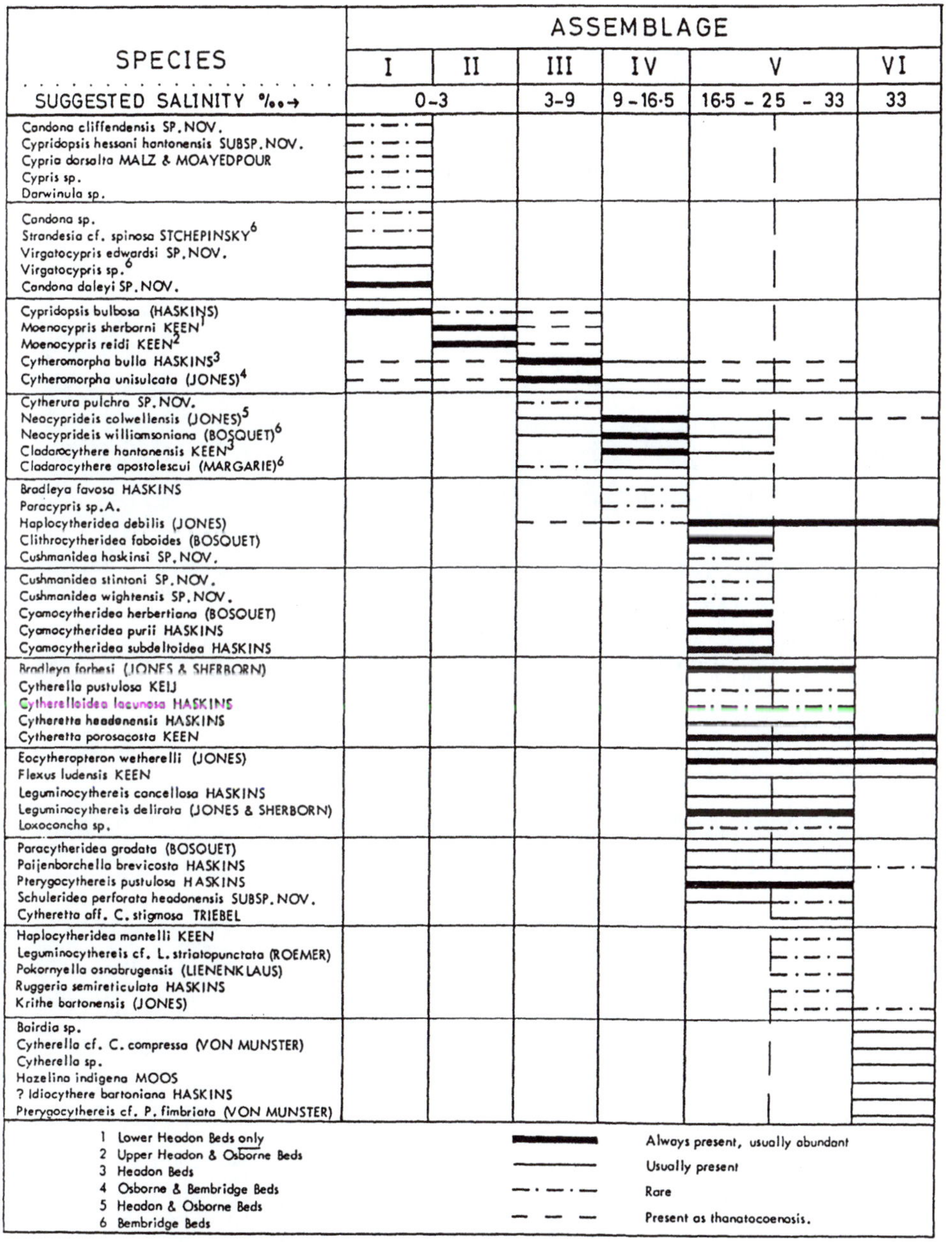

Figure 2.5: Characteristic ostracods from the Headon Beds, Isle of Wight. No.7 and 11 are from Colwell Bay, 13 and 14 from Whitecliff Bay, the remainder from Headon Hill; measurements are of the length in mm. **Fig.1**, *Candona daleyi* Keen 1977, 0.84, x60; **Fig.2**, *Virgatocypris edwardsi* Keen 1977, 1.16, x43; **Fig.3** *Cypridopsis bulbosa* (Haskins 1968), 0.53, x94; **Fig.4** *Moenocypris reidi* Keen 1977, 1.29, x30; **Fig.5**, *Cytheromorpha bulla* Haskins 1971; **Fig.6**, *Neocyprideis colwellensis* (Jones 1857), 0.81, x60; **Fig.7**, *Haplocytheridea debilis* (Jones 1857), 0.60, x83; **Fig.8**, *Leguminocythereis delirata* (Jones and Sherborn 1889),0.82, x61; **Fig.9**, *Pterygocythereis pustulosa* Haskins 1968, 0.78, x64; **Fig.10**, *Cytheretta porosacosta* Keen 1972, 0.75, x70; **Fig.11**, *Eocytheropteron wetherelli* (Jones 1854), 0.72, x70; **Fig.12**, *Forbescythere forbesi* (Jones and Sherborn 1889), 0.75, x67; **Fig.13**, *Hazelina indigena* Moos 1966, 0.54, x93; **Fig.14**, *Idiocythere bartoniana* Haskins 1971, 0.52, x96.

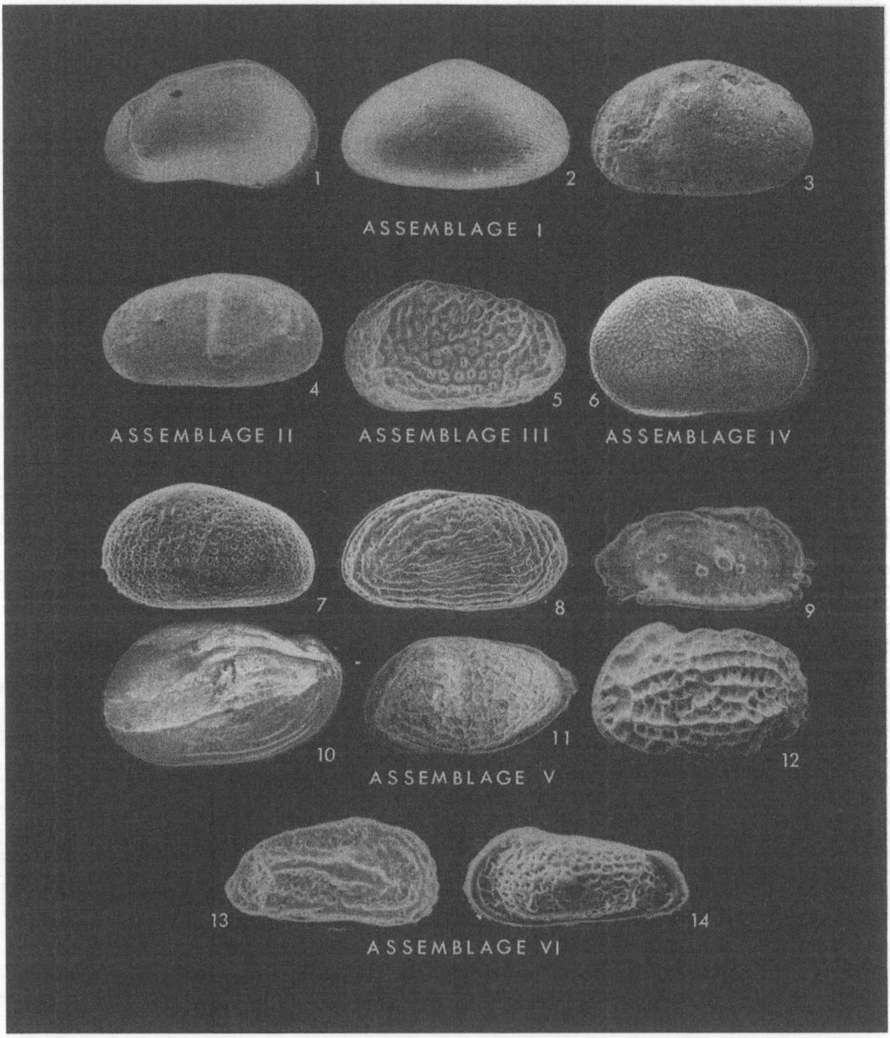

Figure 2.6: Suggested environments of the ostracod assemblages (from Keen, 1977).

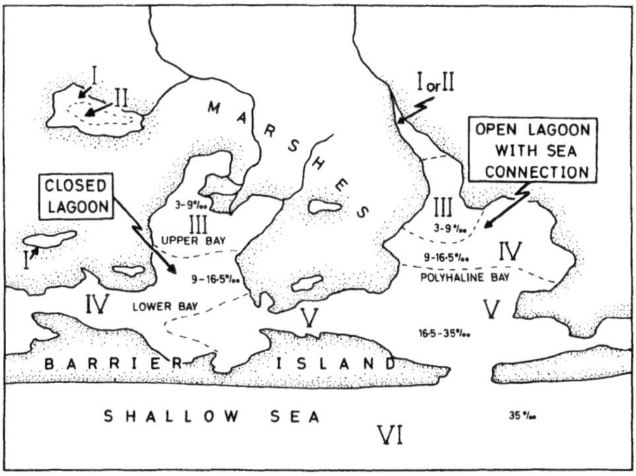


Figure 2.7: The freshwater ostracod assemblages (after Keen, 1975).

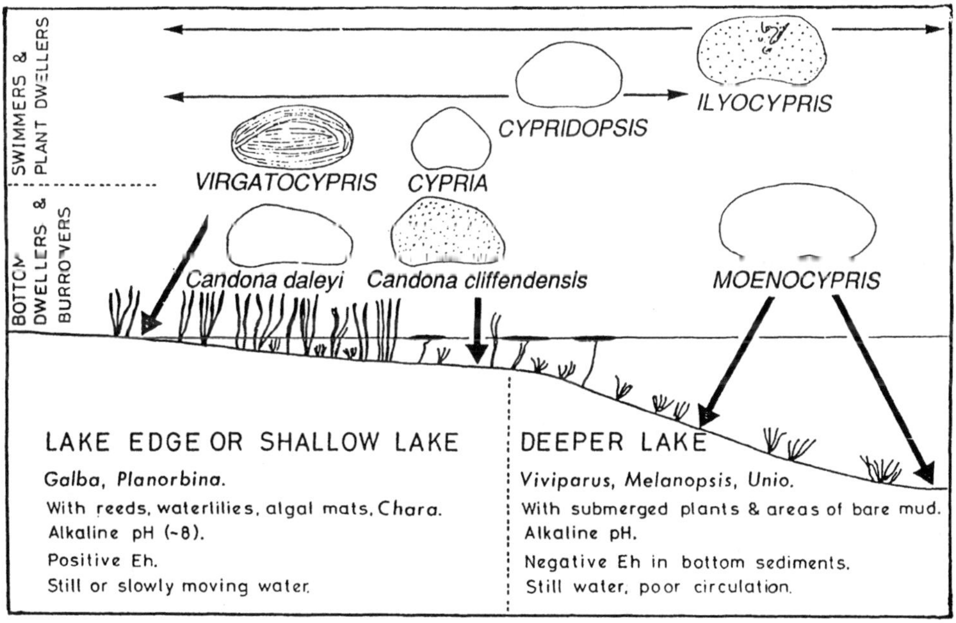


The palaeoecology of *Neocyprideis* has been considered in detail by Keen (1990a). The third of the brackish water assemblages (Assemblage V) is characterised by *Haplocytheridea,* together with a fairly diverse fauna of ostracods found living today in nearshore areas, bays, lagoons, and estuaries. This assemblage is interpreted as an open lagoon or bay fauna, living in shallow polyhaline areas with direct connection with the open sea, while Assemblages III and IV inhabited closed lagoons away from direct connections with the sea (Figure 2.6). The brackish water assemblages occur in sandy clays with a rich fauna of gastropods and bivalves.

Finally, Assemblage VI occurs in the Brockenhurst Bed of Whitecliff Bay, where the presence of genera such as *Bairdia, Hazelina,* and *Idiocythere* suggest truly marine conditions. The fauna is not as rich as most Palaeogene marine faunas so it is interpreted as near-shore, perhaps even littoral, but of normal marine salinity. The impoverishment of the fauna is probably related to the geography of the times (Figure 2.2), where the fauna in the marine embayment is more restricted than that of the open sea.

### 2.2.2 Environmental conditions

The environmental interpretation is shown in Figure 2.6, and this interpretation is applied to the vertical succession in Figure 2.8. The Barton Sands are considered to be deposits of a barrier complex; sedimentological studies (Plint 1984) support the barrier island model.

Figure 2.8: Environmental interpretation showing the lateral migration of the barrier complex. The shoreward side of the barrier is shown by the heavy black line. To the west and north of this lay the coastal floodplain with lagoons, rivers, and lakes; the open sea lay to the east. The transgressive phases are indicated as being very rapid, the regressive phases as being very diachronous (this is probably over exaggerated). The time scale is taken from Haq et al., 1987.

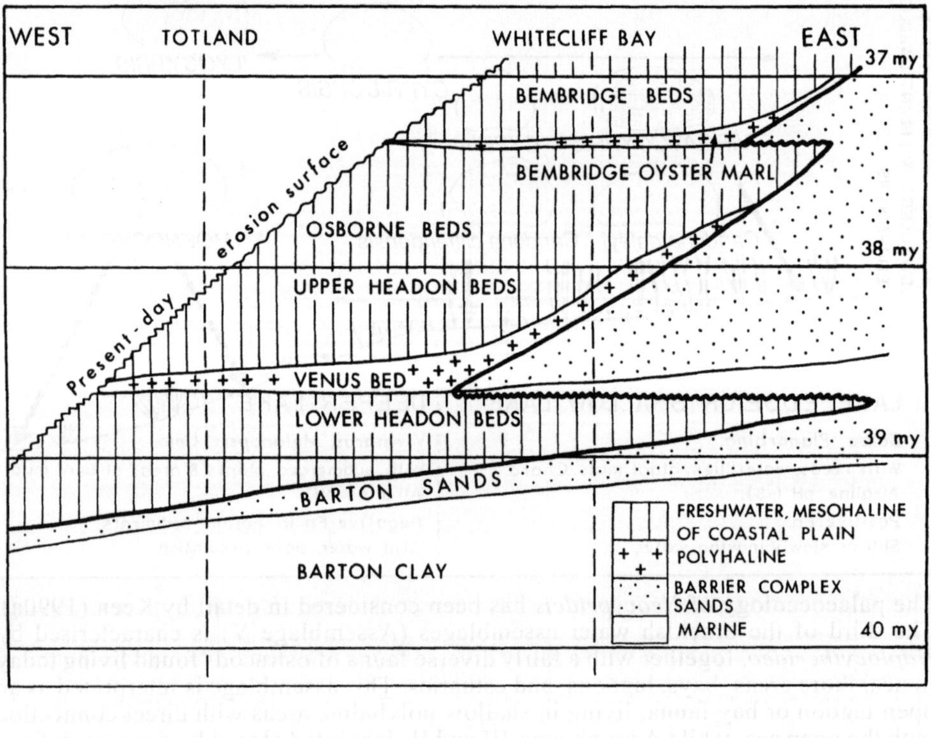

The Lower Headon Beds are believed to have been deposited on a floodplain with rivers, lakes, and lagoons lying behind, i.e. landward of, the coastal barrier. Thus during this time- interval the whole of the Hampshire Basin lay behind the barrier. The rapid marine transgression indicated by the Brockenhurst Bed saw all of these facies move northwestwards. At the height of the transgression Whitecliff Bay and large areas of the New Forest lay seaward of the barrier (= Assemblage VI of the Brockenhurst Bed), while behind the barrier polyhaline lagoons of the Venus Bed facies occurred at Headon Hill and in the western part of the Isle of Wight (Figures 2.6, 2.8). As the sea regressed the barrier complex migrated southeastwards and is represented by the Psammobia Beds (= the "Barren Sands") of Whitecliff Bay. Further regression saw the development of polyhaline lagoons (Venus Bed facies) at Whitecliff Bay, with lower salinity lagoons in the west, which now lay further from direct connections with the sea. The transgression is envisaged as being very rapid, with constant reworking and erosion of sediments, so there is no preservation of the barrier complex, but coastal offlap allowed this facies to be preserved during the subsequent regression. This situation is not unusual; transgressions are frequently represented by an erosive level, while the sediments are deposited during the regressive phase. It follows from this account that the various facies described are diachronous, but can be related to particular events.

The recognition of these assemblages enables more detailed interpretations to be made using the ostracods. Each sample can be classified according to the percentages of freshwater ostracods, *Cytheromorpha, Neocyprideis,* and *Haplocytheridea.* Because *Haplocytheridea* and freshwater ostracods rarely occur together, the samples can be regarded in terms of one, two, or three components and plotted on triangular diagrams (Figure 2.9). These diagrams show the continuous nature of the gradation between the assemblages, indicating considerable overlap between living populations of *Haplocytheridea* and *Neocyprideis, Neocyprideis* and *Cytheromorpha,* and *Cytheromorpha* and freshwater ostracods. If the interpretation of autochthonous / allocthonous specimens indicated on Figure 2.9 is correct, freshwater valves were frequently carried into lagoonal areas, and lower salinity species more easily transported into higher salinity areas than vice versa. This presumably reflects the more dominant role of fluvially derived transportation over tidal transportation in these sheltered bays and lagoons. It is now possible to assign a salinity value to each sample using the various methods and interpretations described in the previous sections, and to construct salinity profiles for the sections studied (Figure 2.10).

### 2.2.3 Salinity profiles, local and international correlation

The salinity profiles are the result of three processes. Firstly there is the effect of localised coastal changes, perhaps caused by storms and hurricanes allowing breaching of the coastal barrier, or by rather longer term processes such as long-shore drift causing the migration of openings in the barrier ridge, or river transported sediment silting up channels and lagoons. Secondly, the relative rates of sedimentation and subsidence will have an effect on overall onlap or offlap. Thirdly, large scale global transgressive - regressive events will have a marked influence on this marginal coastal area. How can the influence of these three factors be separated?

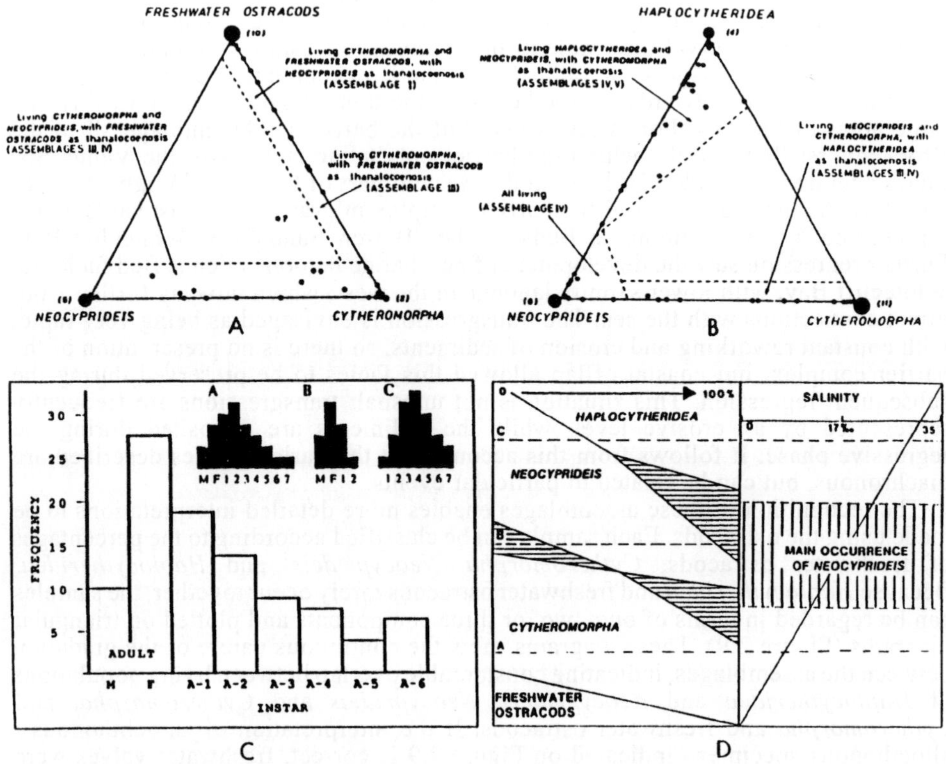


Figure 2.9: Autochtonous and allocthonous ostracod faunas.

**A and B**: Triangular diagrams showing the distribution of samples from the Cyrena pulchra Bed belonging to Assemblages I-V, and illustrating autochthonous (= living) and allocthonous (=thanatocoenosis) components (from Keen, 1977).

**C**: Frequency distribution of adults and instars of *Neocyprideis* from the Cyrena pulchra Bed of Headon Hill. This is a typical structure of an autochtonous species, with adults as well as abundant juvenile instars. The insert diagrams A, B, C represent ideal population structures for A) a low energy biocoenosis, B) a high energy biocoenosis and C) a low energy thanatocoenosis (after Whatley, 1983). These show the effects of postmortem transportation; however further taphonomic processes may distort this ideal structure, often resulting in the elimination of pre-adult instars. Therefore, while it may be possible to demonstrate that a species is autocthonous, the lack of instars does not necessarily indicate an allocthonous species.

**D**: The distribution of salinity related genera from the late Eocene of the Hampshire Basin. The left hand side indicates faunal composition, the right hand side salinity interpretation. Areas with horizontal shading indicate overlap in composition of faunas. Three examples are given: A) 100% *Cytheromorpha*, salinity about 7‰, i.e., low mesohaline; B) mostly *Neocyprideis*, but sometimes with *Cytheromorpha* and/or *Haplocytheridea*, salinity about 18‰, i.e. mesohaline; C) about 70% *Haplocytheridea* and 30% *Neocyprideis*, salinity about 28‰, i.e. polyhaline (after Keen, 1990).

Figure 2.10: Salinity profiles from sections across the Isle of Wight (see Figure 2.1 for locations).

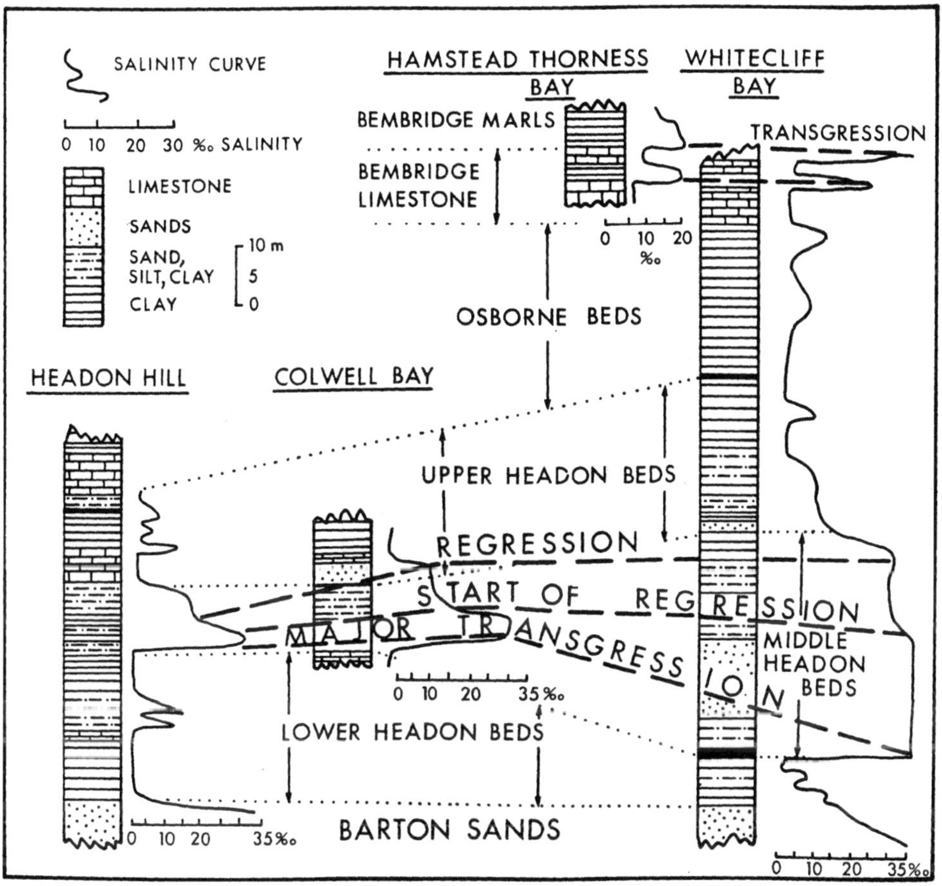


Two examples, where localised coastal changes are believed to be dominant, and which illustrate how susceptible the area was to minor coastal changes will be given. The first is from the Cyrena pulchra Bed in the Lower Headon Beds of Headon Hill, the second from the Upper Headon Beds of the western Isle of Wight. The Cyrena pulchra Bed lies immediately above the lowest of the freshwater limestones at Headon Hill; the ostracods found in the limestone belong to Assemblage I and are clearly autochthonous. The basal green clay of the Cyrena pulchra Bed yields abundant *Neocyprideis* and common *Cytheromorpha* (Figure 2.11), but ascending the succession *Neocyprideis* declines in numbers while *Cytheromorpha* becomes dominant.

Figure 2.11: Section through the Cyrena pulchra Bed, Headon Hill, showing composition of the ostracod fauna and the derived salinity profile (after Keen, 1977).

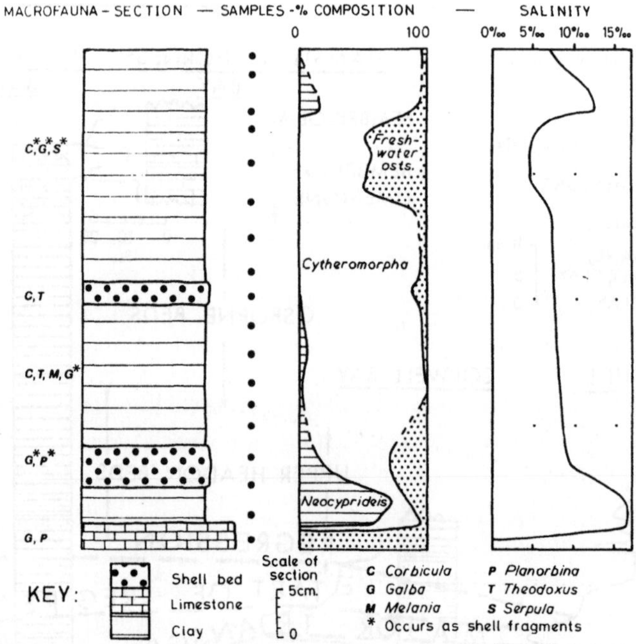


Towards the top of the bed freshwater ostracods become common, but as they are mostly represented by instars and broken valves they are regarded as fluvially transported allocthonous material. The section can be interpreted as a shallow freshwater lake situated near the shoreline of a flat low-lying coastal plain; the sea breached a barrier some distance away allowing the flooding of the lake by saline waters. This rapid rise in salinity as the lake became converted into a lagoon led to the death of the freshwater animals and the introduction of a brackish fauna. The sea connection was short lived, the salinity became reduced, more rapidly at first, but then more gradually, and the lagoon began to silt up. Towards the top of the bed there is an indication of strong fluvial influence followed by a further very short-lived breaching by the sea. The Cyrena pulchra Bed has a very limited geographical distribution which reinforces the interpretation of local coastal change.

The Hatherwood Limestone is the most prominent unit of the Upper Headon Beds at Headon Hill. It consists of 9m of fossiliferous limestone with freshwater gastropods, *Chara,* and vertebrate remains. Pisolites, sinter beds, lignites, conglomerate horizons, and marls occur within the sequence. The ostracods mostly belong to Assemblage I, but one horizon near the top of the limestone contains Assemblage IV. Murray and Wright (1974) recorded very tiny foraminifera from the limestone and interpreted the succession as the deposit of a lagoon of near normal marine salinity. However, the evidence from the molluscs, ostracods, vertebrates and charophytes points overwhelmingly to the predominance of freshwater conditions. The foraminifera do suggest the presence of lagoonal conditions nearby; they may have been windblown into the area. Lagoonal conditions are also indicated by the occurrence of Assemblage IV ostracods within the sequence.

Figure 2.12: The outcrop of the Hatherwood Limestone and Linstone Chine Member around Totland, with a reconstruction of the geography at the time of deposition of the two members.

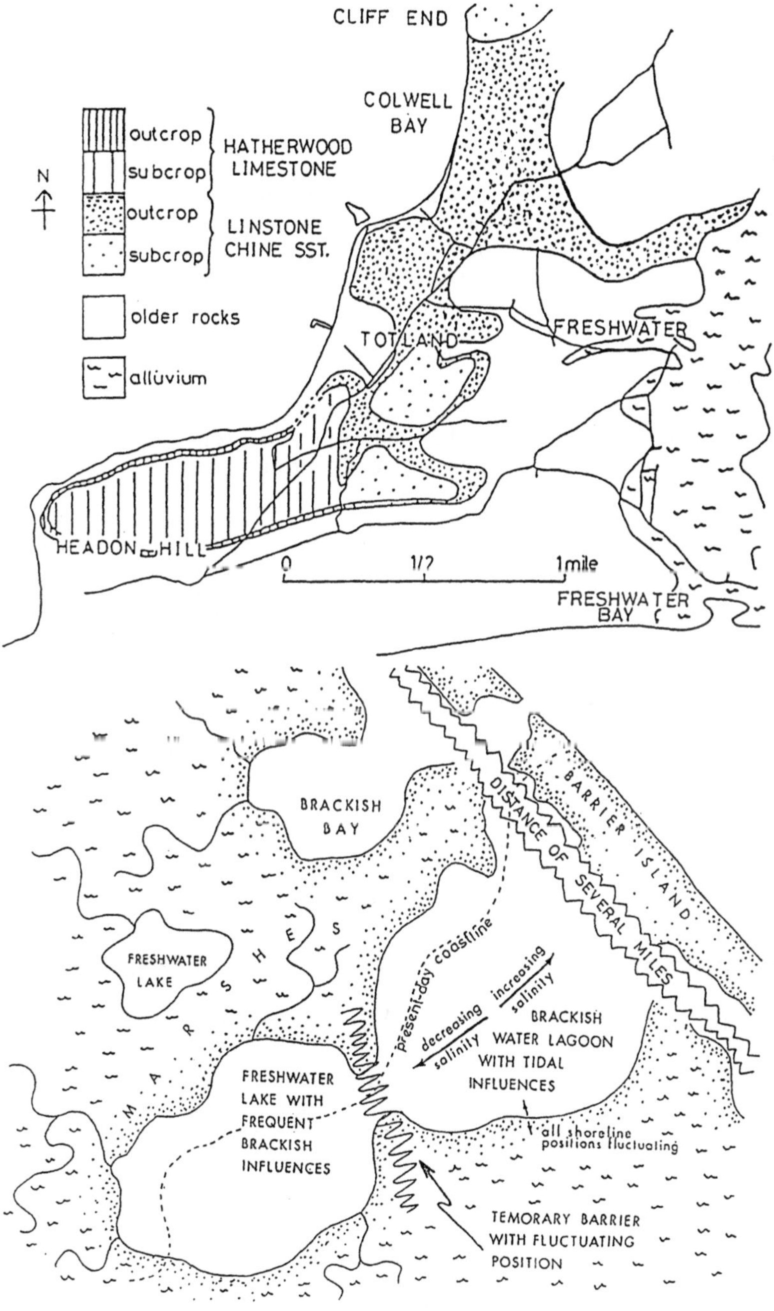

The Hatherwood Limestone is only developed at Headon Hill; it disappears abruptly northwards within a distance of about 100m, and in Totland Bay and Colwell Bay (Figure 2.12) its place appears to be taken by a series of white and chocolate-coloured sands showing ripple drift bedding, mud drapes, and rare seams of brackish water molluscs with ostracods of Assemblages III and IV. These sands have been named the Linstone Chine Member by Insole and Daley (1985), who regard them as a unit underlying the Hatherwood Limestone. The sands are poorly developed at the south of Headon Hill where they are less than 40cm thick; they thicken northwards to 240cm at the northern end of Headon Hill, and rapidly thicken into Totland Bay where they are over 400cm thick. Two interpretations are possible; the sands and the limestones may have been deposited at the same time and are therefore lateral equivalents, or the present stratigraphical order is the result of a combination of erosion and differing rates of sedimentation or non-deposition. Incidentally, a third possibility was suggested by Judd in 1880; he was so impressed by the sudden disappearance of the limestone that he believed it must be due to a fault downthrown to the north. Of the two modern hypotheses, the one of facies change is preferred. It is envisaged that an enclosed lagoon existed to the north and east, with an area in the southwest where lacustrine conditions predominated cut off from this lagoon (Figure 2.12). The occurrence of some horizons within the limestone containing brackish water ostracods strongly supports the presence of nearby lagoonal environments.

Figure 2.13: Correlation of the late Eocene-early Oligocene eustatic curve for the Hampshire Basin with the international sequences of Haq et al., 1987. The transgressive events dated at 38.8 m.y.and 37.5 m.y. are derived from the salinity curves seen in Figure 2.10 and correlate with the Middle Headon Beds and Bembridge oyster Marl transgressions (see also Figure 2.8); later events are derived from younger sequences in the Isle of Wight (from Keen, 1990).

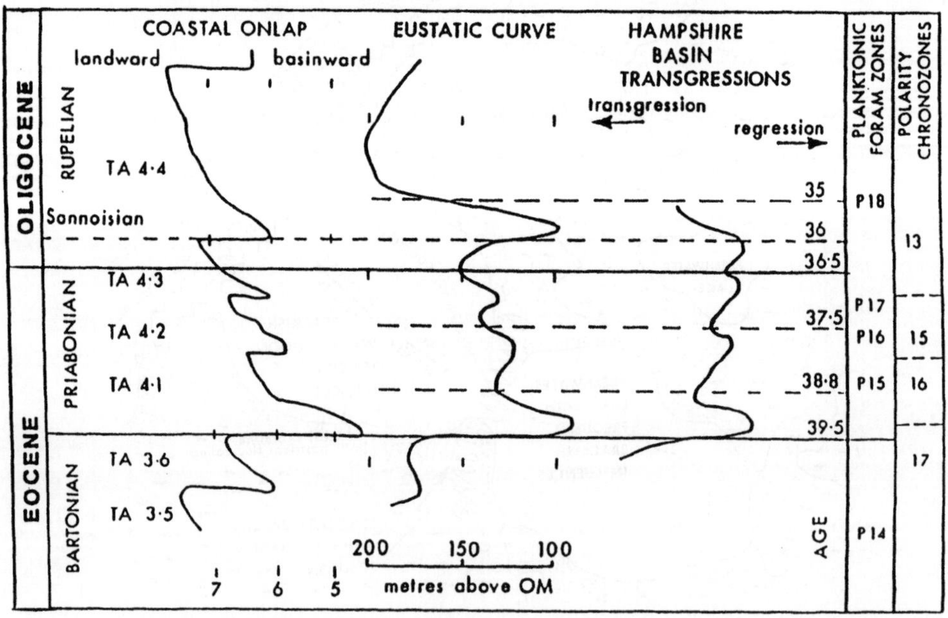

The relationship between sedimentation and subsidence on the one hand, and global sea-level fluctuations on the other, is more difficult to resolve. The persistence of a marginal coastal environment for some 4 - 5 million years suggests that sedimentation and subsidence were more or less in balance. The salinity curves constructed for the Isle of Wight (Figure 2.10) can be converted into a eustatic curve (Figure 2.13) and compared with the global curves given by Haq et al. (1987). Two tie points are recognised (Plint, 1988; Keen, 1990b) with the major regression at 39.5 m.y. and the major transgression at 35.5 m.y. This enables all three intervening transgressive events in the Isle of Wight sequence to be correlated with those of the Haq et al. curve (Figure 2.13). In Figure 2.13 the sequence described corresponds to the lower part of the Priabonian, Cycles TA 4.1 and TA 4.2. In order to complete the picture in the Hampshire Basin the succeeding Bembridge and Hamstead Beds have been included, using similar techniques on the ostracod faunas as described for the lower beds. Two further transgressive events can be recognised (Keen 1990b) at 37.5 m.y. (the Nematura Bed and White Band of the Lower Hamstead Beds) and at 35.5 m.y. (The Upper Hamstead Beds). The curves also show the Priabonian regressive episode sandwiched between two periods of higher sea levels, which is very apparent in the charts of Haq et al.. This correspondence suggests that the transgressive events of the Isle of Wight do in fact represent global events. It is also clear that the Priabonian transgressions of the Isle of Wight became weaker through time, unlike those of the Haq et al. curve. This could be taken to indicate that over the long term, subsidence was not quite keeping up with sedimentation.

This use of global coastal onlap sequences is the key to wider geographical correlation for sequences such as these in the Hampshire Basin. Cycles TA 3.5 - TA 4.4 can be directly correlated with the sequences of the Gulf Coast of the U.S.A., and the ostracod zones of Hazel (1990) can be correlated with those of Europe. It also helps in recognising the positions of important geological boundaries. The Eocene-Oligocene boundary has always been known to occur somewhere within the Headon-Hamstead Beds succession; it has been placed at the base of the Middle Headon Beds, the base of the Lower Hamstead Beds, or the base of the Upper Hamstead Beds. In recent years most workers have settled on the Hamstead Beds as basal Oligocene (Keen 1972, 1990b; Curry et al. 1978). Recently (1989) the International Subcommision on Palaeogene Stratigraphy voted for a boundary stratotype in Italy, at a level coinciding with the extinction of the important Eocene planktonic foraminifer *Hantkenina.* This foraminifer cannot be used in marginal marine areas such as the Hampshire Basin. However, its extinction coincides with the transgressive event at 36.5 m.y. This being so, it is now fairly clear that the base of the Oligocene in England should be drawn at the Nematura Bed, which occurs within the lower part of the Lower Hamstead Beds.

The cycles described here are the third order cycles of Haq et al. (1987), orignially defined as having durations of 1-10 million years. They are amongst the commonest cycles observed in seismic data sets, and these particular cycles can probably be correlated with parasequences. However, the various components of sequence stratigraphy are difficult to recognise in proximal, essentially non-marine successions such as these. The transgressive events are the easiest to recognise and can be assumed to coincide with the maximum flooding surfaces of more distal areas. Ostracods, as well as other microfossils, have an important role to play in the sequence stratigraphy apart from the accurate dating of the rocks (without which much else will fail).

As was stated at the beginning, no group of fossils has provided a satisfactory biozonation of these Late Eocene sediments. The transgressive events can now be seen to provide a means of correlation within the Hampshire Basin as well as with the international time scale. If the transgressive events are taken to represent time planes, the relationship between ostracod biozones and chronozones can be demonstrated (Figure 2.14). In this figure a diagrammatic section is drawn from the Barton Clay to the Bembridge Marls in terms of the facies present. The ostracod zones have been defined in Keen (1978). The *Cytheretta laticosta* Zone is defined as the total range of the nominate species; the base of the *Haplocytheridea debilis* Zone is defined by the last appearance of *C. laticosta,* and its top by the disappearance of *H. debilis.* The Barton Clay at the base of the succession is a marine deposit with a rich fauna, including abundant *C. laticosta.* This is succeeded by the barrier complex sands of the Barton Sands. On the Isle of Wight the only fossils found in these sands are burrows of the trace fossil *Ophiomorpha,* which is usually taken to indicate marine conditions.

The top of the *C. laticosta* Biozone coincides with the boundary between the Barton Clay and Barton Sands. Dinoflagellate studies suggest that deposition of the Barton Sands commenced earlier in the west than in the east (Bujak et al. 1980), and in this account the barrier sands are assumed to have migrated eastwards through time. The sands are also thicker in the east even though they represent a shorter time span; this is probably due to penecontemporaneous erosion of the more landward western succession. Freshwater and brackish water sediments occur above the Barton Sands yielding ostracod Assemblages I - IV as already described. A marine incursion is witnessed by the Middle Headon Beds and the Brockenhurst Bed, which, as already discussed, was probably very rapid and so can be taken to represent a time plane. The descendant of *C. laticosta, C. porosacosta* Keen, is now present, accompanied by *H. debilis.* As the *C. laticosta* Zone is defined as the range zone of the species, where should its top be placed? The strict application of the biozone concept would place it at its last occurrence in the two sections; its equivalent chronozone would have its boundary within the period of deposition of either the Barton Sands or the Lower Headon Beds. The top of the *H. debilis* Zone is similarly confused; in the east, its last occurrence is in the Bembridge oyster Marl at the base of the Bembridge Marls, but in the west its last occurrence was much earlier, in the Middle Headon Beds. Once again the top of the biozone is markedly diachronous with respect to its equivalent chronozone. In the east the whole succession between the Middle Headon Beds and the Bembridge oyster Marl can be included in the *H. debilis* Biozone because they lie within the outer limits of occurrence of the species; in the west only the Middle Headon Beds can be placed in the biozone.

### 2.2.4 Other examples

The conditions of deposition of these well exposed sequences in the Hampshire Basin have been elucidated by a study of their ostracod faunas. Marginal marine deposits such as these are common in the geological record, and because they frequently contain large sand bodies they are of great interest to petroleum geologists. Tertiary examples abound in the Gulf Coast area of the U.S.A., and ostracods have been used in their interpretation. Little of this has been published, but Howe (1971) has given some information. In particular, the presence of genera such as *Haplocytheridea, Hemicyprideis,* and other related genera have been taken as excellent indicators for the presence of marginal marine environments.

Figure 2.14: Chronozones and biozones in the late Eocene of the Isle of Wight (after Keen, 1983).

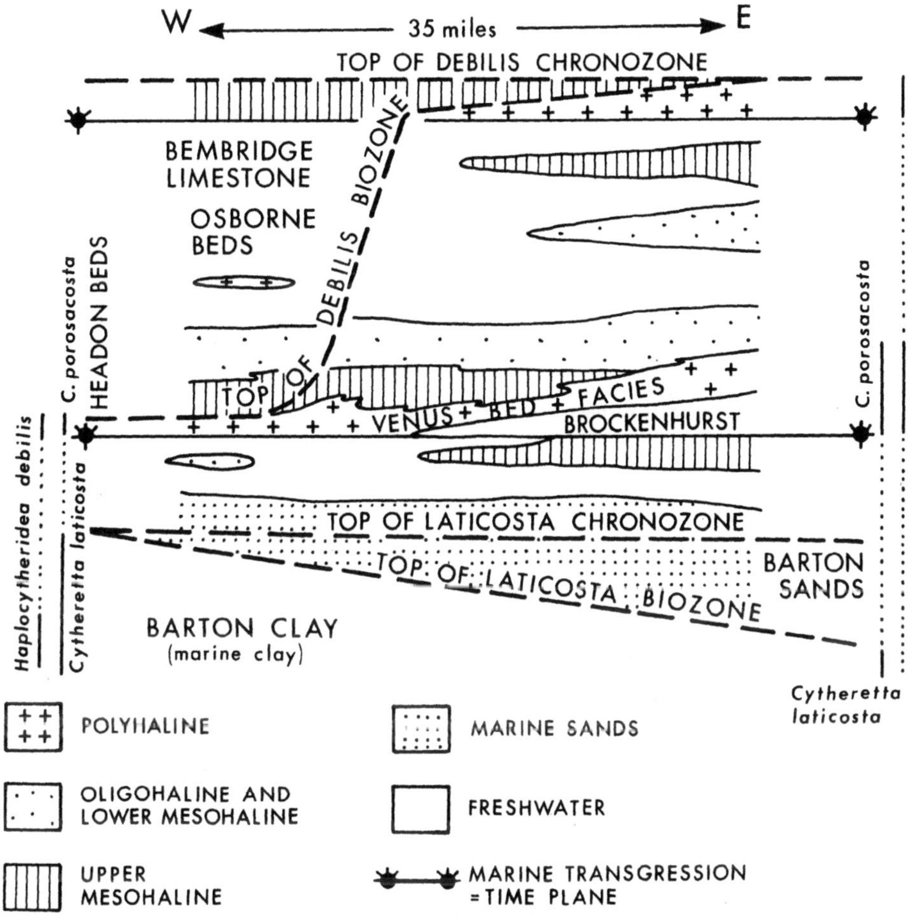


Howe emphasised the use of ecophenotypic tubercles (Figure 2.15) that may be present on the carapace of certain cytherid ostracods. These tubercles generally develop in low salinity waters and so their presence can be used to recognise brackish water environments. Because they will not be present on the same species occurring in a higher salinity environment, they can be utilised to recognise environmental gradients. Living species have been studied to examine this phenomenon, and there is now an extensive literature on the subject (see Keen, 1982 for references). Because ostracods do not store $CaCO_3$ in their body prior to moulting, they have to extract it from the water during the 2 - 3 days they take to form a new calcified carapace after moulting. This means they are very sensitive to the water chemistry, and in fluctuating environmental conditions in marginal marine areas, ecophenotypic features are frequently developed within a species (see Figure 2.15).

Figure 2.15: Examples of ecophenotypic characters developed in Palaeogene ostracods. **Figs. 1 and 4**, *Haplocytheridea debilis*, Middle Headon Beds, Colwell Bay, both x60; **Fig. 1** is a male right valve, **Fig. 4** a female right valve which has developed ecophenotypic tubercles; the development of this feature is more common on right valves than on left valves, and is also more common on the female than on the male; sexual dimorphism is common in ostracods, with the male usually more elongate than the female. **Figs. 2 and 5**, *Hemicyprideis montosa* (Jones and Sherborn 1889), both x70; **Fig. 3**, female left valve, Upper Hamstead Beds, Bouldnor Cliff, Isle of Wight,; **Fig. 4**, female right valve, Lower Hamstead Beds, Bouldnor Cliff, Isle of Wight, showing the development of tubercles; the differences between left and right valves can be seen in these two figures. **Figs. 3 and 6**, *Loxochoncha nystiana* (Bosquet 1852), from the early Oligocene; **Fig. 3**, female left valve, Sables de Berg Bilzen, Belgium, x70; **Fig. 6**, female right valve, Upper Hamstead Beds, Bouldnor Cliff, Isle of Wight, x65. The reticulate form characterises polyhaline-euhaline salinities, the smooth form lower salinities, eg. mesohaline-polyhaline. **Figs. 7-8**, *Neocyprideis;* **Fig. 7**, *N. colwellensis*, Headon Beds of Headon Hill, x75, a juvenile moult showing development of tubercles, each of which is seen to consist of a cluster of smaller tubercles. **Figs. 7 and 8** show the pore canals of different individuals of an undescribed Eocene species, the bar representing 10 µm. These pore canals are of the sieve type, **Fig. 8** is circular, **Fig. 9** irregular in shape; **these may indicate respectively low salinities and hypersaline salinities.**

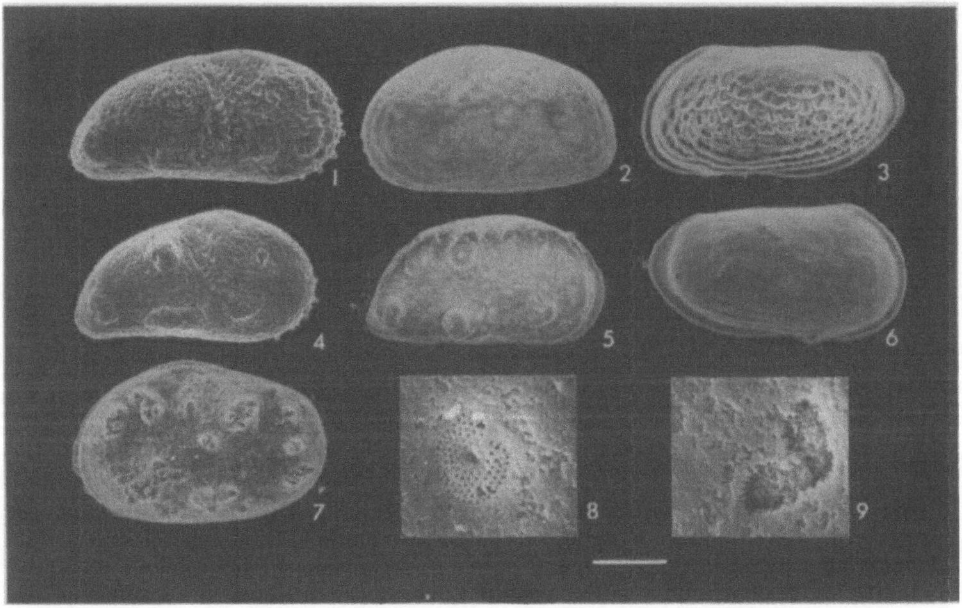

Apart from the tubercles already mentioned, changes in the ornamentation may also occur, with weak ornament being more characteristic of lower salinities. The shape of the normal pore canals has been shown to vary with salinity (Rosenfeld and Vesper, 1977), with circular outlines in low salinities, through elongate shapes to irregular outlines in hypersaline waters (Figure 2.15).

These genera do not persist far back into the Mesozoic, so other genera then have to be used, but a distinctive brackish water ostracod fauna can be recognised from the Middle Jurassic onwards.

## 2.3 Late Jurassic and Early Cretaceous cypridean ostracods

Figure 2.16: The distribution of late Jurassic-early Cretaceous Cypridean ostracod faunas on reconstructed early Cretaceous continental positions (from various sources).

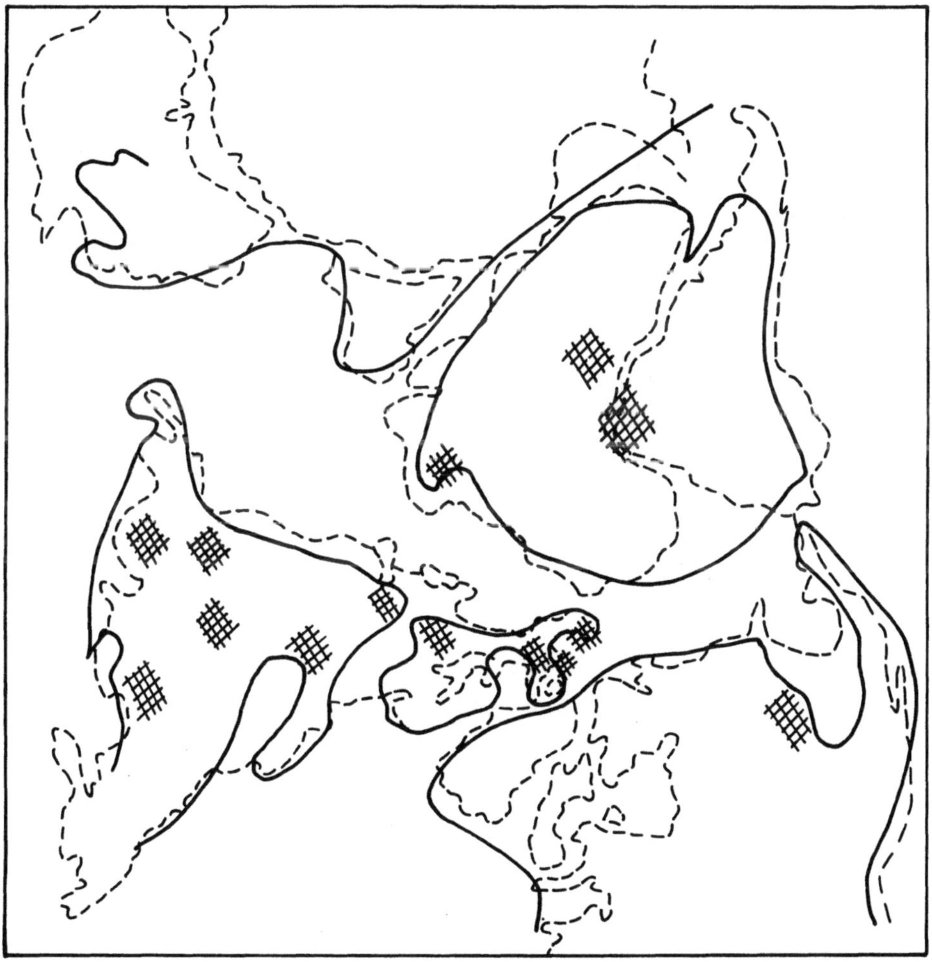

A non-marine "Wealden Facies" frequently developed during the extensive late Jurassic-early Cretaceous regression. This facies is present in many parts of the world, but the absence of typical Mesozoic zonal fossils such as ammonites poses problems for correlation. Rocks of this facies have often been assumed to be of similar age, but it is now clear that they range from the Portlandian to the Albian, i.e. they can be as young as mid Cretaceous. A characteristic ostracod fauna, which is often referred to as the *"Cypridea* Fauna", evolved during this time and it is found in many parts of the world (Figures 2.16, 2.17). These ostracods were first described in the Purbeck and Wealden Beds of southern England, and it is with this area that our account commences. Rocks of this age frequently contain vast numbers of ostracods clearly visible to the naked eye, so early geologists paid a lot of attention to them. The beds they occur in do not yield ammonites because of their non-marine origin, so the ostracods assumed some importance in biostratigraphy. Edward Forbes used the ostracods for zoning the Purbeck Beds in 1851, but it was T.R. Jones in 1885 who produced a zonation of the Purbeck and Wealden Beds which is still valid today, and who recognised their international importance by comparing them with similar faunas from Germany. In more recent years, however, the name most often associated with their study is that of the late F.W. Anderson. His publications span the years between 1939 and 1985; the 1985 paper was published posthumously and contains a summary of all his work together with a complete bibliography.

Figure 2.17. Some species of *Cypridea* from the early Cretaceous of Sussex, southern England; photographs kindly supplied by Prof. John Neale of Hull University, all x55. **Fig. 1.**, *C. begdenensis* Anderson 1967; **Fig. 2** *C. clavata* Anderson (1939); **Fig. 3** *Cypridea* sp. These figures show the range of morphological variation seen in the genus.

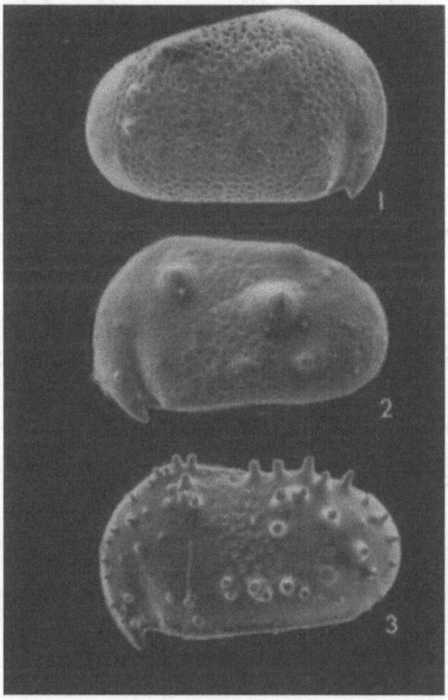

### 2.3.1 Cypridean faunas of southern England and Europe

Anderson (1973, 1985) recognised 12 zones based on the genus *Cypridea* for the Purbeck and Wealden Beds. This spans some 35my., from the late Tithonian (or Portlandian) to the late Barremian. He also established 15 stratigraphically arranged Assemblages, based upon ostracod events such as concurrent ranges, species abundance and species diversity. Anderson's main means of correlation within southern England, however, was by the use of faunicycles. The Purbeck/Wealden Beds were deposited in a wide variety of non-marine conditions, ranging from soil beds through a variety of freshwater deposits to quasi-marine and hypersaline deposits. Salinity has usually been given priority in explaining these ostracod distributions, and Anderson divided the ostracod fauna into two types: the 'C' Phase and the 'S' Phase faunas. The first of these refers to faunas dominated by *Cypridea,* which frequently occurs in vast numbers to the exclusion of all other genera. The second refers to non-Cypridean faunas with marine genera such as *Protocythere, Cytheropteron, Paranotacythere, Parexophthalmocythere,* and *Schuleridea,* together with mesohaline-polyhaline genera such as *Haplocytheridea, Galliaecytheridea, Klieana,* and *Macrodentina,* and problematical genera such as *Fabenella, Mantelliana,* and *Theriosynoecum.*

The 'S' Phase faunas were generally regarded as inhabiting a more saline environment than the 'C' Phase ostracods, and Anderson used these to name 24 marine bands in the succession. A Faunicycle consists of these two phases, an 'S' Phase passing upwards into a 'C' Phase fauna. The 'S' Phase frequently contains one or more minor 'C' Phase episodes within it, and often passes upwards gradually into the main 'C' Phase (Figure 2.18). In general the Lower Purbeck Beds contain thicker 'S' Phase faunas, with a trend towards lower salinities in the Upper Purbeck and Wealden Beds. Using this concept, Anderson has recognised 98 faunicycles (1985) which form the basis for correlation of these deposits throughout southern England, believing them to be climatically controlled, speculating on the possibility of Milankovitch cycles before this became fashionable. However, while the faunicycles themselves, with their alternation of cycles, are a fact, their interpretation is controversial.

Figure 2.18. The Durlston faunicycle (after Anderson, 1971).

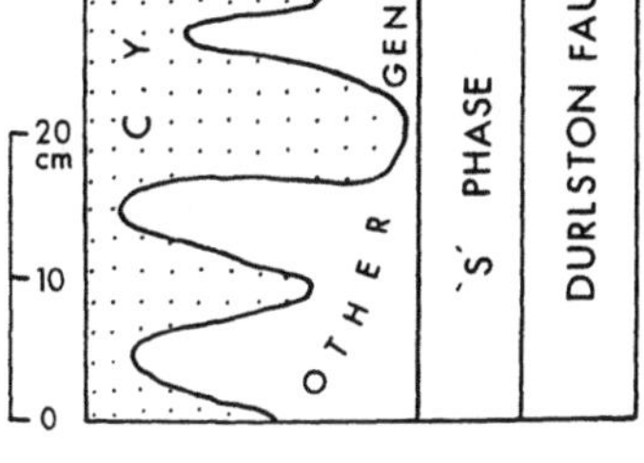

'S' Phase faunas also contain genera such as *Darwinula* which are usually regarded as being freshwater ostracods. Other genera listed above probably tolerated a wide range of salinities. The environmental requirements of *Cypridea* itself probably varied between species. It was a classic r-selective strategist, i.e. adapted to fluctuating environments where opportunistic species with rapidly established populations could thrive. Its parthenogenetic reproduction allowed it to spread rapidly and occupy any suitable environment. The salinity of the water it inhabited, while very low, might not always have been truly freshwater, and the bodies of water may have been semi-permanent, drying out during dry seasons. In this case, the evolution of the cyprid desiccation-resistent egg may have given *Cypridea* an advantage over other freshwater ostracods. Other factors may have assumed importance, such as water depth or oxygen levels. You-gui Li et al. (1988) have related the presence of smooth species to reducing conditions in lakes or swamps, with punctate and bisulcate species in oxidising conditions, while De-quan Ye (1988) has suggested thin-shelled and spinose or reticulate forms characterise deep water, while thick-shelled smooth or nodose forms are characteristic of shallow water lacustrine environments.

*Cypridea* evolved very rapidly in the late Jurassic and early Cretaceous. The number of species present in western Europe lies between 100 and 150 (Kilenyi and Neale, 1978), and while this may be partially due to over splitting, it is clear that a large number of species evolved in a fairly short time. Non-marine ostracods do not usually evolve at such rapid rates as this, and some authors (Sylvester-Bradley, 1947; Kilenyi and Neale, 1978) have suggested polyploid speciation due to constantly changing environments. Polyploidy could also explain the wide range of variation seen in the genus. However, as in all such cases, it is not clear how much of the observed variation is of ecophenotypic origin. Comparison with Recent ostracods living in similar environments indicates the likelihood of phenotypic characters being present. Species of *Cypridea* are differentiated on the basis of lateral outline and number and position of tubercules and spines, and the presence, absence, and strength of reticulation. Variation of these characters is common, and fossil populations often intergrade with one another so that the separation of stratigraphically arranged species is often difficult. At the moment, the question of genotypic versus phenotypic variation cannot be satisfactorily answered, although the majority of workers incline towards the former.

While the environmental interpretation of the faunas is still open to question, their role in stratigraphy is not. They were important in demonstrating that the German "Wealden" is mostly older than the English Wealden; the main part of the German succession correlates with the Upper Purbeck Beds of England, only the uppermost part being of lower Wealden age (Anderson, 1971). Thus the German succession ranges from the Tithonian to the Lower Valanginian, while the English Wealden ranges from the Lower Valanginian to the Upper Barremian. Anderson's Cypridea zones have allowed accurate correlation of the late Jurassic of south eastern France, southwestern France, the Jura, and the Paris Basin (Colin and Oertli 1985), while the Wealden succession of northeastern Spain has been correlated with the English and German sequences by Brenner (1976) using ostracods. They have also been successfully used for offshore correlation in the North Celtic Sea and Fastnet Basins off Ireland, and for the Portlandian and Berriasian (Colin et al. 1981; Ainsworth et al. 1987).

### 2.3.2 Cypridean faunas of China

Late Mesozoic non-marine deposits are widely distributed over large areas of China, where they accumulated to form thick sequences in several depositional basins. The Jurassic and Cretaceous non-marine ostracods of these deposits have been intensively studied in recent years because of their important biostratigraphical role in coal and petroleum exploration. Some useful English language summaries can be found in Hao Yi-chun et al. (1983), Gou Yun-sian and Cao Mei-zhen (1983) and Yu-wen Li (1988). The Chinese workers have recognised a series of biozonations for the various basins, and have been able to suggest correlations with Europe and the U.S.A., as well as developing interesting ideas on the role of these non-marine ostracods in basin development and facies studies. The genus *Cypridea* first appeared in the late Jurassic, and as elsewhere in the world, became abundant in the Lower Cretaceous, but it is also important in the late Cretaceous and persisted into the Palaeocene in China. Few of the species are identical with those of Europe, although several are very similar. Within China there is a certain amount of endemism between the basins, and there is a more abundant fauna in northern China than in southern China for the late Jurassic and early Cretaceous. The Jurassic-Cretaceous boundary has been studied by several authors, and correlation with the Purbeck-Wealden of England and Germany appears to be remarkably good. The most detailed of these correlations is that of Yu-wen Li (1988) (Figure 2.19). Yu-wen Li has used the non-marine ostracods to make a careful biostratigraphical study of the Sichuan Basin in eastern central China, and has been able to show that previous studies placed the boundary much too low in the succession due to the diachronous nature of many of the lithostratigraphical units. The presence of some European species of *Cypridea* has allowed correlation with the Purbeck and Wealden Beds, and the Jurassic-Cretaceous boundary recognised in the Sichuan Basin is based on that recognised in England at the base of the Middle Purbeck Beds. Hao Yi-chun et al. (1983), in a more generalised study, were able to compare faunas from China with the Lower Purbeck Beds of England, the Purbeck and Lower Wealden of Europe, Mongolia and the far east of the former Soviet Union, the Barremian of Mongolia, western Siberia, the Caspian Sea area, and Dakota and Wyoming in the U.S.A, and the Aptian-Cenomanian of the U.S.A., the Lebanon, and the Cocobeach Formation of Gabon in west Africa. You-gui Li et al. (1988) studied the boundary in the Fuxin Basin, northeast of Beijing, where they were also able to make detailed comparisons with the Purbeck and Wealden of Europe.

De-quan Ye (1988) studied the Cenomanian Nenjiang Formation of the Songliao Basin in eastern China. This formation consists of more than 1000m of dark mudstones and sandstones with important oil reservoirs. Sixteen ostracod zones have been established, mostly based on species of *Cypridea*. The diversity and composition of the ostracod faunas, the abundance of specimens, and details of morphology, have been used to work out the depositional characteristics of the Nenjiang Formation. A transgressive episode related to basin subsidence followed by subsequent shallowing can be recognised. An interesting result of this study is the possible use of ostracod shell morphology in the recognition of facies which may be potential source rocks.

Figure 2.19: The correlation of the Jurassic/Cretaceous boundary between England and China according to Yu-wen Li (1988).

| | STAGE | ENGLAND | CHINA |
|---|---|---|---|
| CRETACEOUS | BERRIASIAN | *Cypridea setina* | *Cypridea (Ulwellia)-Simicypris* Assemblage Zone |
| | | *Cypridea vidrana* | |
| | | *Cypridea fasciculata* | *Cypridea dayaoensis - Quinjiania* Assemblage Zone |
| | | *Cypridea granulosa* | |
| JUR | PORTLANDIAN | *Cypridea dunkeri* | *Jingguella (J) acutura - Jingguella (Minheelia) minheesis* Assemblage Zone |

### 2.3.3 Cypridean faunas of Brazil

A final example of the importance of the *Cypridea* fauna to early Cretaceous stratigraphy is taken from South America and West Africa. Krommelbein in his now classic studies of these faunas demonstrated their similarities as evidence for their proximity in the early-mid Cretaceous (see Krommelbein, 1966, for a good summary in English). This has been used as important evidence in dating the break-up of this part of Gondwanaland. However, Krommelbein also demonstrated the importance of these ostracods for biostratigraphy. More recently Moura (1988) has shown how non-marine ostracods can be used in biostratigraphical studies in the Campos Basin, which is situated near Rio de Janeiro on the coast of Brazil. The sequence studied shows a lower "lake sequence" of the Neocomian (probably Barremian) Jiquia local stage consisting of lacustrine-deltaic clastic sediments, passing upwards into a "gulf sequence" of evaporitic and clastic sediments of the Alagoas local stage ( Aptian) (Figure 2.20). Two ostracod zones are recognised in the Jiquia stage; the lower zone of *Petrobasia diversicostata* is divisible into three subzones, from oldest to youngest, the subzones of *Bisulcocyprus postangularis postangularis, Cypridea (Sebasianites) devexa,* and *Cypridea (Pseudocypridina) faveolata;* the upper Jiquia zone is the *Limnocythere troelseni* Zone. The Alagoas stage is characterised by a single zone, the *"Cytheridea"?spp. gr.* 201/218 Zone. The distribution of these zones and subzones has been recorded throughout the Campos Basin from ditch and core samples. They now form the basis for biostratigraphical studies in the Campos Basin, and for stratigraphical correlation within the basin and with other coastal basins further to the north of Brazil. Figure 2.20 shows a typical cross section in the Campos Basin.

Figure 2.20: A cross section in the Campos Basin, Brazil, showing the distribution of ostracod zones (after Moura, 1988).

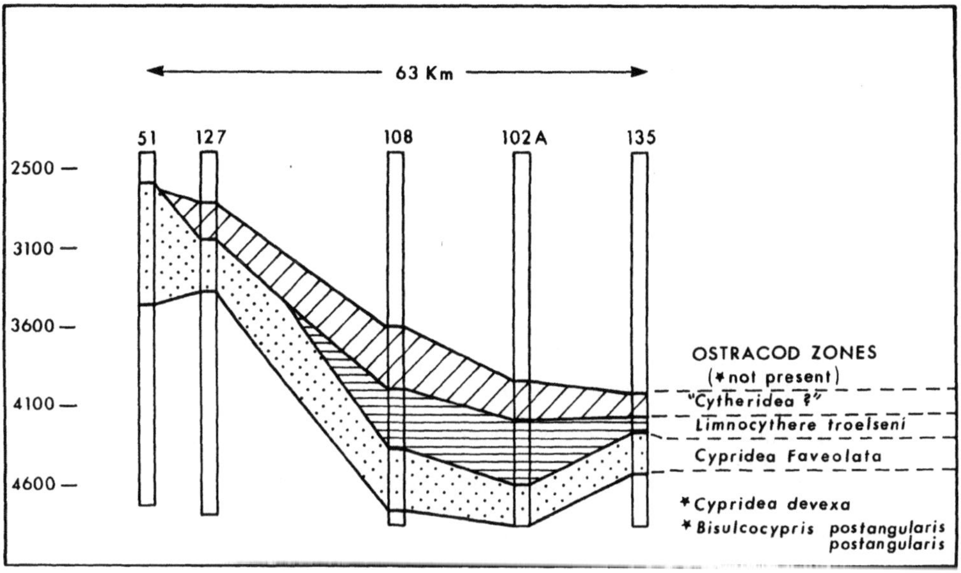


## References


Ainsworth, N.R., O'Neill, M., Rutherford, M.M., Clayton, G., Horton, N.F. and Penney, R.A. 1987. Biostratigraphy of the Lower Cretaceous, Jurassic and uppermost Triassic of the North Celtic Sea and Fastnet Basins. *In:* Brooks, J. and Glennie, K. (eds), *Petroleum Geology of North West Europe,* Graham and Trotman, 611-622.

Anderson, F.W. 1971. The ostracods. In F.W. Anderson and R.A.B. Bazley. *The Purbeck Beds of the Weald (England). Bull. geol. Surv. Gt. Britain,* 34, 27-138.

Anderson, F.W. 1973. The Jurassic-Cretaceous transition: the non-marine ostracod faunas. *In:* Casey, R. and Rawson, P.F. (eds), *The Boreal Lower Cretaceous, Geological Journ., Special Issue 5,* 101-110.

Anderson, F.W. 1985. Ostracod faunas in the Purbeck and Wealden of England. *Journal of Micropalaeontology,* 4 ,1-68.

Bristow, H.W, Reid,C. and Strahan, A., 1889. The Geology of the Isle of Wight (2nd ed.). *Memoir. geol. Surv. U.K., xiv* + 349 pp.

Bujak, J.P., Downie, C., Eaton, G.L. and Williams, G.L. 1980. Dinoflagellate cysts and acritarchs from the Eocene of southern England. *Special Papers in Palaeontology, 24,* 100 pp.

Colin, J.P., Lehmann, R.A. and Morgan, B.E. 1981. Cretaceous and Late Jurassic biostratigraphy of the North Celtic Sea Basin, offshore southern England. *In*: Neale, J. and Brasiet, M. (eds.) *Microfossils from Recent and Fossil shelf seas*, Ellis Horwood, Chichester, 122-155.

Colin, J.P. and Oertli, H.J. 1985. Purbeckien. *In* Oertli, H.J.S. (ed.), *Atlas des ostracodes de France,* Mem. Elf-Aquitaine, 9, 147-161.

Cooper, J. 1976. British Tertiary stratigraphical and rock terms formal and informal, additional to Curry 1958, Lexique Stratigraphique International. *Tertiary Research Special Paper No. 1,* 37 pp.
Curry, D., Adams, C.G., Boulter, M.C., Dilley, F.C., Eames, F.E., Funnel, B.M. and Wells, M.K. 1978. A correlation of Tertiary rocks in the British Isles. *Geological Society of London Special Report* no.12, 78 pp.
Edwards, R.A. and Freshney, E.C. 1987a. Lithostratigraphical classification of the Hampshire Basin Palaeogene deposits (Reading Formation to Headon Formation). *Tertiary Research, 8,* 43-73.
Edwards, R.A. and Freshney, E.C. 1987b. Geology of the country around Southampton. *Memoir of the British Geological Survey,* Sheet 315 (England and Wales), 111 pp.
Forbes, E., 1851. On the succession of strata and distribution of organic remains in the Dorsetshire Purbecks. *Edinburgh Phil. Journ.,49,* 311-313
Forbes, E. 1853. On the fluviomarine Tertiaries of the Isle of Wight. *Jl. geol. Soc. London, 9,* 259-270.
Forbes,E. 1856. On the Tertiary Fluvio-marine Formation of the Isle of Wight. *Mem. geol. Surv.,* U.K., 162 pp.
Gou Yun-sian and Cao Mei-zhen 1983. Stratigraphic and biogeographic distribution of the *Cyprideea*-bearing faunas of China. *In*: Maddocks, R.F. (ed.) *Applications of Ostracoda*, University of Houston, Department of Geosciences, 381-393.
Hao Yi-Chun, Su Deying and Li-You-qui 1983. Late Mesozoic non-marine ostracodes in China. *In*: Maddocks, R.F. (ed.) *Applications of ostracoda*, University of Houston, Department of Geosciences, 377-380.
Haq, B.U., Hardenbol, J. and Vail, P.R. 1987. Chronology of fluctuating sea levels since the Triassic. *Science,* 235, 1156-1167.
Hazel, J.E. 1990. Major changes in Eocene and Oligocene Gulf Coast ostracod assemblages: relationship to global events. *In:* Whatley, R. and Maybury, C. (Eds) *Ostracoda and Global Events,* Chapman and Hall, London, 111-121.
Howe, H.V. 1971. Ecology of American torose Cytherideidae. *Bull. Cent. Rech. Pau-S.N.P.A..,* suppl. 5, 545-557.
Insole, A. and Daley, B. 1985. A revision of the lithostratigraphical nomenclature of the Late Eocene and Early Oligocene strata of the Hampshire Basin, southern England. *Tertiary Research,* 7, 67-100.
Jones, T.R., 1857. A monograph of the Tertiary Entomostraca. *Palaeontographical Society Monograph,* 1-68, 6pls.
Jones, T.R., 1885. On the Ostracoda of the Purbeck Formation with notes on the Wealden. *Quarterly J. geol. Soc. Lon.,* 41, 311-353.
Jones, T.R. and Sherborn, C.D., 1889. A supplemental monograph of the Tertiary Entomostraca of England. *Palaeontographical Society Monograph,* 1-55, 3pls.
Judd, J.W., 1880. On the Oligocene strata of the Hampshire Basin. *Quarterly J. geol. Soc. Lon.,* 36, 137-177.
Judd, J.W., 1882. On the relations of the Eocene and Oligocene strata in the Hampshire Basin. *Quarterly J. geol. Soc. Lon.,* 38, 461-486.
Keen, M.C. 1972. The Sannoisian and some other Upper Palaeogene ostracoda from north west Europe. *Palaeontology, 15,* 267-325, pls. 45-56.
Keen, M.C. 1975. The Palaeobiology of some Upper Palaeogene freshwater ostracods. *Bulletin American Paleontology, 65,* 271-283.
Keen, M.C. 1977. Ostracod assemblages and the depositional environments of the Headon, Osborne and Bembridge Beds (Upper Eocene) of the Hampshire Basin. *Palaeontology, 20,* 405-445, 4pls.
Keen, M.C. 1978. The Tertiary- Palaeogene, *In:* Bate, R.H. and Robinson, E. (Eds), *A Stratigraphical Index of British ostracoda,* Seel House Press, Liverpool, 385-450, 12pls.
Keen, M.C. 1982. Intraspecific variation in Tertiary ostracods, *In:* Bate, R.H., Robinson, E. and Sheppard, L.M. (Eds), *Fossil and Recent Ostracods*, Ellis Horwood, Chichester, 381-405, 5pls.
Keen, M.C. 1983. Ostracods and Tertiary Biostratigraphy, *In:* Maddocks, R.F. (Ed), *Applications of ostracoda,* University of Houston Department of Geosciences,78-95.
Keen, M.C. 1990a. The ecology and evolution of the Palaeogene ostracod *Neocyprideis. Cour. Forsch.-Inst. Senckenberg, 123,* 217-228.
Keen, M.C. 1990b. Ostracoda, sea-level changes and the Eocene-oligocene boundary. *In:* Whatley, R. and Maybury, C. (Eds), *Ostracoda and Global Events,* Chapman and Hall, London, 152-160.
Kilenyi, K. and Neale, J.W. 1978. The Purbeck/Wealden, *In:* Bate R.H., Robinson, E. and Sheppard, L.M. (Eds), *Fossil and Recent Ostracods,* Ellis Horwood, Chichester, 299-325, 6 pls.

Krommelbein, K. 1966. On "Gondwana Wealden" ostracoda from N.E. Brazil and West Africa. *Proc. 2nd West Afr. Micropaleont. Coll. (Ibadan 1965)*, Brill, Leiden, 113-118.

Li, Y.-G., Su, D.-Y. and Zhang, L.-J. 1988. The Cretaceous ostracod faunas from the Fuxin Basin, Liaoning Province. *In:* T. Hanai, N. Ikeya and K. Ishizaki (eds), *Evolutionary biology of Ostracoda,* Elsevier, Amsterdam, 1173-1186.

Li, Y.-W. 1988. The application of Ostracoda to the location of the Non-marine Jurassic-Cretaceous Boundary in the Sichuan Basin of China. *In:* T. Hanai, N. Ikeya and K. Ishizaki (eds), *Evolutionary biology of Ostracoda*, Elsevier, 1245-1260.

Lyell, C. 1824. On a recent formation of freshwater limestone in Forfarshire. *Trans. Geol. Soc. London*, Ser.2, vol.2, 73-96.

Melville, R.V. and Freshney, E.C. 1982. *British Regional Geology. The Hampshire Basin and adjoining areas.* Fourth edition. London: British Geological Survey, viii + 146 pp.

Moura, J.A. 1988. Ostracods from non-marine early Cretaceous sediments of Campos Basin, Brazil, *In:* T. Hanai, N. Ikeya and K. Ishizaki (eds), *Evolutionary biology of Ostracoda*, Elsevier, Amsterdam, 1207-1216.

Murray, J.W. and Wright, C.A. 1974. Palaeogene Foraminiferida and palaeoecology, Hampshire and Paris Basins and the English Channel. *Special Papers in Palaeontology*, **no.14,** 129 pp.

Plint, A.G. 1984. A regressive coastal sequence from the Upper Eocene of Hampshire, southern England. *Sedimentology, 31*, 213-225.

Plint, A.G. 1988. Global eustasy and the Eocene sequence in the Hampshire Basin, England. *Basin Research, 1,* 11-22.

Rosenfeld, A. and Vesper, B. 1977. The variability of the sieve pores in Recent and fossil species of *Cyprideis torosa* (Jones), as an indicator for salinity and paleosalinity. *In:* Loffler, H. and Danielopol, D. (Eds.) *Aspects of Ecology and Zoogeography of Recent and Fossil Ostracoda*, Junk, Amsterdam, 55-67.

Stinton, F.C. 1975. Fish otoliths from the English Eocene. *Monograph palaeontogr. Soc.*, London. (1), 1-56.

Sylvester-Bradley, P.C. 1947. The shell structure and evolution of the Mesozoic ostracod *Cypridea. Quarterly J. geol. Soc. Lon., 103*, viii.

Whatley, R. 1983. The application of Ostracoda to palaeoenvironmental analysis. *In*: Maddocks, R.F. (Ed) *Applications of Ostracoda*, University of Houston, Department of Geosciences, 51-77.

Webster, T., 1814. On the Freshwater Formations in the Isle of Wight, with some observations on the strata over the Chalk in the south-east part of England. *Trans. geol. Soc. Lond., 2*, 161-254.

Ye, D.-Q. 1988. Ostracod evolution and depositional characteristics of the Cretaceous Nenjiang Formation in the Songliao Basin, China. *In:* T. Hanai, N. Ikeya and K. Ishizaki (eds), *Evolutionary biology of Ostracoda,* Elsevier, Amsterdam, 1217-1227.

Michael C. Keen
Dept. of Geology and Applied Geology,
University of Glasgow.
Glasgow G12 8QQ,
Scotland,
U.K.

# 3 PRELIMINARY OBSERVATIONS ON BENTHONIC FORAMININERA ASSOCIATED WITH BIOGENIC GAS SEEP IN THE NORTH SEA

R.W. Jones

**Abstract**

This chapter records the benthonic foraminifera associated with a biogenic gas seep in the North Sea. A sample from an actively seeping site contains a benthonic foraminiferal assemblage statistically distinct from that of a control site. The distinctions are in terms of, firstly, abundance, diversity and dominance, secondly, taxonomic composition, and thirdly, proportions of epifaunal (surface-dwelling) and infaunal (sediment-dwelling) morphotypes.

## 3.1 Introduction

### 3.1.1 Previous studies on the Recent and Sub-Recent benthonic foraminifera of the North Sea

Since the pioneering study of Jarke (1961), there have been a large number of studies on the Recent and sub-Recent benthonic foraminifera of the North Sea. Only those dealing with the west-central area are discussed here.

Gabel (1971) produced an authoritative study on the total (living plus dead) foraminiferal populations of the entire North Sea, defining a total of thirteen biofacies. The area around the present study area falls close to the boundary between his *Cribrononion incertum* [*Elphidium* ex gr. *clavatum*]-*Nonion depressulum, Bulimina* [*Stainforthia*] *fusiformis -B. marginata* and *Cassidulina norcrossi* [*C. laevigata* s.l.] Biofacies (II, IV and VII respectively).

Hughes et al. (in Holmes 1977) subdivided the foraminiferal assemblages of the cored Quaternary of the Central North Sea into six biostratigraphic units. Their Unit F [?Holocene] is dominated by *Bulimina marginata, Hyalinea balthica* and *Trifarina angulosa* (which together account for 75% of the assemblage). Units E, C and A [?Pleistocene, Glacial or Stadial] are dominated by the boreal-arctic and arctic species *Elphidium clavatum* [*E.* ex gr. *clavatum* of this chapter], (35-77%), *Cassidulina obtusa* [*C. obtusa* s.l.] (1-55%), and *Protelphidium orbiculare* and *Haynesina orbicularis*] (<10%), with *Elphidium bartletti* and *Buccella frigida* among the accessory species. Units D and B [?Pleistocene, Interglacial or Interstadial] are again dominated by the boreal-arctic and arctic species *E. clavatum* (56-67%), *C. obtusa* (13-15%) and *P. orbiculare* (2-48%), but they also contain significant proportions of the boreal-lusitanian species *Ammonia batava*, *Bulimina gibba, B. marginata* and *Trifarina angulosa.*

Jansen et al. (1979) examined the foraminiferal assemblages from the cored Holocene Witch Deposits and the Late Pleistocene Fladen and Hills Deposits and Swatchway Beds of Witch Ground and Fladen Ground areas. The Upper Witch Deposits (Holocene, <8400 yr BP) are dominated by the boreal-lusitanian species *Bulimina marginata, Hyalinea balthica, Uvigerina peregrina* and *Trifarina angulosa* and the boreal-arctic *Cribrononion excavatum* forma *clavatum* [*Elphidium* ex gr. *clavatum* of this chapter]. The Lower Witch Deposits (Holocene, 8700 yr BP) are dominated by the boreal-arctic and arctic species *C. excavatum* forma *clavatum* (70%) and *Cassidulina crassa* [*C. obtusa* s.l.] (7-15%). The Fladen Deposits (Pleistocene, 15000 yr BP) and Hills Deposits (Pleistocene, 18000 yr BP) are dominated by *C. excavatum* forma *clavatum, C. crassa* and, locally, *Protelphidium orbiculare* [*Haynesina orbicularis*], with *Virgulina loeblichi* [?*Stainforthia* sp.] and



*D. G. Jenkins (ed.), Applied Micropaleontology,* 69–91.
© 1993 *Kluwer Academic Publishers.*

*Nonion labradorica* [*Nonionellina labradorica*] among the accessory species. The Swatchway Beds (Pleistocene, Interglacial (Eemian) or Interstadial (Weichselian)) contain high and downwardly increasing numbers of the boreal-lusitanian species *Bulimina marginata, Trifarina angulosa, Uvigerina peregrina* and *Hyalinea balthica* (together, 21-85%).

Collison (1980) studied the vertical distribution of the living benthonic foraminifera in cores from off the coast of Northumberland. He found *Fursenkoina fusiformis* [*Stainforthia fusiformis* of this chapter] and *Bulimina marginata* to be dominantly infaunal and *Ammonia batava* to be dominantly surficial (see discussion below; see also Tables 3.6-3.7).

Murray (1985) worked on the living benthonic foraminifera (>63 μ) of the Forties and Ekofisk areas (water depths 70-145m). The living population in both areas is dominated by *Fursenkoina fusiformis* [*Stainforthia fusiformis* of this chapter] (40-95%). Accessory species include *Epistominella vitrea* (0.5-4.1%), *Cassidulina obtusa* (1.2-4.8%) and *Reophax scottii* (0.6-4.5%) at Ekofisk, and *Nonionella* sp. (0.7-17.9%), *Bulimina marginata* (1.1-41.3%), *Cassidulina carinata* [*C. laevigata* s.l.] (0.2-4.5%), *Hyalinea balthica* (0.2-6.5%), *Trifarina angulosa* (0.2-4.7%), *Melonis barleeanus* [*M. affinis*](0.5-4.1%) and *Saccammina sphaerica* (0.5-4.1%) at Forties.

Sejrup et al. (1987) undertook a multidisciplinary study of the Quaternary stratigraphy of two boreholes in the Fladen Ground area (lithostratigraphy, amino-acid stratigraphy, foraminiferal biostratigraphy, magnetostratigraphy and seismic stratigraphy). Unfortunately, they did not record any Holocene benthonic foraminifera.

Powell (1988) provided an invaluable synthesis of the work undertaken previously on the Recent foraminifera of the North Sea. She went on to look at species distributions in relation to various environmental parameters (depth, hydrography, sediment type, temperature, salinity, calcium carbonate availability and nutrition). She concluded that depth is the most important determining factor (and used the depth distributions of twenty-six Recent species to model depth changes through the Cenozoic). Recent species recorded by her as living in depths of <200m (in North Sea Coastal and Estuarine Waters) include *Ammonia beccarii* [*A. batava* of this chapter], *Elphidium excavatum* and *E. incertum* [*E.* ex gr. *clavatum*], *Hyalinea balthica, Bulimina marginata, Cibicides lobatulus* [*C.* ex gr. *lobatulus*], *Uvigerina peregrina, Trifarina angulosa, Fursenkoina fusiformis* [*Stainforthia fusiformis*], *Cassidulina laevigata* and *Islandiella teretis* [*C. laevigata* s.l.], *C. obtusa* [*C. obtusa* s.l.] and *Melonis barleeanus* [*M. affinis*], all of which were also encountered in the present study. Some of these species appear to correlate not only with depth but also with sediment type. *A. beccarii, E. excavatum, H. balthica* and *F. fusiformis* appear to correlate with silty, sandy or gravelly substrates, and *B. marginata, U. peregrina* and *M. barleeanus* appear to correlate with muddy substrates.

Most recently, Murray (1992) described the distribution and population dynamics of benthonic foraminiferal assemblages (>63 μ) from inner shelf areas in the western part of the Southern North Sea (water depths 23-98m). He found live assemblages to be dominated by *Elphidium excavatum* [*E.* ex gr. *clavatum* of this chapter] or *Fursenkoina fusiformis* [*Stainforthla fusiformis*], and dead assemblages to be of similar composition (though with *E. excavatum* more widespread and *F. fusiformis* more restricted in distribution).

3.1.2 Previous studies on benthonic foraminifera associated with seeps

To my knowledge, the only previous study of benthonic foraminifera associated with seeps is that of Kaminski (1988). Kaminski documented the agglutinating (benthonic) foraminifera >63 μ in size in the uppermost 1.5cm of sediment in four box cores from two oil seep and two control sites in the Gulf of Mexico. The sites were in the Green Canyon OCS Lease Blocks 184 and 272 on the Louisiana Continental Slope, in water depths of 532-542m (Block 184 control and seep samples respectively) to 685-696m (Block 272 seep and control samples respectively). The Block 184 seep supports a chemosynthetic community including vestamentifarian worms, whereas the Block 272 seep does not. He found that the agglutinating foraminiferal communities associated with the seeps differed from those of the control samples in being less rich and diverse. The Block 184 seep sample contained only 79 specimens belonging to 15 species, compared with the control sample, which contained 254 specimens belonging to 32 species. The Block 272 seep sample contained 187 specimens belonging to 25 species, compared with the control sample, which contained 292 specimens belonging to 36 species. He also found that the agglutinating foraminiferal communities associated with the seeps contained statistically higher abundances of textulariids (e.g. *Textularia wiesneri*) and trochamminids (e.g. *Trochammina glabra*) and lower abundances of astrorhizids than the control communities.

Kaminski's data indicate that the Block 184 and 272 agglutinating foraminiferal seep communities differ from the corresponding control communities in the relative proportions they contain of epifaunal (surface-dwelling) as opposed to infaunal (sediment-dwelling) species (as defined by Jones & Charnock, 1985). The Block 184 seep community contains 69% epifaunal and 31% infaunal species and the corresponding control sample 60% epifaunal and 40% infaunal, while the Block 272 seep community contains 66% epifaunal and 34% infaunal species and the corresponding control sample 60% epifaunal and 40% infaunal.

3.1.3 Previous studies on other micro-organisms associated with seeps

Riese & Michaels (1991) have recently demonstrated that hydrocarbons (both gaseous and liquid) and associated adsorbed toxins seeping to the surface cause detectable genetic mutations in microbial populations in streams. These mutations result in higher levels of stress- or toxin-tolerance. The precise level of tolerance is proportional to the toxin concentration. It varies in a statistically meaningful fashion at a level below the detection threshold of most standard geochemical techniques.

## 3.2 Present Study

3.2.1 Background

British Petroleum is currently conducting research into the petroleum seeps (both liquid and gaseous) of various areas including the North Sea.

The ultimate objectives of the palaeontological contribution to this research are twofold. The first objective is to ascertain whether it is possible to detect seepage by means of its effects on surface benthonic foraminiferal communities. The second objective is to determine the duration of seepage by high-resolution biostratigraphic and/or isotopic dating of subsurface seep communities. This is important in assisting in determining the timing of petroleum charge in relation to that of trap formation.

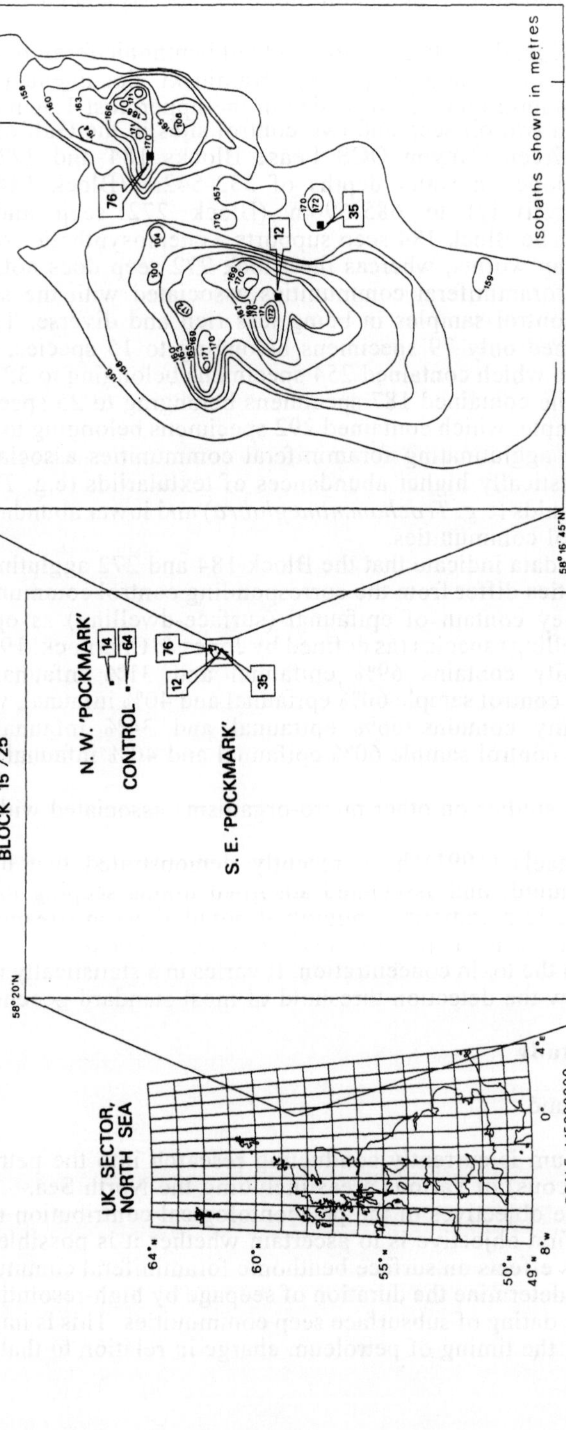


Figure 3.1: Location map

### 3.2.2 Material

Nine samples were available for analysis from in and around the "pockmark" site in 150-180m of water in Block 15/25 in the U.K. Sector of the North Sea (Figure 3.1). Currently active gas seepage at the "pockmark" site is seen on seismic lines across the area and also on side-scan-sonar traces picked up by the NERC R/V *Challenger*. Seepage can also be seen on the video footage shot by the crew of the Max Planck Institute submersible *Jago*, which was lowered from the *Challenger* to investigate the sea-floor around the "pockmark" site. This video footage shows seepage to be localised around comparatively small vents. Although the "pockmark" site is excavated out of post-glacial clay (see also superficial sediment distribution maps in Owens (1977)), the vents are surrounded by carbonate crusts. The carbonates are thought to be formed by organisms consuming methane and producing carbon dioxide; if they could be dated they could provide evidence as to the duration of seep activity. No evidence has yet been established at the "pockmark" site of the development of a Gulf of Mexico-type chemosynthetic community (compare, for instance, Kennicutt et al., 1988; Macdonald et al., 1989; Wade et al., 1989; Brooks et al., 1989; Callender et al., 1990)); indeed, such chemosynthetic communities (and, coincidentally, gas hydrates) appear to occur only in water depths of greater than approximately 500m. However, crustose mats of the sulphide-oxidising bacterium *Beggiotoa* are widespread around the "pockmark" site.

One sample was collected by *Jago* from unconsolidated surface sediment around an active seep site (76A), and the remaining eight were from the tops and bottoms of four 110cm gravity cores collected by the *Challenger*, two from the south-east "pockmark" (Cores 12 and 35), one from the north-west "pockmark" (Core 14), and one from a control site outside the "pockmark" site (Core 64) (Table 3.1).

The sediment methane concentrations in these samples ranged from nil in the control sample (Core 64) to 149 μmol/dm$^3$ in the seep sample (76A) (Table 3.1). Concentrations in samples from Cores 12, 35 and 14 were negligible, indicating no measurable active seepage at the sites where these cores were collected (Table 3.1).

Geochemical analyses on the "pockmark" site hydrocarbons indicate that they are of shallow biogenic origin, and therefore that their occurrence in the Balmoral-Glamis Petroleum Province is entirely coincidental.

### 3.2.3 Methods

Sample preparation consisted of disaggregation in hydrogen peroxide, washing out of the clay fraction, drying in an oven and passage through a nest of sieves (the finest one of which had a 75 μ mesh). The benthonic foraminifera from each sieve fraction of each sample were picked, sorted into species, identified and counted. Statistical data are presented in Appendix 1. Numbers quoted in Appendix 1 are absolute abundances; numbers quoted in parentheses are relative abundances. Taxonomic data are presented in Appendix 2. Taxa listed in Appendix 2 are arranged according to the suprageneric scheme of Haynes (1981). Selected species are illustrated by scanning electron microscopy on Plates 1-2. The benthonic foraminifera were not stained for "live" protoplasm. However, most of the common species were believed to be autochthonous as they had previously been recorded ("live" in some cases) from comparable depths (Powell 1988 and references therein).

Table 3.1: Sample Data

| SAMPLE | DEPTH | YIELD | SEDIMENT METHANE CONCENTRATION | TYPE |
|---|---|---|---|---|
| 76A | 172 m | 505 (21 species) | 149 μmol dm$^{-3}$ | Seep |
| 14 | 165 m | Top 939 (27 species) | Negligible | NW "pockmark" |
| | | Bottom 854 (17 species) | | |
| 35 | 170 m | Top 400 (29 species) | Negligible | SE "pockmark" |
| | | Bottom 452 (25 species) | | |
| 12 | 168 m | Top 541 (25 species) | Negligible | SE "pockmark" |
| | | Bottom 623 (23 species) | | |
| 64 | 152 m | Top 594 (35 species) | Nil | Control |
| | | Bottom 289 (25 species) | | |

## 3.3 Discussion of results (surface samples)

### 3.3.1 Similarity index matrix

All of the samples were compared with one another using specific-level similarity indices (Murray, 1973). The results were plotted out in the form of a similarity index matrix (Table 3.2). Similarity indices of less than a somewhat arbitrarily selected 75% indicate that samples are statistically distinct (Jones, 1984).

The sample from the seep site (76A) is distinct from the sample from the control site (Core 64) (similarity index 61%). The sample from the north-east "pockmark" (Core 14) is indistinguishable from the sample from the seep site (76A) (similarity index 76%). It is distinct from the sample from the control site (Core 64) (similarity index 72%), and also distinct from the samples from the south-east "pockmark" (Cores 12 and 35) (respective similarity indices 71 and 73%). The samples from the south-east "pockmark" (Cores 12 and 35) are distinct from the sample from the seep site (76A) (respective similarity indices 60% and 68%). They are indistinguishable or virtually indistinguishable from samples from the control site (Core 64) (respective similarity indices 88% and 74%), and also indistinguishable from one another (similarity index 76%).

Table 3.2: Similarity index matrix

| | 76A Seep | 14 NW "pockmark" | 35 SE "pockmark" | 12 SE "pockmark" | 64 Control |
|---|---|---|---|---|---|
| 76A Seep | | | | | |
| 14 NW "pockmark" | 76 | | | | |
| 35 SE "pockmark" | 68 | 73 | | | |
| 12 SE "pockmark" | 60 | 71 | 76 | | |
| 64 Control | 61 | 72 | 74 | 88 | |

## 3.3.2 Abundance, diversity and dominance

Benthonic foraminiferal abundance ranged from 289-939 specimens per 20cc sample. Diversity ranged from 17-35 species per sample. Dominance ranged from 19-38%.

The seep sample (76A) is characterised by lower abundance, lower diversity and higher dominance than the control sample (Core 64) (Table 3.3).

Table 3.3: Abundance, diversity and dominance

| SAMPLE / ABUNDANCE, DIVERSITY AND DOMINANCE | SEEP | CONTROL |
|---|---|---|
| ABUNDANCE | 505 | 594 |
| DIVERSITY | 21 | 35 |
| | Fisher diversity 4 | Fisher diversity 8 |
| DOMINANCE | 23% | 21% |

Similar differences have been noted by the author between seep and control samples from thermogenic gas seep sites elsewhere in the North Sea and in middle-lower bathyal depths in the North-East Atlantic and Gulf of Mexico.

### 3.3.3 Taxonomic Composition (Ordinal Level)

The seep sample is characterised by lower absolute and relative abundances of the Orders Lituolida, Nodosariida and Buliminida, and by higher absolute and relative abundances of the Order Rotaliida than the control sample (Table 3.4). Miliolida occur in the same proportions in both samples. Astrorhizida and Robertinida are absent.

Table 3.4: Taxonomic composition (ordinal level)

| SAMPLE / ORDER | SEEP | CONTROL |
|---|---|---|
| ASTRORHIZIDA<br>LITUOLIDA | 0% | 1% |
| MILIOLIDA | 1% | 1% |
| NODOSARIIDA | 1% | 3% |
| BULIMINIDA | 50% | 62% |
| ROBERTINIDA | 0% | 0% |
| ROTALIIDA | 48% | 34% |

### 3.3.4 Taxonomic composition (specific level)

Of the commonly occurring species (relative abundance >5% in one or more samples), *Bulimina marginata* and *Trifarina angulosa* (Buliminida) occur in lower absolute and relative abundances in the seep than in the control sample, and are negatively correlated with seepage, while *Uvigerina peregrina* and *Cassidulina laevigata* s.l. (Buliminida), *Hyalinea balthica* and *Elphidium* ex gr. *clavatum* (Rotaliida) occur in higher absolute abundances (other than *H. balthica*) and relative abundances in the seep than in the control sample, and are positively correlated with seepage (Table 3.5). Note in this context that the species that are positively correlated with seepage cannot at the present time be incontrovertibly stated to be chemosynthetic: only carbon isotope analyses would reveal whether they are actually metabolising seeping hydrocarbons. Note also that *Elphidium* ex gr. *clavatum* is positively correlated not only with seepage but also with coarse-grained substrates (see above).

Table 3.5: Taxonomic composition (specific level)

| RELATIVE ABUNDANCE / SPECIES | SEEP | CONTROL |
|---|---|---|
| *Bulimina marginata* | 9% | 21% |
| *Trifarina angulosa* | 5% | 10% |
| **Uvigerina peregrina* | 18% | 10% |
| **Cassidulina laevigata* s. l. | 16% | 10% |
| **Hyalinea balthica* | 23% | 20% |
| **Elphidium* ex gr. *clavatum* | 18% | 2% |

### 3.3.5 Proportions of epifaunal and infaunal morphotypes

Epifaunal morphotypes are those that live on the sediment or in the "flocculent layer" at the sediment-water interface; infaunal morphotypes are those that live in the sediment (Collison, 1980; Haynes, 1981; Corliss, 1985, 1991; Jones & Charnock, 1985; Gooday, 1986; Mackensen & Douglas, 1989; Corliss & Fois, 1991; Murray, 1991, 1992; Nagy, in prep.; see also Tables 3.6 and 3.7).

Table 3.6: Depth distribution of species within the sediment

| SPECIES | DEPTH DISTRIBUTION |
|---|---|
| *Ammonia* ex gr. *beccarii* [*A. batav-a/-us*] | Predominantly 0-2 cm subsurface |
| *Bulimina marginata* | 0-10 cm subsurface |
| *Cassidulina laevigata* s.l. | Predominantly 0-1 cm subsurface |
| *Cibicides* ex gr. *lobatulus* | Epifaunal |
| *Elphidium* ex gr. *clavatum* | 0-10 cm subsurface |
| *Melonis affinis* [*M. barleeanum/us*] | 0-5 cm subsurface |
| *Stainforthia* [*Fursenkoina*] *fusiformis* | Predominantly deeply infaunal (6-10 cm subsurface) |
| *Uvigerina peregrina* | 0-2 cm subsurface |

Table 3.7: Categorization of epifaunal and infaunal morphotypes

| | |
|---|---|
| EPIFAUNAL AND SURFICIAL | SACCAMMINIDAE<br>LITUOLIDAE<br>MILIOLIDAE<br>NODOSARIIDAE p.p. *(Lagena, Lenticulina)*<br>GLANDULINIDAE<br>CASSIDULINIDAE<br>SPIRILLINIDAE<br>DISCORBIDAE<br>EPONIDIDAE<br>PSEUDOPARRELLIDAE<br>ANOMALINIDAE<br>CHILOSTOMELLIDAE *(Pullenia)*<br>NONIONIDAE<br>ELPHIDIIDAE<br>ROTALIIDAE |
| INFAUNAL (DEEPLY INFAUNAL*) | VALVULINIDAE<br>NODOSARIIDAE p.p. *(Amphicoryna, Dentalina, Vaginulina)*<br>BULIMINIDAE*<br>TURRILINIDAE*<br>UVIGERINIDAE*<br>BOLIVINITIDAE*<br>EOUVIGERINIDAE* *(Procerolagena)* |

The seep sample is characterised by higher relative abundances of epifaunal morphotypes (and correspondingly lower relative abundances of infaunal morphotypes) than the control sample (Table 3.8). Similar, though more exaggerated, differences have been noted by the author between seep and control samples from thermogenic gas seep sites elsewhere in the North Sea and in the North-East Atlantic and Gulf of Mexico.

Table 3.8: Proportions of epifaunal and infaunal morphotypes

| MORPHOTYPE \ SAMPLE | SEEP | CONTROL |
|---|---|---|
| EPIFAUNAL & SURFICIAL | 66% | 48% |
| INFAUNAL (DEEPLY INFAUNAL*) | 34% (34%) | 52% (51%) |

The disproportionately low relative abundances of infaunal morphotypes in the seep-affected samples probably indicate inimical conditions within the sediment. Whether these conditions were caused by seeping hydrocarbons or by associated adsorbed toxins is as yet unclear. However, the results of elemental analyses of seep-affected samples from the North-East Atlantic have not revealed any unusually high concentrations of those heavy metals shown by, among others, Sharifi et al. (1991) to have deleterious effects on benthonic foraminiferal populations.

### 3.4 Discussion of results (subsurface samples)

Significantly higher total sulphide contents in the subsurface than in the surface from Cores 35 and 64 suggest possible past seepage at the presently probably inactive sites where these cores were collected (Figure 3.2). Perhaps significantly, the subsurface sample from Core 64 is distinct from the surface sample from the same site (similarity index 73%). However, it is also distinct from the surface sample from the seep site (76A) (similarity index 65%). The subsurface sample from Core 35 is indistinguishable from the surface sample from the same site (similarity index 77%). It is also indistinguishable from the surface sample from the control site (Core 64) (similarity index 78%), but distinct from the surface sample from the seep site (76A) (similarity index 70%). In view of the apparent incompatibility of the geochemical and foraminiferal evidence for past seepage, it has proved impossible to determine the duration of seepage at the North Sea site.

However, unreleased data from a recent BP oil exploration well in the Danang Basin, offshore Vietnam, has shown that foraminifera can be used to determine the duration of seepage. In the Vietnamese well, foraminiferal "seep communities" occur over some 20m of a continuously cored deep-water (bathyal) shale sequence of Middle or Late Miocene age. This 20m is thought to represent something of the order of 200 000 years. The "seep communities" are similar to those of the North-East Atlantic, but are locally characterised by a superabundance (>50%) of *Gyroidina* spp.

Figure 3.2: Total Sulphide Contents in Surface and Subsurface Samples.

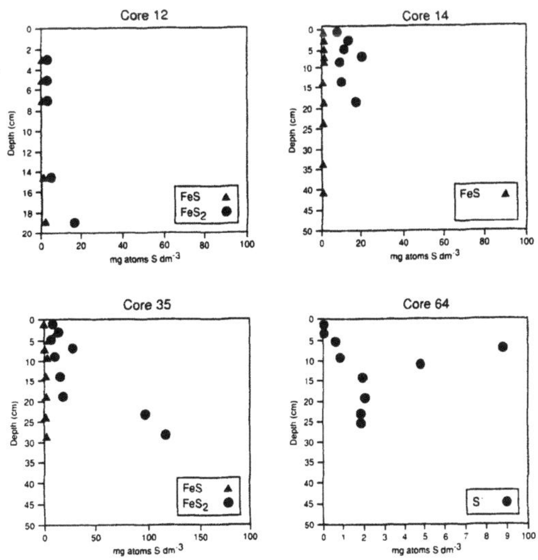

## 3.5 Conclusions

### 3.5.1 Surface Samples

It is apparently possible to detect seepage in the North Sea by means of its effects on surface benthonic foraminiferal communities, though the representativeness of the data sets and duplicability of results remain somewhat in question.
The seep sample contains communities that are quantitatively (statistically) distinct from those of the control sample. The distinctions are in terms of:
- lower abundance and diversity and higher dominance;
- a lower relative abundance of the Orders Astrorhizida, Lituolida, Nodosariida and Buliminida and a higher relative abundance of the Order Rotaliida;
- a lower relative abundance of the species *Bulimina marginata* and *Trifarina angulosa* (Buliminida) and a higher relative abundance of the species *Uvigerina peregrina* and *Cassidulina laevigata* s.l. (Buliminida) and *Hyalinea balthica* and *Elphidium* ex gr. *clavatum* (Rotaliida);
- higher relative abundances of epifaunal as opposed to infaunal morphotypes.

### 3.5.2 Subsurface Samples

It was not possible to determine the duration of seepage at the North Sea site by dating the subsurface seep communities.

## Acknowledgements

I wish to acknowledge the assistance of Chris Bird, Chris Clayton, Tim Dodd, Chris Ebdon, Steve Hay, Mark Houchen, Alan Higgins, Mike Lines, Brian Mitchener and Jane Thrasher of BP and Paul Dando of the Marine Biological Association of the United Kingdom, Plymouth.

BP granted permission to publish.

### References

Brooks, J.M., Kennicutt, M.C., II, Macdonald, I.R., Wilkinson, D.L., Guinasso, N.L., Jr. & Bidigare, R.R., 1989. Gulf of Mexico Hydrocarbon Seep Communities - Part IV. Description of Known Chemosynthetic Communities. Proceedings of the 21st Offshore Technology Conference: 663-667.

Callender, W.R., Staff, G.M., Powell, E.N. & Macdonald, I.R., 1990. Gulf of Mexico Hydrocarbon Seep Communities - V. Biofacies and Shell Orientation of Autochthonous Shell Beds below Storm Wave Base. Palaios, 5: 2-14.

Collison, P., 1980. Vertical Distributon of Foraminifera off the Coast of Northumberland, England. Journal of Foraminiferal Research, 10(1): 75-78.

Corliss, B.H., 1985. Microhabitats of Benthic Foraminifera within Deep-Sea Sediments. Nature, 314(4): 435-438.

Corliss, B.H., 1991. Morphology and Microhabitat Preferences of Benthic Foraminifera from the Northwest Atlantic Ocean. Marine Micropaleontology, 17(3/4): 195-236.

Corliss, B.H. & Fois, E., 1991. Morphotype analysis of deep-sea benthic foraminifera from the Gulf of Mexico. Palaios, 5(6): 589-605.

Gabel, B., 1971. Die Foraminiferen der Nordsee. Helgolander Wissenschaftliche Meeresuntersuchungen, 22: 1-65.

Gooday, A.J., 1986. Meiofaunal Foraminiferans from the Porcupine Seabight (North-East Atlantic): Size Structure, Standing Stock, Taxonomic Composition, Species Diversity and Vertical Distribution in the Sediment. Deep-Sea Research, 33(10): 1345-1373.

Haynes, J.R., 1981. Foraminifera. London; MacMillan, 433 pp.
Holmes, R., 1977. Quaternary Deposits of the Central North Sea, Institute of Geological Sciences Report, 77/14.
Jansen, J.H.F., Doppert, J.W.C., Hoogendoorn-Toering, K., de Jong, J. & Spaink, G., 1979. Late Pleistocene and Holocene Deposits in the Witch and Fladen Ground Area, Northern North Sea. Netherlands Journal of Sea Research, 13(1): 1-39.
Jarke, J., 1961. Die Beziehungen Zwischen Hydrographischen Verhaltnissen, Faziesentwicklung und Foraminiferen-Verbreitung in der Heutigen Nordsee als Vorbild fur die Verhaltnisse Wahrend der Miozan-Zeit. Meyniana, 10: 21-36.
Jones, R.W., 1984(MS). Late Quaternary Benthonic Foraminifera from Deep-Water Sites in the North-East Atlantic. Unpublished Doctoral Dissertation, University College of Wales, Aberystwyth.
Jones, R.W. & Charnock, M.A., 1985. "Morphogroups" of Agglutinating Foraminifera, their Life Positions and Feeding Habits and Potential Applicability in (Paleo)Ecological Studies. Revue de Micropaleontologie, 4(2): 311-320.
Kaminski, M.A., 1988(MS). Cenozoic Deep-Water Agglutinated Foraminifera in the North Atlantic. Unpublished Doctoral Dissertation, Massachussets Institute of Technology/Woods Hole Oceanographic Institution.
Kennicutt, M.C., II, Brooks, J.M., Bidigare, R.R. & Denoux, G.J., 1988. Gulf of Mexico Hydrocarbon Seep Communities - I. Regional Distribution of Hydrocarbon Seepage and Associated Fauna. Deep-Sea Research, 35(9): 1639-1651.
Macdonald, I.R., Boland, G.S., Baker, J.S., Brooks, J.M., Kennicutt, M.C., II & Bidigare, R.R., 1989. Gulf of Mexico Hydrocarbon Seep Communities - II. Spatial Distribution of Seep Organisms and Hydrocarbons at Bush Hill. Marine Biology, 101: 235-247.
Mackensen, A. & Douglas, R.G., 1989. Down-Core Distribution of Live and Dead Deep-Water Benthic Foraminifera in Box Cores from the Weddell Sea and the California Continental Margin. Deep-Sea Research, 36(6): 879-900.
Murray, J.W., 1973. Distribution and Ecology of Living Benthic Foraminiferids. London, Heinemann Educational Books, 274 pp.
Murray, J.W., 1985. Recent Foraminifera from the North Sea (Forties and Ekofisk Areas) and the Continental Shelf West of Scotland. Journal of Micropalaeontology, 4(2): 111-125.
Murray, J.W., 1991. Ecology and Palaeoecology of Benthic Foraminifera. London, Longman Scientific & Technical, 397 pp.
Murray, J.W., 1992. Distribution and Population Dynamics of Benthic Foraminifera from the Southern North Sea. Journal of Foraminiferal Research, 22(2): 114-128.
Nagy, J., in preparation. Morphogroup Patterns of Foraminifera in Delta-Influenced Environments of the Jurassic North Sea.
Owens, R., 1977. Quaternary Deposits of the Central North Sea, 4. Institute of Geological Sciences Report, 77/13.
Powell, A.D.J., 1988(MS). Palaeobathymetric Analysis of Tertiary Sediments in the Northern North Sea and North-East North Atlantic Ocean. Unpublished Doctoral Dissertation, University of Exeter.
Riese, W.C. & Michaels, G.B., 1991. Microbiological Indicators of Subsurface Hydrocarbon Accumulation. American Association of Petroleum Geologists Bulletin, 75(3): 662 (abstract only).
Sejrup, H.P., Aarseth, I., Ellingsen, K.L., Reither, E., Jansen, E., Lovlie, R., Bent, A., Brigham-Grette, J., Larsen, E. & Stoker, M., 1987. Quaternary Stratigraphy of the Fladen Area, Central North Sea: a Multidisciplinary Study. Journal of Quaternary Science, 2: 35-58.
Sharifi, A.R., Croudace, I.U. & Austin, R.L., 1991. Benthic Foraminiferids as Pollution Indicators in Southampton Water, Southern England, U.K. Journal of Micropalaeontology, 10(1): 109-113.
Wade, T.L., Kennicutt, M.C., II & Brooks, J.M., 1989. Gulf of Mexico Hydrocarbon Seep Communities - Part III. Aromatic Hydrocarbon Concentrations in Organisms, Sediment and Water. Marine Environmental Research, 27: 19-30.

R.W. Jones
BP Exploration (Frontier & International)
4-5 Long Walk, Stockley Business Park
Uxbridge, Middlesex, UB11 1BP.

Appendix 1: Statistical Data

Sample 76A (172m)

Species

*Hyalinea balthica* 116 (23%)
*Uvigerina peregrina* 93 (18%)
*Elphidium* ex gr. *clavatum* 89 (18%)
*Cassidulina laevigata* s.l. 81 (16%)
*Bulimina marginata* 45 (9%)
*Trifarina angulosa* 27 (5%)
*Melonis affinis* 18 (4%)
*Cibicides* ex gr. *lobatulus* 7 (1%)
*Ammonia batava* 6 (1%)
*Sigmoilopsis schlumbergeri* 4 (1%)
*Haynesina orbicularis* 3 (1%)
*Brizalina* spp. 2 (<1%)
*Elphidium groenlandicum* 2 (<1%)
*Fissurina* spp. 2 (<1%)
*Cassidulina obtusa* s.l. 1 (<1%)
*Dentalina baggi* 1 (<1%)
*Discorbinella* (?) sp. 1 (<1%)
Miliolidae Indet. 1 (<1%)
Nonionidae Indet. 1 (<1%)
*Pyrgo elongata* 1 (<1%)

Orders

Buliminida 50%
Rotaliida 48%
Miliolida 1%
Nodosariida 1%

Epifaunal and Infaunal Morphotypes

Epifaunal Morphotypes 66%
Infaunal Morphotypes 34% (deeply infaunal morphotypes 34%)

Core 12 (168m) (top)

Species

*Bulimina marginata* 134 (25%)
*Hyalinea balthica* 101 (19%)
*Cassidulina laevigata* s.l. 52 (10%)
*Uvigerina peregrina* 51 (9%)
*Stainforthia fusiformis* 49 (9%)
*Melonis affinis* 34 (6%)
*Trifarina angulosa* 26 (5%)
*Fissurina* spp. 23 (4%)
*Elphidium* ex gr. *clavatum* 16 (3%)
*Nonionella turgida* 11 (2%)
*Brizalina* spp. 5 (1%)
*Cassidulina obtusa* s.l. 4 (1%)
*Pullenia osloensis* 4 (1%)
*Quinqueloculina seminulum* 4 (1%)
*Amphicoryna scalaris* 3 (1%)
*Lagena sulcata* 3 (1%)
*Martinottiella communis* 3 (1%)
*Nonionella* spp. 2 (<1%)
*Vaginulina trondheimensis* 2 (<1%)
*Ammonia batava* 1 (<1%)
*Lenticulina* sp. 1 (<1%)
*Procerolagena gracilis* 1 (<1%)
*Stainforthia concava* 1 (<1%)

Orders

Buliminida 60%
Rotaliida 31%
Nodosariida 1%
Miliolida 1%
Lituolida 1%

Epifaunal and Infaunal Morphotypes

Epifaunal Morphotypes 48%
Infaunal Morphotypes 52% (deeply infaunal morphotypes 51%)

Core 12 (168m) (bottom)

Species

*Hyalinea balthica* 237 (38%)
*Bulimina marginata* 134 (22%)
*Stainforthia fusiformis* 83 (13%)
*Cassidulina laevigata* s.l. 43 (7%)
*Uvigerina peregrina* 33 (5%)
*Melonis affinis* 24 (4%)
*Trifarina angulosa* 23 (4%)
*Fissurina* spp. 9 (1%)
*Elphidium* ex gr. *clavatum* 9 (1%)
*Nonionella turgida* 6 (1%)
*Procerolagena gracilis* 4 (<1%)
*Lagena sulcata* 3 (<1%)
*Pullenia osloensis* 3 (<1%)
*Ammonia batava* 2 (<1%)
*Brizalina* spp. 2 (<1%)
*Nonionella* spp. 2 (<1%)
*Rosalina* spp. 2 (<1%)
*Amphicoryna scalaris* 1 (<1%)
*Cassidulina obtusa* s.l. 1 (<1%)
*Haynesina orbicularis* 1 (<1%)
*Lenticulina* sp. 1 (<1%)

Orders

Buliminida 52%
Rotaliida 46%
Nodosariida 2%

Epifaunal and Infaunal Morphotypes

Epifaunal Morphotypes 55%
Infaunal Morphotypes 45% (deeply infaunal morphotypes 45%)

Core 35 (170m) (top)

Species

*Cassidulina laevigata* s.l. 76 (19%)
*Hyalinea balthica* 76 (19%)
*Uvigerina peregrina* 68 (17%)
*Bulimina marginata* 43 (11%)
*Melonis affinis* 31 (8%)
*Stainforthia fusiformis* 29 (7%)
*Trifarina angulosa* 21 (5%)
*Elphidium* ex gr. *clavatum* 10 (3%)
*Fissurina* spp. 8 (2%)
*Pyrgo elongata* 5 (1%)
*Quinqueloculina seminulum* 5 (1%)
*Cassidulina obtusa* s.l. 3 (1%)
*Veleroninoides crassimargo* 3 (1%)
*Martinottiella communis* 2 (<1%)
*Cibicides* ex gr. *lobatulus* 2 (<1%)
*Ammonia batava* 1 (<1%)
*Elphidium ustulatum* 1 (<1%)
*Epistominella vitrea* 1 (<1%)
*Haynesina orbicularis* 1 (<1%)
*Lenticulina* sp. 1 (<1%)
Miliolidae Indet. 1 (<1%)
*Nonionella* sp. 1 (<1%)
*Procerolagena gracilis* 1 (<1%)
*Psammosphaera fusca* 1 (<1%)
*Saccammina sphaerica* 1 (<1%)
*Vaginulina trondheimensis* 1 (<1%)

Orders

Buliminida 60%
Rotaliida 33%
Nodosariida 4%
Miliolida 4%
Astrorhizida/Lituolida 2%

Epifaunal and Infaunal Horphotypes

Epifaunal Morphotypes 59%
Infaunal Morphotypes 41% (deeply infaunal morphotypes 40%)

Core 35 (170m) (bottom)

Species

*Hyalinea balthica* 109 (21%)
*Bulimina marginata* 97 (21%)
*Cassidulina laevigata* s.l. 88 (19%)
*Uvigerina peregrina* 45 (10%)
*Trifarina angulosa* 21 (5%)
*Stainforthia fusiformis* 17 (4%)
*Elphidium* ex gr. *clavatum* 13 (3%)
*Melonis affinis* 11 (2%)
*Fissurina* spp. 10 (2%)
*Procerolagena gracilis* 6 (1%)
*Oolina* spp. 5 (1%)
*Lagena sulcata* 4 (1%)
*Vaginulina trondheimensis* 4 (1%)
*Amphicoryna scalaris* 3 (1%)
*Brizalina* spp. 3 (1%)
*Cibicides* ex gr. *lobatulus* 3 (1%)
*Nonionella turgida* 3 (1%)
*Cassidulina obtusa* s.l. 2 (<1%)
*Oolina hexagona* 2 (<1%)
*Ammonia batava* 1 (<1%)
*Dentalina flintii* 1 (<1%)
Discorbidae Indet. 1 (<1%)
Miliolidae Indet. 1 (<1%)
*Pullenia osloensis* 1 (<1%)
*Spirillina* sp. 1 (<1%)

Orders

Buliminida 62%
Rotaliida 32%
Nodosariida 6%
Miliolida <1%

Epifaunal and Infaunal Morphotypes

Epifaunal Morphotypes 56%
Infaunal Morphotypes 44% (deeply infaunal morphotypes 42%)

Core 14 (165m) (top)

Species

*Hyalinea balthica* 280 (30%)
*Bulimina marginata* 191 (20%)
*Uvigerina peregrina* 157 (17%)
*Cassidulina laevigata* s.l. 145 (15%)
*Trifarina angulosa* 63 (6%)
*Melonis affinis* 32 (3%)
*Elphidium* ex gr. *clavatum* 33 (3%)
*Ammonia batava* 6 (<1%)
*Stainforthia fusiformis* 5 (<1%)
*Cassidulina obtusa* s.l. 3 (<1%)
*Cibicides* ex gr. *lobatulus* 3 (<1%)
*Fissurina* spp. 3 (<1%)
*Cassidulina subglobosa* 2 (<1%)
*Cibicidoides pachyderma* 2 (<1%)
*Nonionella turgida* 2 (<1%)
*Vaginulina trondheimensis* 2 (<1%)
*Amphicoryna scalaris* 1 (1%)
*Brizalina* sp. 1 (<1%)
*Discorbinella* (?) sp. 1 (<1%)
*Galwayella* sp. 1 (<1%)
*Gyroidina* sp. 1 (<1%)
*Lagena* sp. 1 (<1%)
*Martinottiella communis* 1 (<1%)
Nodosariidae Indet. (<1%)
*Rosalina* sp. 1 (<1%)
*Veleroninoides crassimargo* 1 (<1%)

Orders

Buliminida 60%
Rotaliida 38%
Nodosariida 1%
Lituolida <1%

Epifaunal and Infaunal Morphotypes

Epifaunal Morphotypes 55%
Infaunal Morphotypes 45% (deeply infaunal morphotypes 44%)

Core 14 (165m) (bottom)

Species

*Bulimina marginata* 211 (25%)
*Uvigerina peregrina* 196 (23%)
*Cassidulina laevigata* s.l. 153 (18%)
*Hyalina balthica* 142 (17%)
*Elphidium* ex gr. *clavatum* 62 (7%)
*Melonis affinis* 37 (4%)
*Trifarina angulosa* 27 (3%)
*Cibicidoides pachyderma* 10 (1%)
*Fissurina* spp. 5 (1%)
*Cassidulina obtusa* s.l. 2 (<1%)
*Oolina* spp. 2 (<1%)
*Quinqueloculina seminulum* 2 (<1%)
*Amphicoryna scalaris* 1 (<1%)
*Brizalina* sp. 1 (<1%)
*Lagena sulcata* 1 (<1%)
*Lenticulina* sp. 1 (<1%)
*Procerolagena gracilis* 1 (<1%)

Orders

Buliminida 69%
Rotaliida 29%
Nodosariida 1%
Miliolida <1%

Epifaunal and Infaunal Morphotypes

Epifaunal Morphotypes 49%
Infaunal Morphotypes 51% (deeply infaunal morphotypes 51%)

Core 64 (152m) (top)

Species

*Bulimina marginata* 125 (21%)
*Hyalinea balthica* 117 (20%)
*Trifarina angulosa* 61 (10%)
*Cassidulina laevigata* s.l. 60 (10%)
*Uvigerina peregrina* 58 (10%)
*Stainforthia fusiformis* 51 (9%)
*Melonis affinis* 50 (8%)
*Elphidium* ex gr. *clavatum* 12 (2%)
*Fissurina* spp. 9 (2%)
*Nonionella turgida* 7 (2%)
*Amphicoryna scalaris* 5 (1%)
*Epistominella vitrea* 5 (1%)
*Martinottiella communis* 5 (1%)
*Brizalina* spp. 3 (<1%)
*Gyroidina* spp. 3 (<1%)
*Nonionella* spp. 3 (<1%)
*Cassidulina obtusa* s.l. 2 (<1%)
*Cibicidoides pachyderma* 2 (<1%)
Glandulinidae Indet. 2 (<1%)
*Veleroninoides crassimargo* 2 (<1%)
*Oolina hexagona* 2 (<1%)
*Procerolagena gracilis* 2 (<1%)
*Pyrgo elongata* 2 (<1%)
*Biloculinella* spp. 1 (<1%)
*Bulimina gibba* 1 (<1%)
*Buliminella* sp. 1 (<1%)
*Cassidulina subglobosa* 1 (<1%)
*Cibicides* ex gr. *lobatulus* 1 (<1%)
Elphidiidae Indet. 1 (<1%)
*Lagena laevis* 1 (<1%)
*Lagena sulcata* 1 (<1%)
*Recurvoidella bradyi* 1 (<1%)
*Stainforthia concava* 1 (<1%)
*Triloculina trigonula* 1 (<1%)

Orders

Buliminida 62%
Rotaliida 34%
Nodosariida 3%
Lituolida 1%
Miliolida 1%

Epifaunal and Infaunal Morphotypes

Epifaunal Morphotypes 48%
Infaunal Morphotypes 52% (deeply infaunal morphotypes 51%)

Core 64 (152m) (bottom)

Species

*Cassidulina laevigata* s.l. 58 (20%)
*Hyalinea balthica* 51 (18%)
*Uvigerina peregrina* 30 (10%)
*Trifarina angulosa* 28 (10%)
*Bulimina marginata* 24 (8%)
*Stainforthia fusiformis* 23 (8%)
*Melonis affinis* 13 (5%)
*Brizalina* spp. 11 (4%)
*Fissurina* spp. 9 (3%)
*Vaginulina trondheimensis* 8 (3%)
*Nonionella turgida* 5 (2%)
*Cibicides* ex gr. *lobatulus* 4 (1%)
*Elphidium* ex gr. *clavatum* 4 (1%)
*Ammonia batava* 3 (1%)
*Amphicoryna scalaris* 3 (1%)
Miliolidae Indet. 2 (<1%)
Nodosariidae Indet. 2 (<1%)
*Oolina* spp. 2 (<1%)
*Procerolagena gracilis* 2 (<1%)
*Quinqueloculina seminulum* 2 (<1%)
Glandulinidae Indet. 1 (<1%)
*Lagena* sp. 1 (<1%)
*Lenticulina* sp. 1 (<1%)
*Nonionella* sp. 1 (<1%)
*Pyrgo elongata* 1 (<1%)

Orders

Buliminida 61%
Rotaliida 28%
Nodosariida 9%
Miliolida 1%

Epifaunal and Infaunal Morphotypes

Epifaunal Morphotypes 56%
Infaunal Morphotypes 44% (deeply infaunal morphotypes 40%)

APPENDIX 2: TAXONOMIC DATA

Order ASTRORHIZIDA

Family Saccamminidae

*Psammosphaera fusca* Schulze
*Saccammina sphaerica* Brady

Order LITUOLIDA

Family Lituolidae

*Recurvoidella bradyi* (Robertson)
*Veleroninoides crassimargo* (Norman)

Family Valvulinidae

*Martinottiella communis* (d'Orbigny)

Order MILIOLIDA

Family Miliolidae

*Biloculinella* sp.
*Pyrgo elongata* (d'Orbigny)
*Quinqueloculina seminulum* (Linne)
*Sigmoilopsis schlumbergeri* (Silvestri)
*Triloculina trigonula* (Lamarck)
Miliolidae Indet.

Order NODOSARIIDA

Family Nodosariidae

*Amphicoryna scalaris* (Batsch)
*Dentalina baggi* Galloway & Wissler
*Dentalina flintii* (Cushman)
*Lagena laevis* (Montagu)
*Lagena sulcata* (Walker & Jacob)
*Lagena* spp.
*Lenticulina* spp.
*Vaginulina trondheimensis* (Feyling-Hanssen)
Nodosariidae Indet.

Family Glandulinidae

*Fissurina* spp.
*Galwayella* sp.
*Oolina hexagona* (Williamson)
*Oolina* spp.
Glandulinidae Indet.

Order BULIMINIDA

Family Buliminidae

*Bulimina gibba* Fornasini
*Bulimina marginata* d'Orbigny
*Stainforthia concava* (Hoglund)
*Stainforthia fusiformis* (Williamson)

Family Turrilinidae

*Buliminella* sp.

Family Uvigerinidae

*Trifarina angulosa* (Williamson)
*Uvigerina peregrina* (Cushman)

Family Bolivinitidae

*Brizalina* spp.

Family Eouvigerinidae

*Procerolagena gracilis* (Williamson)

Family Cassidulinidae s.l.

*Cassidulina laevigata* d'Orbigny s.l.
*Cassidulina obtusa* Williamson s.l.
*Cassidulina subglobosa* Brady

Order ROTALIIDA

Family Spirillinidae

*Spirillina* sp.

Family Discorbidae

*Discorbinella* (?) sp.
*Rosalina* spp.
Discorbidae Indet.

Family Eponididae

*Gyroidina* sp.

Family Pseudoparrellidae

*Epistominella vitrea* Parr

Family Anomalinidae

*Cibicides* ex gr. *lobatulus* (Walker & Jacob)
*Cibicidoides pachyderma* (Rzehak)
*Hyalinea balthica* (Schroter)

Family Chilostomellidae

*Pullenia osloensis* Feyling-Hanssen

Family Nonionidae

*Melonis affinis* (Reuss)
*Nonionella turgida* (Williamson)
*Nonionella* sp.
Nonionidae Indet.

Family Elphidiidae

*Elphidium* (*Elphidium*) ex gr. *clavatum* (Williamson)
*Elphidium* (*Elphidiella*) *groenlandicum* Cushman
*Elphidium* (*Toddinella*) *ustulatum* Todd
*Haynesina orbicularis* (Brady)
Elphidiidae Indet.

Family Rotaliidae

*Ammonia batava* (Hofker)

PLATE 3.1: ASTRORHIZIDA (Figs. 1-2), LITUOLIDA (Figs. 3-5), MILIOLIDA (Figs. 6-10), NODOSARIIDA (Figs. 11-19) and BULIMINIDA (Figs. 20-30)

Fig. 1 *Psammosphaera fusca* Schulze x53.
Fig. 2 *Saccammina sphaerica* Brady x35.
Fig. 3 *Veleroninoides crassimargo* (Norman) x71.
Fig. 4 *Recurvoidella bradyi* (Robertson) x248.
Fig. 5 *Martinottiella communis* (d'Orbigny) x28.
Fig. 6 *Biloculinella* sp. x57.
Fig. 7 *Pyrgo elongata* (d'Orbigny) x71.
Fig. 8 *Quinqueloculina seminulum* (Linne) x106.
Fig. 9 *Sigmoilopsis schlumbergeri* (Silvestri) x85.
Fig. 10 *Triloculina trigonula* (Lamarck) x85.
Fig. 11 *Amphicoryna scalaris* (Batsch) (macrospheric) x71.
Fig. 12 *Amphicoryna scalaris* (Batsch) (microspheric) x78.
Fig. 13 *Dentalina baggi* Galloway & Wissler x23.
Fig. 14 *Dentalina flintii* (Cushman) x23.
Fig. 15 *Lagena laevis* (Montagu) x141.
Fig. 16 *Lagena sulcata* (Walker & Jacob) x127.
Fig. 17 *Vagulina trondheimensis* (Feyling-Hanssen) x85.
Fig. 18 *Galwayella* sp. x233.
Fig. 19 *Oolina hexagona* (Williamson) x283.
Fig. 20 *Bulimina gibba* Fornasini x177.
Fig. 21 *Bulimina marginata* d'Orbigny x106.
Fig. 22 *Stainforthia concava* (Hoglund) x177.
Fig. 23 *Stainforthia fusiformis* (Williamson) x141.
Fig. 24 *Buliminella* sp. x233.
Fig. 25 *Trifarina angulosa* (Williamson) x78.
Fig. 26 *Uvigerina peregrina* (Cushman) x71.
Fig. 27 *Procerolagena gracilis* (Williamson) x99.
Fig. 28 *Cassidulina laevigata* d'Orbigny s.l. x106.
Fig. 29 *Cassidulina obtusa* Williamson s.l. x106.
Fig. 30 *Cassidulina subglobosa* Brady x212.

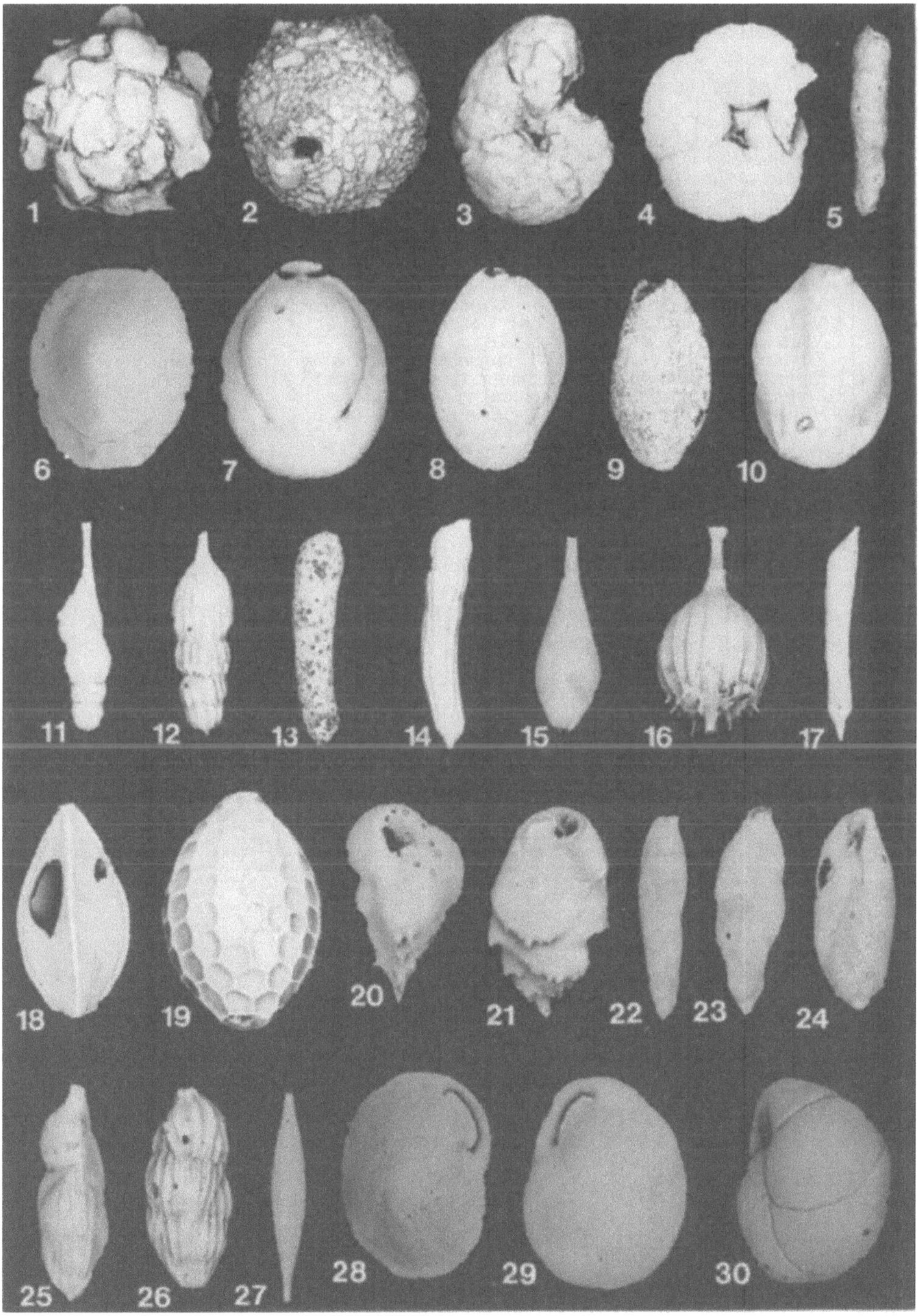
1
2
3
4
5
6
7
8
9
10
11
12
13
14
15
16
17
18
19
20
21
22
23
24
25
26
27
28
29
30

PLATE 3.2: ROTALIIDA

Fig. 1 *Spirillina* sp. General view x424.
Fig. 2 *Discorbinella* (?) sp. Spiral view x141.
Fig. 3 *Discorbinella* (?) sp. Umbilical view x141.
Fig. 4 *Gyroidina* sp. Umbilical view x354.
Fig. 5 *Epistominella vitrea* Parr. Spiral view x354.
Fig. 6 *Epistominella vitrea* Parr. Umbilical view x283.
Fig. 7 *Cibicides* ex gr. *lobatulus* (Walker & Jacob). Umbilical view x35.
Fig. 8 *Cibicidoides pachyderma* (Rzehak). Spiral view x127.
Fig. 9 *Cibicidoides pachyderma* (Rzehak). Umbilical view x92.
Fig. 10 *Hyalinea balthica* (Schroter). General view x78.
Fig. 11 *Pullenia osloensis* Feyling-Hanssen. General view x318.
Fig. 12 *Melonis affinis* (Reuss). General view x106.
Fig. 13 *Melonis affinis* (Reuss) (*M. pompilioides* type). General view x78.
Fig. 14 *Nonionella turgida* (Williamson). General view x177.
Fig. 15 *Nonionella* sp. General view x354.
Fig. 16 *Elphidium* ex gr. *clavatum* (Williamson) (*E. excavatum* type). General view x53.
Fig. 17 *Elphidium groenlandicum* Cushman. General view x53.
Fig. 18 *Elphidium ustulatum* Todd. General view x120.
Fig. 19 *Haynesina orbicularis* (Brady). General view x106.
Fig. 20 *Ammonia batava* (Hofker). Spiral view x99.
Fig. 21 *Ammonia batava* (Hofker). Umbilical view x92.
Fig. 22 *Ammonia batava* (Hofker). Apertural view x127.

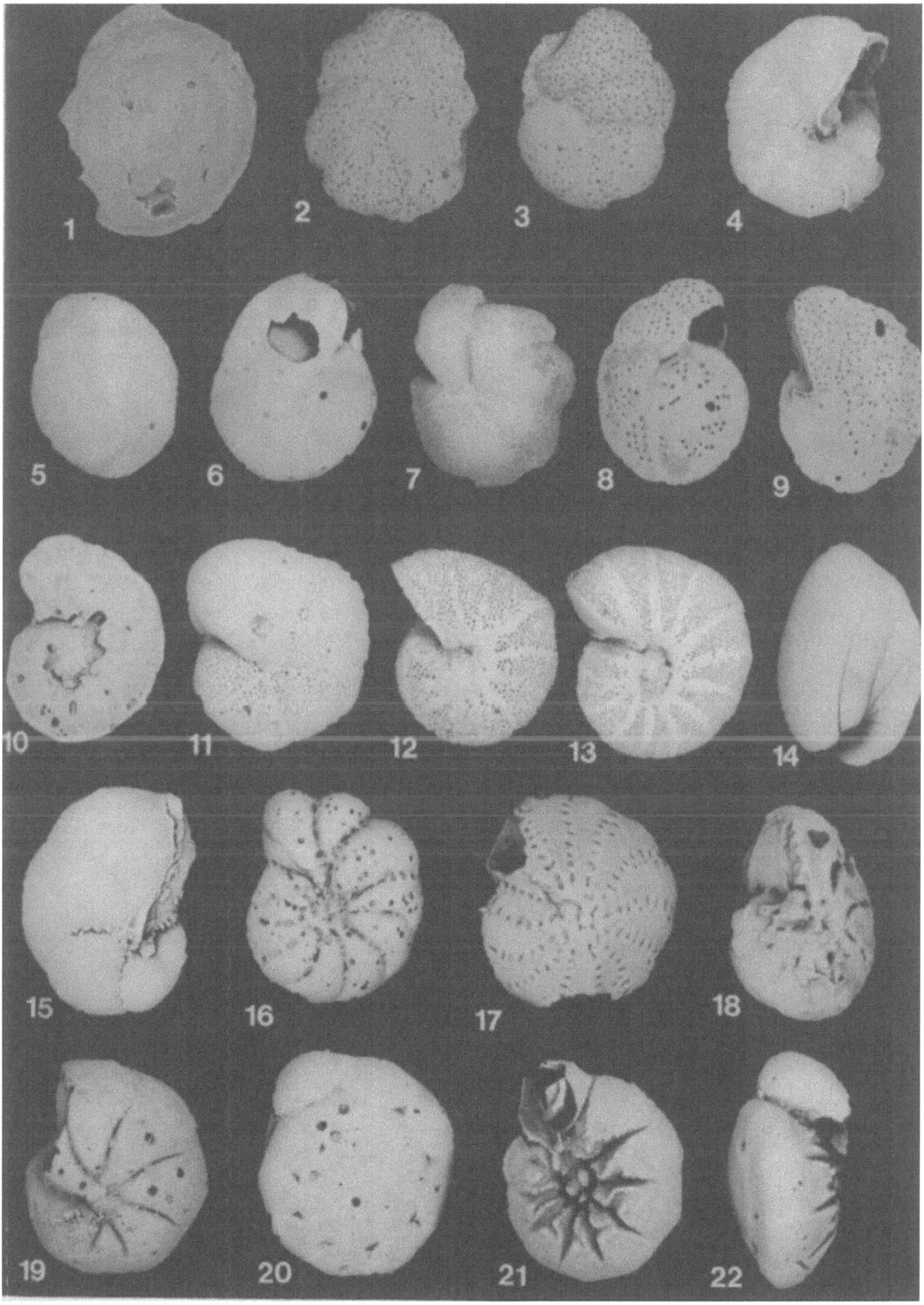
1
2
3
4
5
6
7
8
9
10
11
12
13
14
15
16
17
18
19
20
21
22

# 4 APPLICATION OF MICROPALAEONTOLOGY TO HYDROCARBON EXPLORATION IN THE NORTH SEA BASIN

Philip Copestake

## Abstract

The role of micropalaeontology in the hydrocarbon exploration of the North Sea is reviewed. The small size and abundance of microfossils in drill samples and their rapid evolution makes them valuable for correlation, age determination, unconformity identification, depositional sequence characterization, lithostratigraphic characterization and palaeoenvironmental interpretation in the subsurface.

The application of micropalaeontology is crucial to exploration, appraisal and field development studies and impacts on drilling problems (such as coring point selection and terminal depth decisions), assessment of reservoir distribution (and estimation of reserves), trap evaluation and source rock evaluation. These applications are illustrated by reference to both published and previously unpublished case histories.

The relative value of particular microfossil groups is discussed in relation to the range of known North Sea exploration plays. Micropalaeontology has evolved in parallel with, and in response to the requirements of the oil industry for more detailed correlation studies and reservoir zonations. As the North Sea hydrocarbon province passes into its mature phase, micropalaeontology is seen as one of many technologies with a vital part to play in prolonging the life of the province, by maximizing the extraction of hydrocarbons from increasingly small accumulations.

## 4.1 Introduction

The aim of this chapter is to review the applied role of micropalaeontology in hydrocarbon exploration and production in the North Sea Basin. The development of the discipline, in tandem with the pursuit of particular plays, is discussed and illustrated with selected case histories, several of which have not previously been published. Emphasis is placed on the principles of using the discipline in an applied context, rather than in a purely biostratigraphic way. The science of micropalaeontology has historically been supported and driven during its development because of its applied usefulness, and the oil industry continues to provide employment for many micropalaeontologists for this reason. Despite this, papers describing the applied utility of micropalaeontology are relatively few; this is mainly due to the confidential nature of the many industry studies in which micropalaeontology has been applied successfully to exploration problem solving. Most case studies that have been published or are due to be published are based on individual hydrocarbon fields, and do not emphasize the principles of applied micropalaeontology or discuss the full range of applied uses in the field of hydrocarbon exploration and production. This chapter was thus written in an attempt to provide an overview of the range of techniques and principles involved in the application of the science. The review is necessarily brief in its coverage of some areas, but an attempt has been made to mention all the significant applications of the science and of the important publications on the North Sea, to which the interested reader can refer for more detail. The emphasis in the review may seem to be unduly biased to the United Kingdom sector of the North Sea region. However, because of the greater maturity of exploration in the UK sector relative to other sectors of the North Sea, more papers have been published on this area than, for instance, on the Norwegian, Danish or Dutch sectors.



*D. G. Jenkins (ed.), Applied Micropaleontology*, 93–152.
© 1993 *Kluwer Academic Publishers.*

Coverage is restricted to the North Sea, because of its importance as an exploration province. A number of papers have also been published on offshore European areas such as West of Shetland, the Celtic Sea and Fastnet Basin, but these have not been included in this review. No plates of microfossils are included because a number of publications are now available which cover this aspect (see section 4.4 for a review of important publications).

The case histories discussed cover the exploration phase and subsequent appraisal and development of discoveries. The term micropalaeontology is used in its broad sense to include the study of all microfossil groups whether of plant affinity (e.g. spores, pollen, dinoflagellate cysts, diatoms) or animal affinity (e.g. radiolaria, foraminifera).

Since the initial gas and oil discoveries were made in the 1960s and 1970s in the North Sea, the basin has developed unusually rapidly to a mature stage and has proved to be one of the most prolific hydrocarbon provinces in the world. In just over twenty years, over 120 named oil and gas fields have been discovered in the UK, Norwegian, Danish and Dutch sectors, in addition to at least as many smaller accumulations that may be developed at a future date (Brennand, Van Hoorn and James, 1990) (Figure 4.1). Over this period micropalaeontology has developed as the only consistently reliable means of age dating the subsurface successions and of correlating between wells. The discipline has developed in response to the needs of the industry for increasingly detailed biozonations (including reservoir zonations) as discoveries are appraised, reservoir distributions are mapped and fields put into production (with the attendant requirement for high resolution biozonations).

The particular advantages of using microfossils for subsurface biostratigraphy are:-

(a) small size.

The small size of microfossil groups enables them to survive destruction by standard drill bits (average diameters: dinocysts 0.05mm, calcareous nannoplankton 0.015mm, foraminifera 0.4mm). As the bulk of samples comprise accumulations of rock fragments (or "ditch cuttings") derived by fracturing of rocks by drill bits, most macrofossils are destroyed by the drilling process.

(b) abundance.

Microfossils can achieve astonishing abundance, some samples comprising almost 100% fossil oozes. In these cases the microfossils can be of rock forming density. An example from the North Sea is of Lower Aptian and Lower Barremian limestones (Tuxen Formation, Hesjedal and Hamar, 1983) packed with planktonic and benthonic foraminifera. In such occurrences, several thousands of specimens can occur in a 10ml volume sample. In an equivalent sample one would be fortunate to find a single fragment of a macrofossil. Changes in the nature of these assemblages are used by the exploration micropalaeontologist to interpret the age and environment of deposition of the rock and correlate it with sequences in other wells.

Figure 4.1 Distribution of oil and gas fields in the North Sea.

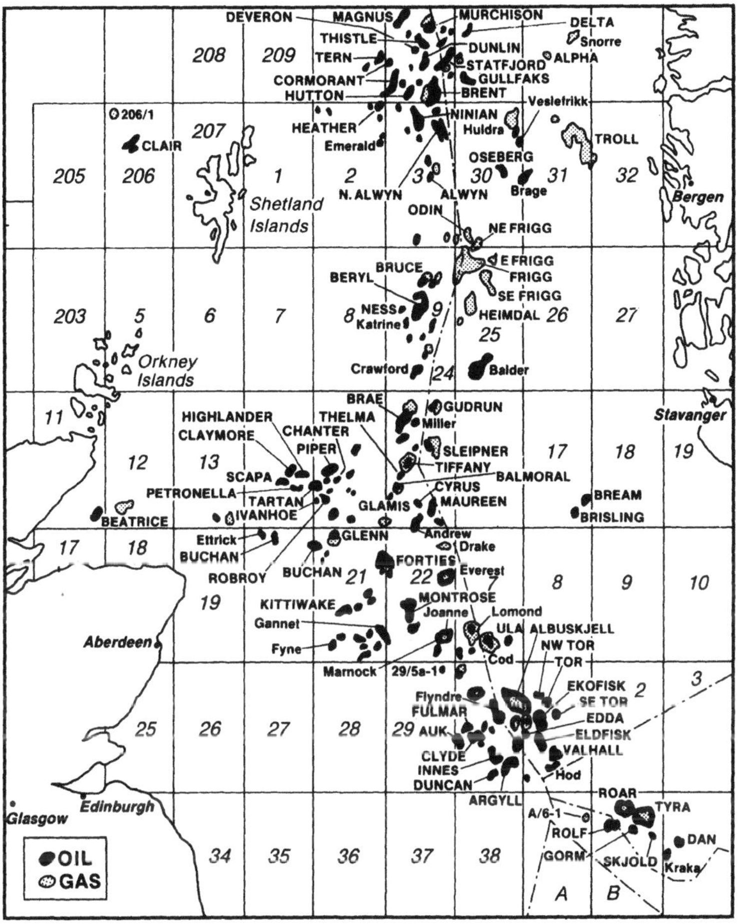


## 4.2 Review of techniques

Microfossil assemblages can be used in a variety of ways to make geological interpretations and deduce subsurface relationships which are important to exploration and production.

4.2.1. Correlation

This is the micropalaeontological technique with most impact in exploration. The recognition of stratigraphically significant and distinctive assemblages allows rock units to be correlated between wells. Changes in these assemblages, defined by stratigraphic "tops" (uphole disappearances, extinctions, last appearance datums, LADs, or first downhole occurrences, FDOs), "bases" (uphole appearances, inceptions, first appearance datums, FADs or last downhole occurrences, LDOs) or abundance changes of index species can be tied between wells as correlation lines. Experience within a basin indicates which of these changes, or bioevents, are the most reliable for correlation. These bioevents are considered to approximate to time lines, allowing for inaccuracies of sample spacing, sample processing and detrimental drilling factors (see below). Between the correlation lines variations in lithofacies can be mapped to determine, for instance, the nature and extent of reservoir distribution, such as sandstones and carbonates.

It is essential to compare this biostratigraphic correlation framework with the wireline logs to assess which log changes (log "breaks" or "picks") are the most useful for correlation and, related to this, which lithostratigraphic changes appear to be synchronous within the basin.

### 4.2.2 Age determination

Knowledge of how microfossil ranges compare with standard chronostratigraphic divisions of the geological time scale (e.g. Cretaceous Period, Hauterivian Stage, *inversum* ammonite Biozone) (Figure 4.2) means that these subdivisions can be recognized in well samples. For the Cenozoic, the standard biozones are based on the microfossil groups of calcareous nannoplankton (NP and NN Zones of Martini, 1971) and planktonic foraminifera (P Zones of Blow 1969, 1979; Berggren and Miller, 1988). Thus where these microfossil groups occur in well samples, standard biozones can be recognized directly. For most of the Phanerozoic, however, standard biozones are based on macrofossil groups such as ammonites (Jurassic, Cretaceous), belemnites (Cretaceous) and bivalves (Cretaceous). For correlation from wells to these standard biozones, it is thus essential that microfossil distributions in standard onshore sequences are studied in detail and calibrated against the standard biozones. The extrapolation of onshore age ranges is not without difficulty, however, because some North Sea taxa are known to have different ranges offshore from those in onshore sections.

An interpretation of absolute geological time (geochronological age) is more tenuous. There are relatively few, chronostratigraphically widely spaced, points from which K-Ar dates have been obtained and thus the age of intervening strata has to be estimated. Due to experimental inaccuracies, combined with varying opinion on the reliability of different radiometric dating techniques, several published geochronological schemes exist with different ages for stage and period boundaries (Harland et al., 1982; Kent and Gradstein, 1985; Snelling, 1985; Haq, Hardenbol and Vail, 1987; Harland et al., 1990). This can introduce significant errors into burial history and geochemical modelling studies based on well data.

It is important to be able to "date" a fossil assemblage as precisely as possible (ideally in terms of standard biozones). Accurate basin evaluation, including assessment of such factors as source rock hydrocarbon generation potential necessitates reconstruction of burial depth and duration and the age magnitude of missing sections at unconformities. Accurate reconstruction of tectonic history and timing of trap creation (including time of deposition of sealing sediment) relative to timing of hydrocarbon generation and migration also rely on the ages derived from microfossil age interpretations for dating.

Accurate comparison of identified depositional sequences for dating boundaries with published schemes (e.g. Vail, Mitchum and Thompson, 1977; Haq et al., 1987) also requires very accurate biostratigraphic dating. One problem with the latter authors' approach is that they tend to name sequence boundaries with age assignations in millions of years. This is unfortunate, because the agreed absolute time scales

Figure 4.2 Stratigraphic ranges of major microfossil groups. Ranges compiled from following sources: radiolaria, Kling (1978), Anderson (1983); diatoms, Burckle (1978); silicoflagellates, Haq (1978), Perch-Neilsen (1985a); foraminifera, Ross and Haman (1989); calcareous nannoplankton, Perch-Neilsen (1985b); ostracodes, Pokorny (1978); calpionellids, Remane (1985); calcareous algae, Wray (1978); conodonts, Higgins and Austin (1985), Muller (1978); acritarchs, Williams (1978), Knoll and Swett (1985); dinoflagellates, Williams and Bujak (1985), Williams (1978); chitinozoa, Jansonius and Jenkins (1978); spores and pollen, Heusser (1978).

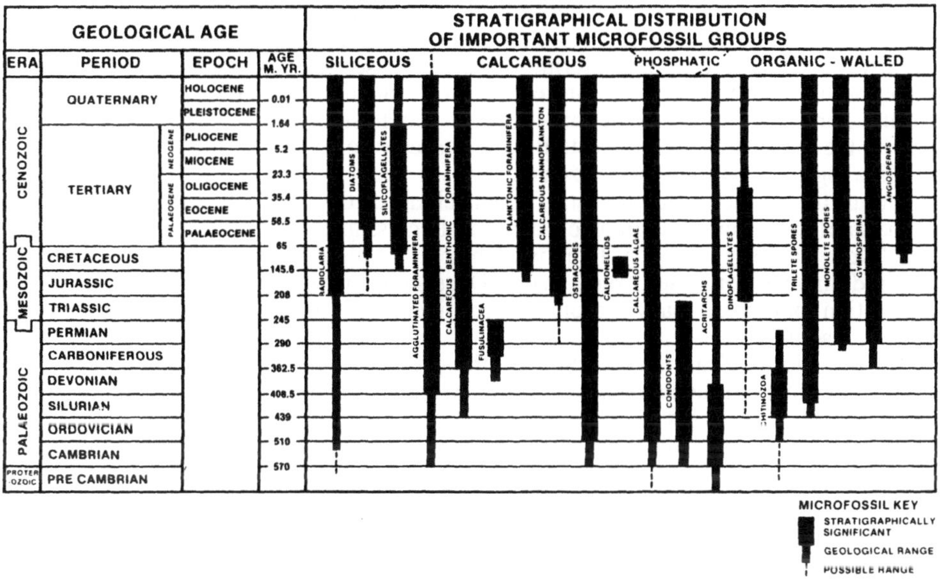


change with alarming regularity according to the latest developments in geochronological techniques and these assignations may quickly become out of date (e.g. the 195 million years Late Sinemurian sequence boundary of Haq et al., 1987, which was formerly referred to as the 184 million years sequence boundary by Vail, Hardenbol and Todd, 1984)

### 4.2.3 Unconformity Identification

The identification of missing sections, whether in the subsurface or at outcrop can only be reliably demonstrated by biostratigraphy. In well sequences such hiatuses are indicated by the absence of one or more biozones (assuming that marker species absence is not due to poor sample coverage or detrimental drilling parameters, see below). Good quality biostratigraphic data will allow the time duration of the missing section to be estimated by correlation with standard biozones and thence to an absolute time scale.

It is important to date as precisely as possible the unconformity at the location where the magnitude of the break is the smallest. This allows the geological event which caused the unconformity to be dated at the point where the strata around the break pass laterally into conformability ("correlative conformity" of Vail et al., 1977). Thus, the unconformity is dated between the oldest rocks overlying it and the youngest rocks underlying it.

Unconformity identification is of great importance to stratigraphic trap evaluation, as the pinch out of a reservoir sandstone onto or beneath an unconformity surface will provide a trapping mechanism where overlying sediments provide a viable seal.

In a tectonically active basin such as the North Sea, many regional unconformities have been generated by the interplay of tectonic changes and eustatic sea level changes. These factors combine to create regional sequence boundaries which are often, though not always, developed as unconformities in a well section. Accurate identification of unconformities is of great importance to basin evaluations, as these correspond to the seismic horizons mapped by the oil company geophysicists. Thus if a biostratigrapher can, by the integration of the biostratigraphic succession with the lithostratigraphy and wireline log patterns, suggest horizons of missing section in a well, this provides a most important tie for the explorationist. This matching of the well data with the seismic is the basis of the chronostratigraphic notation often assigned to regional seismic markers (e.g. "Near Base Cretaceous"). Even with the current high resolution seismic data being acquired in the North Sea, such that 3-D seismic surveys are now being shot for exploration purposes, many unconformities identified from wells are often beyond seismic resolution.

In the North Sea, unconformities are probably the major single reason for prospective intervals, containing reservoirs, being absent from a particular well location. However, if the missing section is identified from biostratigraphic data in a well, it may be possible, by correlation with other wells, to predict a location (possibly downdip) at which the reservoir may reappear. Further, owing to the tectonic cause of many unconformities, reservoir sands are often located on unconformity surfaces. If these can be identified and mapped (in wells and seismically), this can provide predictive power to the explorationist.

### 4.2.4 Application to sequence stratigraphy analysis

#### 4.2.4.1 Sequence stratigraphy models

The depositional sequence concept was first established by Sloss (1963) but was later modified and more clearly defined by the Exxon school, particularly Vail et al. (1977), Vail et al. (1984) and Van Wagoner et al. (1989). These authors observed that sedimentary successions are naturally divisible into discrete depositional sequences. A depositional sequence is defined as "a stratigraphic unit composed of a relatively conformable succession of genetically related strata and bounded at its top and base by unconformities or their correlative conformities" (Mitchum et al., 1977). Since its promotion by Vail, Mitchum and Exxon colleagues in 1977, the depositional sequence concept has revolutionized interpretations of subsurface stratigraphy, particularly because such sequences can be recognized on seismic sections as seismic sequences. The sequences are recognized by reflector terminations at surfaces known as sequence boundaries; these surfaces are of chronostratigraphic significance in that all strata above the surface are younger than all the strata below, even though the age of the rocks above or below the surface is variable. It was claimed (Vail et al., 1977) that these depositional sequences represent a succession of changing environments determined by eustatic changes of sea level and that an individual sequence represents the rocks deposited between a sea level rise and a sea level drop. While this assertion of eustatic control has not found widespread acceptance (e.g. Miall, 1986, 1992; Summerhayes,

1986; discussions at Dijon Conference, May, 1992), the sequence approach is of great value to petroleum exploration. It has been observed that particular parts of sequences contain reservoir rocks and other parts sealing shales. For example, in a slope or basinal setting, relative sea level fall at the base of a sequence can result in the deposition of a submarine fan facies due to the erosion of shelf sediments and their redeposition beyond the shelf edge (lowstand systems tract). Subsequent sea level rise (transgressive systems tract) invariably results in blanketing and sealing of the fan by shales (sometimes of source rock facies), forming a regional condensed section. During the final part of the sequence, where sea level reaches and pauses at a maximum (highstand systems tract), regressive deposits can build out into the basin as deltas and coastal barrier complexes, or, in tropical areas, as prograding reef facies. This full sequence is not developed at a single location, the submarine fans being deposited on the slope or basin plain, the deltaic sediments being deposited on the shelf. With this knowledge of the distribution of reservoir, source and seal facies within sequences, their recognition on seismic in undrilled areas becomes a powerful predictive tool.

Once the seismic sequences are tied into well sections, where the well ties are confident, it is found that the sequences can be characterized by micro-palaeontological events (tops, bases and species abundances). Of particular significance is the fact that the most transgressive parts of sequences are frequently characterized by flood abundances of oceanic or deep water microfossils such as radiolaria or planktonic foraminifera. The microfossil abundances are frequently associated with peaks on the gamma ray electric logs reflecting the radioactive, organic-rich content of the shale horizons. These shales mark the levels of significant onlap (transgression) onto structural highs and clearly reflect a relative increase in sea level (whether of eustatic or tectonic origin). Such rock units are condensed deposits, and the fossil concentration and organic rich content are due, in part, to the lack of land-derived clastic sediment in the deposit as a result of the transgression suppressing the significant offshore movement of coarse clastics. These major condensed sections are interpreted to contain a surface, the maximum flooding surface (Galloway, 1989; surface of maximum flooding of Haq et al., 1987). In the subsurface it is relatively easy to recognize these maximum flooding surfaces on palaeontological, lithological and log character and to identify depositional packages or sequences between them. These packages correspond to the depositional episodes or genetic stratigraphic sequences of Galloway (1989). In the North Sea basin, many of the major sequence breaks (unconformities) appear to coincide with condensed sections in well successions, and thus correspond with the level of the hiatal surface generated at the time of transgression, as has also been documented by Galloway (1989) in the Gulf Coast, USA. This contrasts with the Exxon model, which does not consider this type of unconformity at all, but lays emphasis on the sequence boundary, theoretically generated at a time of relative drop in sea level and resulting in a reduction in accommodation space and associated subaerial exposure, and located between maximum flooding surfaces (e.g. Posamentier and Vail, 1988; Van Wagoner et al., 1989, 1990). The Exxon type sequence boundaries are much more subtle to identify than the regional condensed sections, and often cannot be proven to be present at all in wells; this may be due to the apparent lack of evidence of subaerial exposure in many well sections. In reality, both sequence boundaries sensu Exxon and genetic sequence boundaries sensu Galloway are present and both may equate to an unconformity. However, the fact remains that in North Sea well sections many of the observable unconformities fall at the levels of the regional condensed sections. This matches the observations of Kaufmann (1988), who noted that the unconformities which occur in offshore basinal sequences in the Upper Cretaceous of the Western Interior Basin, USA at around eustatic highstands, form regionally within very narrow time intervals and represent bypass and condensation intervals of short duration reflecting sediment starvation.

The presence of a third unconformity surface, the transgressive surface, may also confuse the sequence boundary placement, particularly as, conceptually, this surface, otherwise known as the ravinement surface (Nummedal and Swift, 1987), can be diachronous. This surface is located above the sequence boundary, but beneath the maximum flooding surface; however, it may merge with either of the other two surfaces. It will merge with the sequence boundary where the lowstand systems tract sediments are not preserved and it will merge with the maximum flooding surface where the transgressive systems tract sediments are not developed. In a cored section any of these three surfaces may be seen as an erosion surface; thus identification of an erosion surface with any one of them is difficult in a single well study or without refined biostratigraphic data.

The consistently stated view of all stratigraphers undertaking sequence stratigraphic analysis (e.g. Vail and Wornardt, 1990), is that the maximum flooding surface condensed section is the easiest feature to identify in well sections. Ease of practicality may thus favor this approach, rather than that of Exxon, as the primary means of sequence stratigraphic subdivision.

Location of the sequence boundary sensu Exxon may be valuable to attempt; however, the actual placement of it in a well section is frequently uncertain in conformable sections. In sand prone shelf successions (such as the Piper Formation in the North Sea), a typical sequence may comprise a highstand prograding, coarsening upwards sand unit overlain by a transgressive, fining upwards sand unit. The sequence boundary sensu Exxon may be located at the boundary between the two sands, that is within a lithostratigraphically defined sand member or formation. Shallow marine Upper Jurassic sand units in the North Sea which have been referred to the Piper, Sgiath and Fulmar Formations are of this type (Harker Gustav and Riley, 1987; Johnson, Mackay and Stewart, 1986) and, when cored, it is sometimes possible to locate pebble horizons within the sands which may correspond to Exxon type 1 sequence boundaries, at which subaerial exposure may have occurred at the basin margins. If this is correct, it may be predicted that in the basin, lowstand facies, of possible reservoir facies, may be present. Such an observation may be of great value in indicating additional exploration plays which may still be present in the deep, unexplored part of the basin, such as the Central Graben. In basinal claystone-dominated sections where no sands are developed, the sequence boundary can only be suggested to occur at a very slight log feature (at a gamma ray minimum and sonic velocity maximum), but it should be noted that this is often not where the actual stratigraphic break is observed from biostratigraphic data, the latter being at the regional condensed section below the position of the sequence boundary (at a gamma ray log maximum and sonic log velocity minimum).

The problem of which sequence model to follow is exacerbated by the fact that North Sea seismic data cannot at present resolve the two unconformities which may be present (maximum flooding surface and sequence boundary sensu Exxon).

Over the past few years much effort has been expended by North Sea operating companies to apply sequence stratigraphic concepts to subsurface exploration. At the 4th Conference on the Petroleum Geology of North West Europe (April, 1992, proceedings due to be published in 1993), a number of papers were given describing the application of sequence stratigraphy to the Jurassic and Palaeocene of the North Sea and West of Shetland areas. These included a paper by Galloway (in press) in which he described a sequence stratigraphic framework for the North Sea Tertiary; he ascribes sequence boundaries between 15 "depositional episodes" over the Palaeocene-Miocene interval to various causes including basin margin flooding, shoaling and wave base lowering, regional sediment starvation, reorganization in the pattern of sediment supply and major changes in marine energy fluxes. Other papers describing the application of sequence stratigraphy to the North Sea Palaeogene were given by Vining et al. (in press) (Esso), Armentrout et al. (in press) (Mobil), den Hartog Jager et al. (in press) (Shell), Anderton (in press) (BP) and Morton et al. (in press) (British Geological Survey and Unocal). An additional paper was presented by Mobil on the sequence stratigraphy of the Palaeogene of the Faeroes Basin, West of Shetland (Mitchell et al., in press). Papers describing Upper Jurassic sequence stratigraphy of the North Sea at the conference were fewer (Donovan et al., in press, Partington et al., in press, Partington, Copestake and Mitchener, in press). All these authors agreed that the sequence studies could not have been performed successfully without the availability of high resolution biostratigraphy.

A major project is currently underway to produce a sequence stratigraphic scheme, similar in format to that summarized by Haq et al. (1987), for the European basins (Project Mesozoic and Cenozoic Sequence Stratigraphy of European Basins). The aim of this project is to document sequence stratigraphic schemes for Europe and to publish results by late 1993. Many papers were delivered at the International Symposium on Mesozoic and Cenozoic Sequence Stratigraphy of European Basins (Dijon, France, May, 1992). Of these, those by Neal (Palaeogene, Central North Sea), Michelsen et al. and Heilmann-Clausen et al. (Danish North Sea), Kiorboe (Tertiary, Faeroe Basin), Stephen et al. (Jurassic, Inner Moray Firth), Underhill and Partington (Middle Jurassic, North Sea) were pertinent to the North Sea, and all relied heavily on biostratigraphy as a vital aid to sequence analysis.

The essential role of micropalaeontological stratigraphy in sequence analysis is well illustrated by Mitchener et al. (1992), who describe the sequence scheme applied by BP to a regional assessment of the Brent Group (Middle Jurassic) of the North Viking Graben region, spanning a large part of the UK and Norwegian sectors between 58-62 degrees latitude. This sequence stratigraphic approach offers a more realistic model for basin analysis and evaluation of remaining exploration potential within the basin than a traditional lithostratigraphic scheme, and it also incorporates the possibility of being able to predict areas of improved reservoir development. This particular sequence model is very dependent on the palynological biostratigraphy to aid the recognition of sequences within well sections, because the sequences are often beyond seismic resolution. However, in a non to marginal marine setting such as the Brent Group and its equivalents, such as the Bruce and Beryl Sands and Sleipner Formation, there is an obvious possible palaeoenvironmental control upon isolated stratigraphic tops of rare dinoflagellate cysts, for instance. Whittaker et al. (1992) has discussed some of the pitfalls in attempting to establish a chronostratigraphically valid palynological correlation in the Brent Group.

A part of the new research on sequence stratigraphy is the classification of sedimentary cycles. All stratigraphers have observed the cyclicity of sedimentary rock successions. Duval, Cramez and Vail (in press) describe a classification of stratigraphic cycles into first order or continental encroachment cycles (with a duration of >50 million years), second order or major transgressive-regressive cycles (duration of 3my to 50my) and third order or sequence cycles (of 0.5my to 3 my duration). The last cycles can be subdivided into systems tracts and are those which most explorationists utilize for basin analysis. First, second and third order cycles are those which are shown on the Haq et al. (1987) chart. Fourth to sixth order cycles, termed either parasequence cycles (Van Wagoner et al., 1990) or Milankovitch cycles (with a duration of 0.01my to 0.5my) are the building blocks of systems tracts. It is found in the North Sea that high resolution biostratigraphy can adequately resolve first to third order cycles, but often the fourth order cyclicity occurs within regionally based biozonations schemes.

Stratigraphic sequences (third order cycles) frequently show characteristic wireline log signatures, dependent on setting (i.e. basin or shelf). Some typical well log signatures of sequences have been figured by Rioult et al. (1991) and Vail and Wornardt (1990), but these are based primarily on the Gulf Coast USA. The log signatures for a succession of North Sea Jurassic sequences from an Inner Moray Firth well are shown in Figure 4.3.

Despite the recent explosion of literature and research on sequence stratigraphy, there is very little published regarding the use of biostratigraphy in sequence analysis. Vail and Wornardt (1990) and Wornardt (in press) have described how biostratigraphic data can be used in particular to help the identification of condensed sections. Faunal abundances can be used to attempt to separate maximum flooding surface condensed sections (which contain high abundance and high diversity of microfauna, plus evidence of abrupt deepening) from minor condensed sections. The latter yield less significant abundance peaks of lower diversity. Such minor condensed sections occur at the top of basin floor fans, slope fans and lowstand prograding complexes. On well logs, they will display high gamma signatures similar to regional condensed sections at maximum flooding surfaces; a pitfall of sequence stratigraphy from well data is the confusion of local (or minor) and regional condensed sections (containing maximum flooding surfaces). Wornardt (in press) describes a number of different types of minor condensed sections and discusses how these can be separated from each other and from the maximum flooding surface condensed section. His work is based on the Tertiary of the Gulf Coast USA, in which calcareous benthic and planktonic foraminifera and calcareous nannofossils are continuously present. This contrasts with the North Sea in which agglutinated foraminifera, radiolaria and dinoflagellate cysts are predominant in marine successions. Aspects of the technique can, however, be applied to the North Sea, and particular major condensed sections contain characteristic influxes of planktonic microfossils, such as planktonic foraminifera and nasselarian radiolaria. Abundances of dinocysts are more difficult to interpret, as certain types appear to be abundant in lowstand systems tracts while others are abundant in transgressive intervals.

The three fold calibration of seismic data, biostratigraphy and wireline logs is essential in sequence analysis because of problems of seismic resolution. The cross check provided by micropalaeontological and log characterizations of sequences in wells provides vital control points between which seismic interpretation can be difficult due to complex tectonics, variable seismic data quality or simply resolution. Where sequences become thin, as on structural highs or in basinal settings, their recognition may be beyond seismic resolution, but they may still be recognizable by micropalaeontology and log signature. Calibration of the logs with micropalaeontological biozones is also essential as different sequences can show very similar log signatures and may only be separable using micropalaeontological criteria.

Figure 4.3 Depositional sequence interpretation of part of the Upper Jurassic of well 12/21-2.

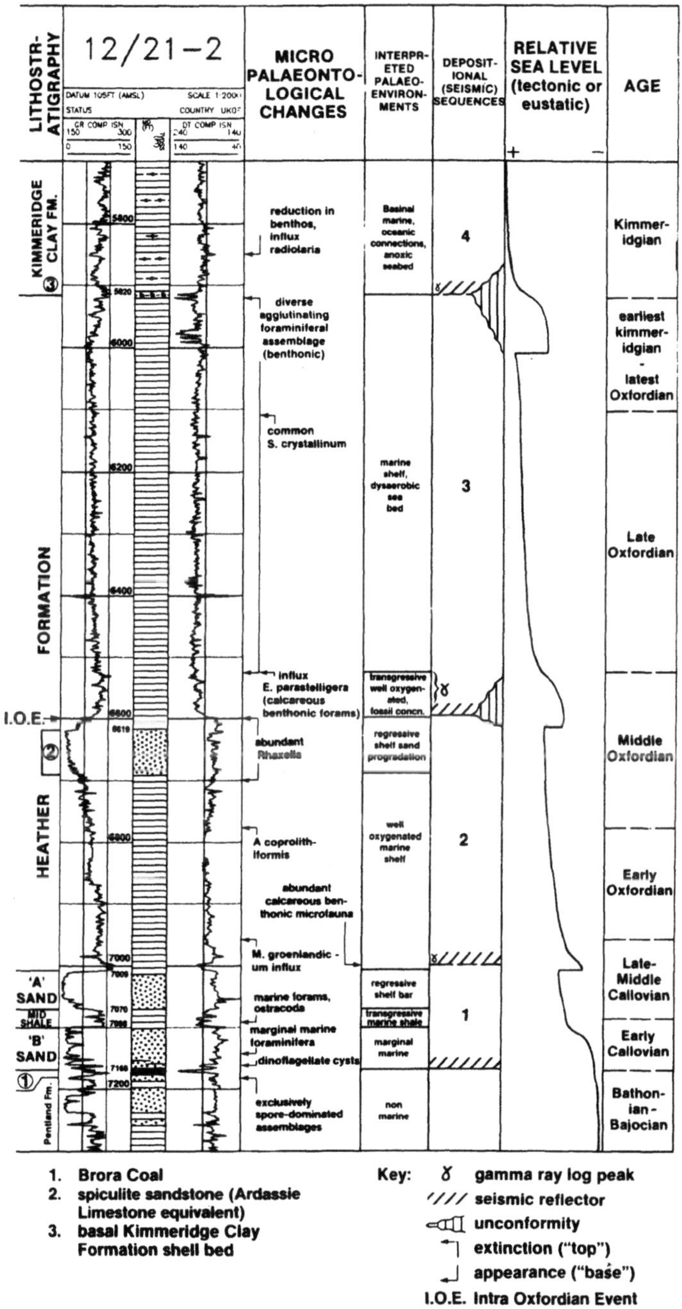

4.2.4.2 Sequence analysis of well 12/21-2, Inner Moray Firth

Figure 4.3 illustrates the sequence analysis of well 12/21-2 in the Inner Moray Firth basin of the Central North Sea. The interval displayed is of Middle to Late Jurassic age and ranges from Bajocian-Bathonian to Early Kimmeridgian, with an apparently complete development of the Callovian-Oxfordian. The interval is dated on the basis of changes in the dinoflagellate cysts and foraminifera populations extracted from the well, which can be correlated with ammonite-dated onshore sections around the Moray Firth coast. The well lies close to the Beatrice Field (block 11/30) which displays a very similar stratigraphy. Stratigraphic data from the area have been published by Linsley et al. (1980), Johnson (1984) and Andrews and Brown (1987). Vail and Todd (1981), Underhill (1991), Stevens (1991) and Stephen et al. (in press) have discussed a sequence analysis of the Jurassic in the Beatrice area and MacLennan and Trewin (1989) have published palynological and sedimentological data from Beatrice. The Inner Moray Firth basin is exceptional in offering the best seismic resolution of Upper Jurassic sequences of any part of the North Sea, owing to the relatively shallow burial depth, the development of thick Jurassic sections, and the lack of a thick Tertiary clastic and Upper Cretaceous Chalk Group cover.

The 12/21-2 well section can be subdivided into four depositional sequences. The boundaries between these sequences are placed approximately at the horizons of major seismic sequence boundaries in the succession. At each of these levels, unconformities have been documented within the Inner Moray Firth basin, which can be identified on seismic sections and in wells. In 12/21-2, however, the sequence boundaries appear to be conformable, within the resolution of the biostratigraphy. This is thus a valuable section in which to accurately place the sequence boundaries and maximum flooding surfaces into a chronostratigraphic scheme. The boundaries between these sequences are placed at log markers which are considered to correspond to maximum flooding surface condensed sections. This view is based on the fact that equivalent log markers, characterized by their distinctive biostratigraphic signature, can be correlated throughout the North Sea basin. In the interval discussed, four major regional condensed sections can be identified, and three of these fall close to horizons of the major seismic events in the succession. One of the regional condensed sections, at the top of the Lower Callovian, is beyond seismic resolution.

The well section is believed to be predominantly shallow marine throughout. No lowstand systems tracts such as coastal plain intervals or deep marine sands are developed. It is also notable that intervals ascribed to transgressive systems tracts are very thin and are represented by claystones. These factors have resulted in the Exxon type sequence boundaries and the genetic sequence boundaries (maximum flooding surfaces) falling close together within the succession, and they cannot be separated seismically. The seismic sequence boundary depths have been taken from the composite log for the well and are placed at velocity changes on the sonic log. Lithostratigraphic terminology is taken from published sources, but the reservoir terminology of the Beatrice Field has not been formally described. The identifiable sequences, and their characterization, are as follows:-

Sequence 1 (J1.4 seismic sequence of Underhill, 1991)

Age: Early-Late Callovian. This is based on the appearance of characteristic dinoflagellate cysts at the base of the sequence, including an influx of *Chytroesphaeridia hyalina* (Raynaud, 1978) Lentin and Williams, 1981 within the Mid Shale, indicative of the Early Callovian, from the base to the mid part of the sequence.

This agrees with the age interpretation of the equivalent lithostratigraphic units in well 11/30-2 by MacLennan and Trewin (1989) who assign an intra Early Callovian (*calloviense* ammonite zone) age to a dinocyst association 11 ft above the base of the B Sand in that well. The occurrence of top common *Nannoceratopsis pellucida* Deflandre, 1938 within the A Sand (7050 ft) indicates a Mid Callovian age for the lower A Sand and upper Mid Shale. This agrees with the age interpretation placed on the equivalent section in the Beatrice Field (MacLennan and Trewin, 1989). The upper part of the A Sand is typically impoverished biostratigraphically, but the basal Heather Formation marine claystone overlying the sand contains an abundance of *Mendicodinium groenlandicum* (Pockock and Sargeant, 1972) Davey, 1979, indicative of a latest Callovian or older age. The A Sand is thus typically ascribed a Mid to Late Callovian age in the Beatrice area.

Lithostratigraphy: represented by the A Sand and B Sand, separated by the Mid Shale (Johnson, 1984; Stevens, 1991) ( Beatrice Formation of Andrews and Brown 1987).

Palaeoenvironment: the base of the B Sand (7167 ft-7160 ft) appears initially to be of non marine origin, indicated by the absence of marine fossils. This agrees with the observations of MacLennan and Trewin (1989) in the Beatrice Field. From 7160 ft, sediments can be interpreted as being of marine origin, indicated by the appearance of marine microplankton (dinoflagellate cysts) and coarse-grained agglutinating foraminifera of marine or marginal marine affinity. The B Sand is overlain by the marine Mid Shale containing abundant bivalves, an abundance increase of dinoflagellate cysts and the first incoming low diversity microfauna of calcareous foraminifera and ostracoda. This is overlain by a marine shelf sand (?barrier bar) (A Sand), containing marine microplankton (dinoflagellate cysts).

Sequence interpretation: the base of the sequence equates to the base of the Callovian, which, together with the appearance of marine (albeit initially marginal marine) deposits above the non-marine Pentland Formation (capped by the Brora Coal), clearly represents a marine transgression. The basal sand of the Callovian, the B Sand, is fine to medium-grained and carbonaceous, with abundant shell fragments, and probably represents a transgressive sand, which can be referred to the transgressive systems tract. This Early Callovian transgression is a regional event seen throughout the North Sea Basin and also in the Hebrides Basin of Western Scotland. In the Inner Moray Firth Basin, it is also the level of a significant unconformity, with truncation beneath and onlap above (see Andrews and Brown, 1987, Figure 4.4; Vail and Todd, 1981). This is termed the "Bathonian/Callovian Unconformity" by Stevens (1991). A seismic marker is also identifiable at this sequence boundary. The base of the Callovian is a clear sequence boundary probably related to sea level rise. Sequence stratigraphic theory suggests that sequence boundaries should be placed below coals, which represent sediment starvation at the beginning of transgression. In fact, Andrews and Brown (1987) document an unconformity in the Inner Moray Firth at the base of the Brora Coal. This is probably correct, although traditionally in this area the unconformity has been placed at the top of the coal (e.g. Stevens, 1991). The view that the transgression began below the level of the Brora Coal is supported by the recovery of marine dinoflagellate cysts in the shale immediately beneath the coal in well 11/30-2 from the Beatrice Field (MacLennan and Trewin, 1989).

The B Sand passes upwards into a calcareous marine, black shale, containing abundant bivalves (Mid Shale), which probably represents the most transgressive part of the sequence. This is interpreted here as a maximum flooding surface (see Figure 4.3), as it can be correlated over a wide area of the North Sea, particularly through the Viking Graben, within the lower part of the Heather Formation. Vail and Todd (1981) also recognize that this event can be correlated to the Viking Graben, but interpret this as a mid Callovian hiatus. There is no biostratigraphic evidence of a break in the Inner Moray Firth, although a break can sometimes be identified in Viking Graben wells at this level. This maximum flooding surface could be used to identify a regional genetic sequence boundary within this local Inner Moray Firth sequence. On correlating away from the Inner Moray Firth, the lower, transgressive systems tract sand section (B Sand) is often absent from South Viking Graben sections. However, it may correlate with part of the Tarbert Formation in the North Viking Graben.

The sequence is capped by a regressive shelf sand unit, the A Sand, representing the final, highstand part of the cycle (see MacLennan and Trewin, 1989). This sand displays a well developed coarsening upwards log signature in its lower part, above the high gamma log response of the Mid Shale, typical of highstand shelf sands. The mid part of the sand is aggradational (from 7050 ft), with a flat gamma ray response, and the uppermost part (from 7028 ft) exhibits a retrogradational, fining upwards signature with an increasing gamma log response. The Exxon methodology (see log signatures in Rioult et al., 1991) may indicate the position of a candidate sequence boundary at the base of the aggradational log unit, and a transgressive surface at the base of the retrogradational unit. The fining upwards profile culminates in a prominent unit of high gamma log values. This is the maximum flooding surface condensed section. The actual maximum flooding surface is picked at 7000 ft, at a prominent low sonic velocity trough. This is well dated biostratigraphically as intra Late Callovian.

The top of this sequence is marked by a major seismic reflector (see Linsley et al., 1980, Figure 4.8). Vail and Todd (1981) have observed downlap above this sequence boundary. The latter authors recognize this sequence in the Beatrice area as their cycle J2.3 and Underhill (1991) as sequence J1.4. Of note is the fact that these authors place the seismic sequence boundary at the top of the sand. This is very close to the level of the maximum flooding surface, but occurs above the transgressive surface and significantly above the theoretical level of the Exxon type sequence boundary.

Sequence 2 (J1.5 seismic sequence of Underhill, 1991)

Age: latest Callovian to late Middle Oxfordian. The basal shale contains ammonites, together with an influx of the dinocyst *M. groenlandicum*, indicative of a continuing latest Callovian age. The overlying shale in part represents the Early Oxfordian, based on a succession of diagnostic dinocyst species, including the first downhole occurrences of common *Acanthaulax scarburghense* (6870 ft) and of *Wanea fimbriata* (6920 ft), together with the first downhole occurrence of the foraminiferid *Ammobaculites coprolithiformis* (Schwager, 1867) correlatable with the onshore outcrops. The top of the sequence occurs within an abundance of the foraminiferid *Epistomina parastelligera* (Hofker, 1954), the top of which is diagnostic of the latest Middle Oxfordian in onshore sections.

Lithostratigraphy: represented by the marine shales of the lower Heather Formation (Deegan and Scull, 1977) (equivalent to the Uppat Formation of Andrews and Brown, 1987), and capped by an interval rich in *Rhaxella* sponge spicules. The latter always forms a distinctive unit on the gamma and sonic logs, and may be developed as a spiculite sand (as in 12/21-2, see Figure 4.3 Alness Spiculite Formation of Andrews and Brown 1987), or as a siliceous horizon, as in the Beatrice Field (Linsley et al., 1980; Andrews and Brown, 1987).

Palaeoenvironment: most of the interval is represented by shales rich in ammonites and diverse dinocyst and calcareous foraminifera assemblages diagnostic of a well oxygenated, shelf marine depositional setting. The spiculite deposit at the top is of shallow marine shelf origin.

Sequence interpretation: the basal boundary of the sequence is discussed above. From the basal shale upwards, the rest of the section constitutes a broad coarsening upwards sequence, well displayed on the gamma and sonic logs. The sequence culminates in a clean, well sorted, bioclastic, *Rhaxella* rich sand (Alness Spiculite) which displays a well developed, cleaning upwards, profile of upwards decreasing gamma ray values and increased sonic velocity. The base of this profile appears to be a gamma ray maximum at 6720 ft, which may correspond to a maximum flooding surface. The progradational sand is interpreted as a shelf highstand deposit. The top of the sand unit is a prominent gamma log break, which marks the base of an increasing gamma log profile, culminating in a unit of high gamma values between 6570 ft and 9530 ft. This interval contains an abundance of microfossils and is interpreted as a regional condensed section containing a maximum flooding surface. The surface is placed at the maximum gamma peak at 6530 ft. The interval between the latter and the top of the sand is interpreted as the transgressive systems tract. The top of the highstand coarsening upwards profile, at 6619 ft, is the position at which the Exxon type sequence boundary may be interpreted. In this instance it appears to be coincident with the transgressive surface (which defines the base of the transgressive systems tract).

The change at the sonic break corresponds to the event known in the Inner Moray Firth area as the Intra Oxfordian Event (Andrews and Brown, 1987; Underhill, 1991), and marks the level of a regional hiatus and a prominent seismic marker. In sequence stratigraphic terms, the seismic sequence boundary appears to be falling within the transgressive systems tract, not at the Exxon type sequence boundary. The log break probably correlates with the top of a clastic transgressive sand unit. In 12/21-2, thin sands are recorded on the composite log within this interval (6602 ft-6619 ft). In other wells in the area, better developed sands are seen within this transgressive interval (e.g. 12/23-1, see figure 4 in Andrews and Brown, 1987). The sequence equates to the J1.5 sequence of Underhill (1991) and the lower part of Vail and Todd's (1981) J2.1 sequence.

Within the sequence, a high gamma claystone occurs, which by correlation with other wells in the Inner Moray Firth, appears to correlate between wells, and is occasionally the horizon of an unconformity (see Andrews and Brown, 1987, figure 4). This is a candidate for a local subsequence boundary, which is beyond seismic resolution in the Inner Moray Firth. Elsewhere in the North Sea, however, this is the horizon of a seismic sequence boundary defining the top of the J40 sequence of Rattey and Hayward (in press).

Sequence 3 (J2.1 Sequence of Underhill, 1991)

Age: Middle Oxfordian to earliest Kimmeridgian. This is based on good Middle Oxfordian dates at the base (*E. parastelligera* abundance) overlain by shales rich in Late Oxfordian microfossils (including common *Scriniodinium crystallinum* (Deflandre, 1938, Klement, 1960). The top of the sequence is of earliest Kimmeridgian (*baylei* ammonite zone) age, based on the uphole termination of a rich agglutinated foraminiferal assemblage, plus the extinction of *S. crystallinum*.

Lithostratigraphy: the interval is developed entirely in marine shale referable to the upper part of the Heather Formation (Deegan and Scull, 1977), equivalent to the Lower Warm Shale of Andrews and Brown (1987).

Palaeoenvironment: the basal part of the sequence contains abundant calcareous benthonic foraminifera indicative of well oxygenated marine shelf conditions. Above this occurs a sharp change (and associated log break) marked by the uphole disappearance of the dominantly calcareous microfauna, which is replaced by a diverse agglutinating foraminiferal assemblage. This is indicative of reduced marine circulation and reduced sea bed oxygenation. This microfauna terminates at the top of the sequence.

Sequence interpretation: the marine shale with abundant calcareous microfauna at the base of the sequence is interpreted as a transgressive shale (see above for discussion). This section has characteristic upward sloping gamma and sonic log responses and is considered to represent the basal, transgressive part of the sequence. The microfossil concentration is succeeded by shales with high gamma ray values at around 6530-6550 ft; these features are diagnostic of a marine flooding surface. The base of the sequence is placed at the Intra Oxfordian Event, a regional log feature, seismic marker and level of local unconformity (see above). These features appear to coincide with the onset of major faulting in the Moray Firth region (Andrews and Brown, 1987) and thus a tectonic rather than eustatic control on this sequence boundary is indicated. The top of the sequence is placed at the top of an upward increase in gamma ray values at 5920 ft, marking the base of the overlying Kimmeridge Clay Formation (Deegan and Scull, 1977). This is also the level of a regional seismic marker, plus a local and regional hiatus. It is interpreted as a maximum flooding surface condensed section, correlated approximately with the top of the *baylei* ammonite zone. In the microfossil assemblages, a sharp uphole change is seen across this condensed section, with the disappearance of the diverse agglutinating foraminiferal assemblages of the upper Heather Formation, succeeded by the uphole appearance of a radiolaria-dominated microfauna. The latter is characteristic of a deeper marine setting. It occurs shortly above a local bivalve shell bed which can be identified in the microfaunal preparations. This shell concentration is interpreted to be part of the condensed section. These palaeontological changes are a very widespread feature and clearly reflect major basin changes probably involving deepening and relative sea level rise, though it is uncertain whether the origin is tectonic or eustatic.

Sequence 3 correlates with the upper part of Vail and Todd's (1981) J2.1 sequence and with the J2.1 sequence of Underhill (1991).

Regional correlation and biostratigraphic data from the well suggests that the sequence as identified in the Inner Moray Firth contains two further maximum flooding surfaces, in addition to those at the base and top. One of these can be placed at around 6292 ft (correlating with the base of common *Haplophragmoides canui*), and is believed to possibly correlate with the *serratum* ammonite zone. A second maximum flooding surface condensed section, placed at 6906 ft, is fingerprinted by the top occurrence of *Spiroplectammina "oxfordiana"* (unpublished), within an abundance of agglutinated foraminifera, the top occurrence of common *S. crystallinum* (6110 ft) and the top of common *H. canui* (6150 ft). These micropalaeontological features are considered to correlate approximately with the top of the Upper Oxfordian (*rosenkrantzi* ammonite zone). The four flooding surfaces identified in seismic sequence 3 are of significance for correlation eastwards into the Outer Moray Firth area, into the successions of the Piper (block 15/17), Ivanhoe , Rob Roy and Scott Fields (the latter three fields being in blocks 15/21-15/22). In these fields, shallow marine reservoir sands are developed within the sequence, which have been referred to the Piper and Sgiath Formations (Harker et al., 1987) or "Main Piper Sandstone Unit" and "Supra Piper Sandstone Unit" (Boldy and Brealey, 1990). In these fields, the reservoir sands are punctuated by marine claystones (I Shale of Harker et al., 1987; Mid Shale Unit of Boldy and Brealey, 1990) which correlate with the maximum flooding surface condensed sections seen also in the claystone-dominated Inner Moray Firth successions (e.g. 12/21-2). In the Outer Moray Firth fields, the correct identification and correlation of these four regional condensed sections is the key to reservoir subdivision and correlation, and is of direct relevance to recent discussions between the operators of the fields regarding reservoir distribution and the extent of oil pools.

Mitchum et al. (1977) have advocated the technique of recognition of unconformities and sequences in wells and correlation between wells by log signature alone. In the North Sea Basin, correlation between wells by wireline logs only is often difficult and cannot be relied upon without calibration by biostratigraphy. This is due to the fact that many different formations, members and sequences have very similar log signatures. Thus, the reliable recognition of sequences on logs requires a biostratigraphic framework to distinguish comparable log signatures of different sequences. Mitchum et al. (1977) restricted the use of biostratigraphy to merely date depositional sequences, without appreciating the importance of biostratigraphy to characterize, identify and correlate sequences and key components of sequences.

As can be seen from the discussion above, biostratigraphy plays a crucial role in the correct application of sequence stratigraphy, not only in the North Sea, but in any basin.

### 4.2.5 Characterization of formations ("fingerprinting")

The principle of utilizing the fossil content of a rock unit to characterize it is not widely appreciated in exploration geology. The approach is, however, an accepted aid to lithostratigraphic definition, as stated by Hedberg (1976): "Fossils may be important in the recognition of a lithostratigraphic unit either as a minor but distinctive physical constituent or because of their rock-forming character...". Experience has shown this to be the case with microfossils, a fact well exemplified in the North Sea. In this exploration province most of the marine formations, that is from the Jurassic to Pleistocene, are typified by diagnostic microfossil associations. This is a very important observation which sometimes allows these formations to be recognized on fossil assemblage alone (i.e. "fingerprinted"). This is particularly useful where formations are thinly developed between stratigraphic breaks and the distinctive log signatures of the complete formations are not developed, or where the available logs are of poor quality. The top of the Kimmeridge Clay Formation (Deegan and Scull, 1977) (and age equivalent Mandal Formation and Draupne Formation, Vollset and Dore, 1984) is often a very important boundary to pinpoint, because it contains organic-rich claystones, which are the primary source rock in the basin, together with sands of known reservoir quality. The downhole penetration of the formation is easily identifiable in well samples by sudden changes in microfossil content, there being a sharp appearance of a diverse association of dinoflagellate cysts (with distinctive amorphous kerogen), radiolaria and agglutinating foraminifera. This change is accompanied by gamma and sonic log breaks and a lithology change and is usually well expressed on seismic sections (e.g. as the "near base Cretaceous event").

Another important lithostratigraphic boundary is the passage from the Middle Jurassic Brent Group into the underlying Lower Jurassic Dunlin Group (terminology after Deegan and Scull, 1977). The Brent Group is a major reservoir in several northern

Figure 4.4 Schematic cross section of the Brent Field (from Bowen 1975) showing the configuration of structure (tilted fault block), reservoir (Statfjord Fm. and Brent Group, separated by the argillaceous Dunlin Group) and seal (Upper Jurassic and Upper Cretaceous shales) typical of many fields in the north Viking Graben ("Brent Province"). This favorable combination of exploration factors is termed a "play" (see Parsley, 1990 for a review

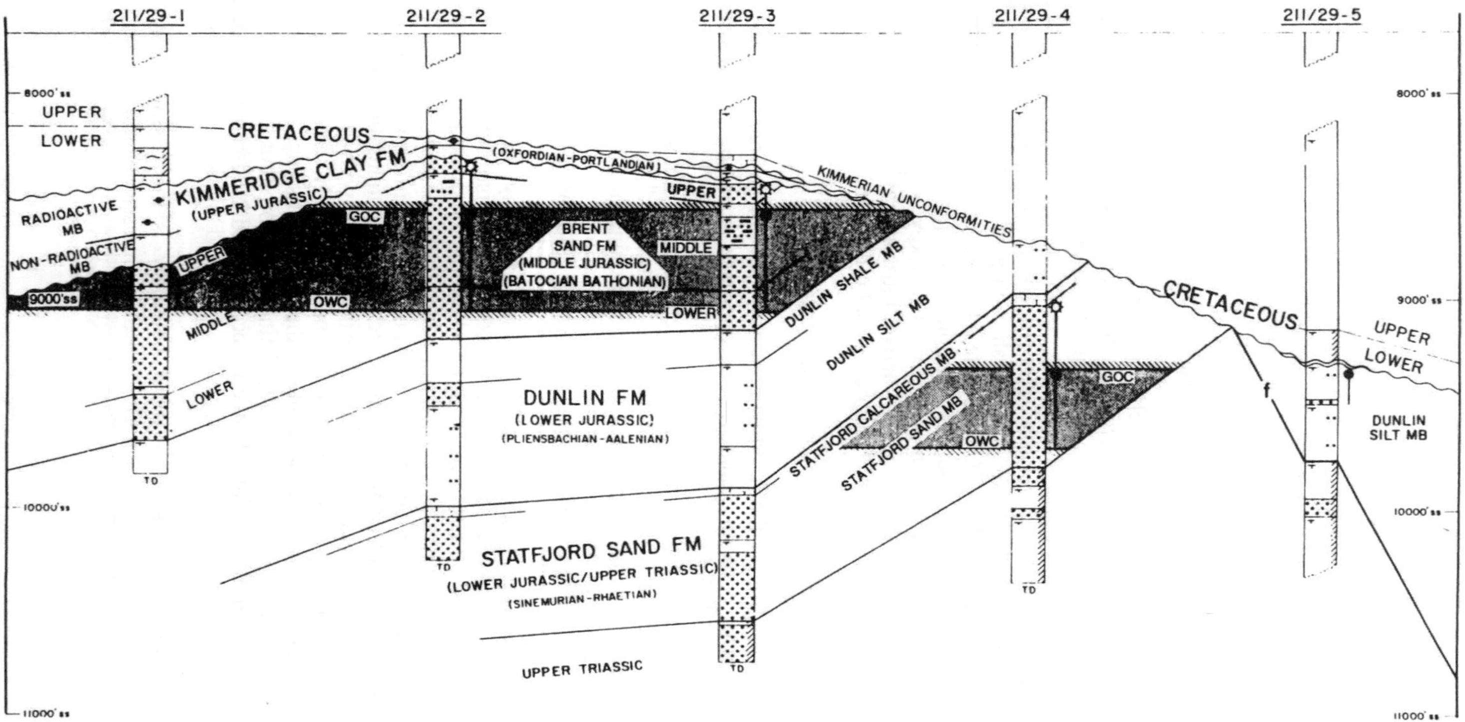

North Sea fields such as Brent (Bowen, 1975; Struijk and Green, 1991), Thistle (Hallett, 1981; Williams and Milne, 1991) and Statfjord (Kirk, 1980). The Dunlin Group is a predominantly shaly unit separating the Brent reservoir from a deeper, secondary reservoir, the Statfjord Formation (Deegan and Scull, 1977) (Figure 4.4).

Traditional lithostratigraphic schemes erected for the North Sea (such as Deegan and Scull, 1977; Vollset and Doré, 1984) have relied only on the pure lithostratigraphic character of rock units to distinguish groups and formations. With the requirement for increased resolution the biostratigraphic character is becoming more widely used in the definition and separation of lithostratigraphic units. Recent papers by Crittenden, Cole and Harlow (1991) who proposed a new more detailed lithostratigraphic scheme for the Lower Cretaceous, and Mudge and Copestake (1992a, b) who revised the lower Palaeogene stratigraphy, have used the biostratigraphic character as a fundamental aid in lithostratigraphic subdivision. The advantage of this method is that the resulting lithostratigraphic boundaries have chronostratigraphic significance and that they parallel schemes based on sequence stratigraphic principles. Indeed, the major lithostratigraphic boundaries so identified often correspond to genetic sequence boundaries.

The following example illustrates the potential pitfalls of undertaking a purely lithostratigraphic/log based correlation without using the biostratigraphic character of the sediments to aid the identification of lithostratigraphic units. Figure 4.5 shows a three well correlation from the Don Field (Block 211/18, see Morrison, Bennet and Byat, 1991 for field description) at around the base of the Brent Group. In each well two sands with an intervening shale overlie a major shale unit. The upper part of Figure 4.5 shows the "layer-cake" correlation produced using logs only, with the major shale unit being referred to the Dunlin Group and the section from the lower sand upwards referred to the Brent Group. Detailed micropalaeontological work in these formations indicates that each of these lithostratigraphic units has a characteristic microfossil "fingerprint". Application of biostratigraphic data to the wells (Figure 4.5b) shows that the "log correlation" was incorrect, that there are three sands developed and that the lower sand in the 24 and 14 wells does not correlate with the lower sand in the 23 well. The latter is an older sand which is sporadically developed within the Cook Formation (Dunlin Group) in the northern North Sea. The upper sand is an upper Rannoch Formation sand ("Rannoch Sand") and does correlate between the three wells, but the lower sand in the 24 and 14 wells is absent from the 23 well, due either to an unconformity or fault.

Figure 4.5 Application of biostratigraphy to correlation of three wells in the Don Field, over the lower Brent - top Statfjord interval. 5a: correlation based on log character alone, equating the lower two sands in each well, with an intervening shale, in a simple "layercake" fashion. 5b: correlation using biostratigraphy, demonstrating that the basal Brent Group, Broom Fm. sand is absent from the 23 well, in which the lower sand is an intra Dunlin Group, Cook Formation sand. This proves the absence of the lower part of the Brent reservoir in the 23 well, due either to an unconformity or fault.

a

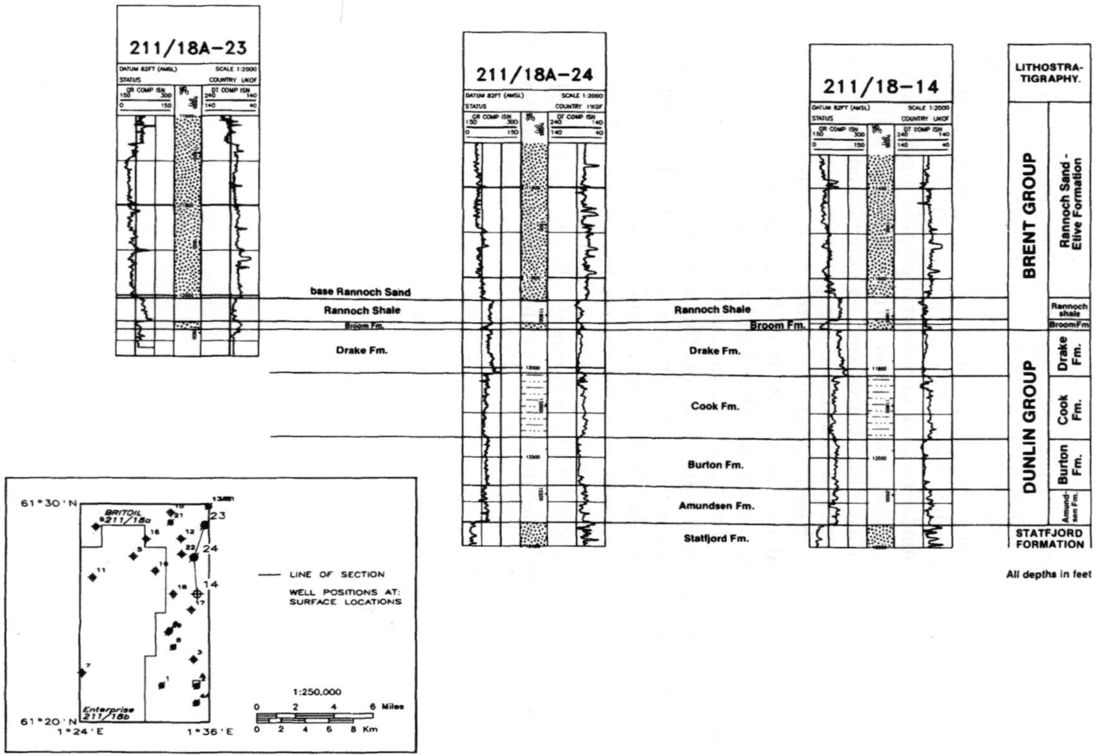

b

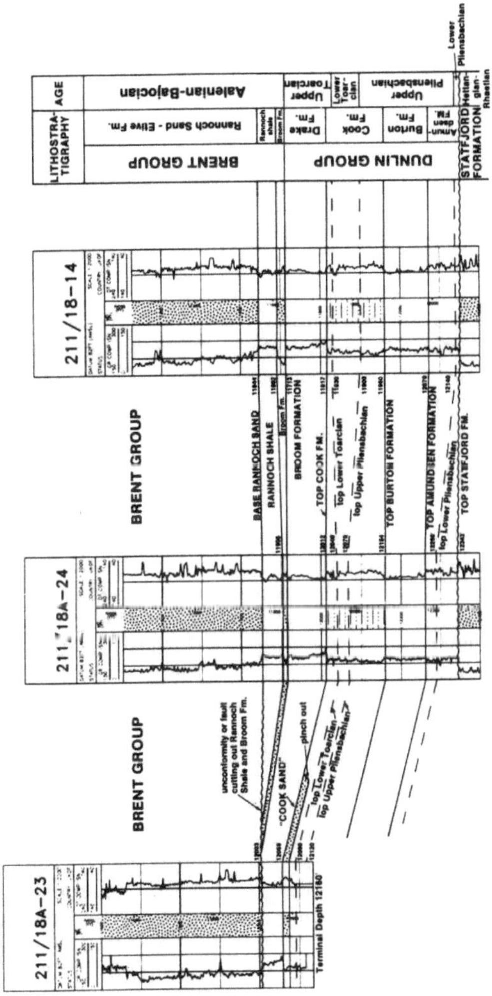
211/18A-23
Terminal Depth 12180
BRENT GROUP
unconformity or fault cutting out Rannoch Shale and Broom Fm.
"COOK SAND"
pinch out
top Lower Toarcian
top Upper Pliensbachian
211/18A-24
BRENT GROUP
BASE RANNOCH SAND
RANNOCH SHALE
Broom Fm.
BROOM FORMATION
TOP COOK FM.
top Lower Toarcian
top Upper Pliensbachian
TOP BURTON FORMATION
TOP AMUNDSEN FORMATION
top Lower Pliensbachian
TOP STATFJORD FM.
211/18-14
LITHOSTRATIGRAPHY
AGE
BRENT GROUP
Rannoch Sand - Etive Fm.
Rannoch shale
Aalenian-Bajocian
DUNLIN GROUP
Drake Fm.
Cook Fm.
Burton Fm.
Amundsen Fm.
Upper Toarcian
Lower Toarcian
Upper Pliensbachian
Lower Pliensbachian
STATFJORD FORMATION
Hettangian-Rhaetian

This interpretation rests on the fact that the Drake and Cook Formations of the Dunlin Group contain characteristic microfossil associations, which confirmed the original top Dunlin pick in the 14 and 24 wells but shows that the shale below the deepest sands in the 23 well is not at the top of the Dunlin Group. The latter lies beneath the upper sand, the Rannoch Shale having been cut out by an unconformity in the 23 wells. Of note is the fact that this unconformity corresponds to the level of a regional sequence boundary (top J10 sequence of Rattey and Hayward, 1992). This illustrates that significant biostratigraphic changes often parallel sequence boundaries.

This technique is thus of importance as:

(a) a basic means of lithostratigraphic evaluation.
(b) a means of reservoir identification and recognition of missing parts of the reservoir. This impacts also on reservoir rock volume and hydrocarbon reserves calculations.

4.2.6 Palaeoenvironmental interpretation

Palaeoenvironmental interpretations based on microfossils are possible due to the preference of organisms for particular habitats. Animals and plants inhabit environments ranging from terrestrial to deep marine and an understanding of their ecological requirements (e.g. temperature, salinity, oxygen levels, turbidity and water depth) in the modern day enables fossil associations to be interpreted palaeoenvironmentally. Individual organism types have known environmental ranges, e.g. calcareous nannofossils are restricted to the marine, outer shelf to bathyal realm, while dinoflagellate cysts range from marginal to bathyal marine environments. The recovery of several fossil groups from a sediment allows the palaeoenvironment to be more tightly constrained. For example, the presence of abundant radiolaria, with dinocysts and nannofossils restricts the environment to upper bathyal marine (see Figure 4.6).

Such simple interpretations of palaeoenvironmental setting can be further refined by the recognition of environmentally specialized genera or species within fossil assemblages. For instance, particular species of benthonic foraminifera are known to have occupied particular depth ranges, e.g. *Gavelinella beccariiformis* (White, 1928) and *Spiroplectammina spectabilis* (Grzybowski, 1898), which are present in the Upper Cretaceous to Palaeocene and Palaeocene to Eocene respectively in the North Sea; they have an interpreted upper depth limit of 500-700m (mid-upper bathyal) (Van Morkhoven, Berggren and Edwards, 1986). Within the crustacean Class Ostracoda, whole families have very restricted environmental tolerances, enabling very precise palaeoenvironmental interpretations to be made (e.g. Whatley, 1988; Neale, 1988).

A major area of micropalaeontology currently undergoing active development is palynofacies. This is the evaluation of palaeoenvironments utilizing a method of classification of assemblages of organic particles. This technique is particularly powerful in field reservoir development studies. Papers outlining applications of the methodology to North Sea cases are documented by Van der Zwan (1990) on the Troll and Draugen Fields of offshore Norway, Whittaker (1984) also on the Troll Field, Whittaker et al. (1992) and Williams (1992) on the Brent Group of the UK North Sea and Boulter and Riddick (1986) on the Paleocene Forties Member of the Forties Field of the Central North Sea.

Figure 4.6 Generalized palaeoenvironmental distribution of major microfossil groups in the North Sea Basin (Triassic-Tertiary) and relationship to occurrences of main reservoir and source rock facies. Microfossil ranges interpretative and subject to modification by offshore transportation processes, variation in depth of carbonate compensation depth (CCCD) with climate, reduced fossilization potential of spores and pollen in terrestrial settings and general lack of non marine aquatic environments.

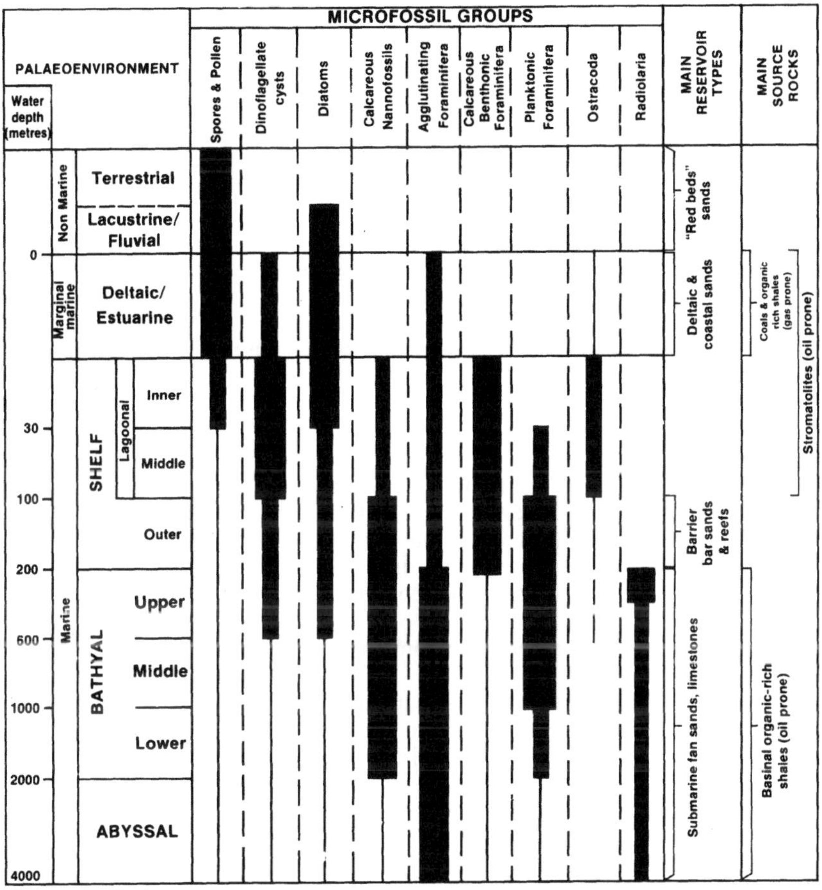


The value of palaeoenvironmental interpretations to oil exploration is in the characterization of facies which may be of reservoir, source or seal potential.

Reservoir: In marine clastic settings, reservoir sands can be palaeontologically characterized as marginal (deltaic, coastal barrier complex), offshore bar or submarine fan. This is important for indicating likely sand body geometry, continuity and sand quality. In carbonate environments, lagoonal, reef or fore-reef settings can be characterized. This can be important in siting wells to locate reef reservoirs where previous wells may have penetrated lagoonal or fore-reef facies.

Source: Source rock facies are developed in particular environments e.g. carbonate lagoons (stromatolitic facies) basinal marine (organic-rich carbonates or anoxic shales). These environments are distinguished by particular microfossil associations. The main North Sea source rock, the Kimmeridge Clay Formation, for example, is a deep marine (bathyal) dysaerobic to anoxic shale facies characterized by abundant dinoflagellate cysts, chlorophycaean algae and radiolaria.

Seal: Sealing shales in clastic hydrocarbon provinces are often transgressive, deep marine sediments rich in deep water markers such as planktonic foraminifera and radiolaria. Examples from the North Sea include Brent fields (e.g. Brent, Thistle) where the Middle Jurassic sands are sealed by Upper Jurassic-Cretaceous marine shales, the Clyde Field where the Fulmar Sand Formation (of Kimmeridgian age) is sealed by later Kimmeridgian-Early Volgian shales (Kimmeridge Clay Formation) and the Maureen Field, where the Maureen Formation sands (Early-early Late Palaeocene) are sealed by an uppermost Maureen Formation transgressive shale (rich in radiolaria). These marine claystones often correspond to maximum flooding surface condensed sections at their bases.

## 4.3 Role of micropalaeontology in hydrocarbon exploration and development programmes

### 4.3.1 Introduction

Petroleum exploration is subdivided into the three phases of exploration (for prospects), appraisal (of discoveries) and development (of fields). Micropalaeontology plays a fundamental role in each of these phases once wells have been drilled, and each phase is typified by different problems requiring different biostratigraphic approaches. Each of these phases and their type of micropalaeontological work will be described in turn.

### 4.3.2 Requirements for hydrocarbon accumulation

For any geologist (micropalaeontologists included) working in the oil industry, a constant awareness of the three major controlling factors on hydrocarbon accumulation is needed. These are: reservoir, trap (including seal) and source. Important information regarding these crucial factors can be derived from several geological subdisciplines, including micropalaeontology, as follows:

| Requirement | Information from (*"fingerprinting";+correlation and biozonation) |
|---|---|
| Reservoir | biostratigraphy*+, sedimentology |
| Trap | biostratigraphy+, seismic mapping |
| Seal | lithostratigraphy, biostratigraphy* |
| Source | geochemistry, biostratigraphy* |

These factors are related in cartoon fashion in Figure 4.7, illustrating some typical North Sea play configurations. Trap/play type 1, on the upthrown (footwall) side of tilted fault blocks is typical of Middle Jurassic Brent Group fields such as Brent, Statfjord and Hutton in the North Viking Graben. Trap/play types 2 and 3 on the downthrown (hanging wall) side of fault blocks are typical of Upper Jurassic submarine fan type reservoirs in the South Viking Graben-Outer Moray Firth area such as Brae and Claymore.

Figure 4.7 Requirements for hydrocarbon accumulation illustrated by schematic reservoir, trap and source rock interrelationships.

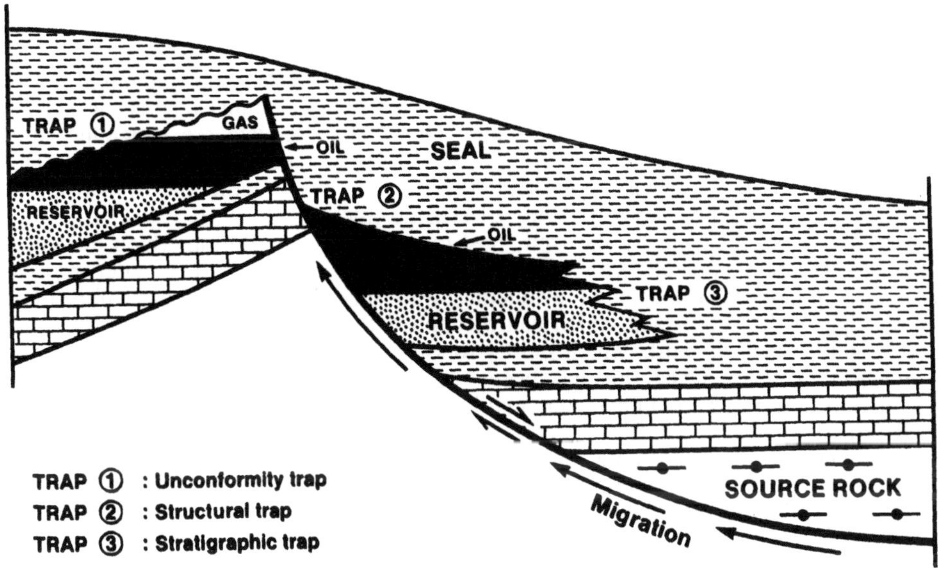


### 4.3.3 Exploration for prospects

In the initial wildcat phase of exploration, the simple technique of recognition of a large structure on seismic sections and drilling it (without particular consideration of stratigraphy) is prevalent within the industry. This has to be the case in previously undrilled regions where the ages of reservoir and source rocks are unknown. Even when drilling in an area adjacent to a relatively mature hydrocarbon province, the prospects and play types (i.e. the favorable combination of reservoir, source and trap) may be found to be significantly different. This is well exemplified in the North Sea area where the play types vary between the southern, central and northern North Sea, northern offshore Norway and West of Shetland (see Parsley 1990 and numerous papers in Brooks and Glennie, 1987 , Abbots, 1991). The "find a large structure and drill it" approach typified the discoveries in the early 1970's of well known giant North Sea oil fields such as Ekofisk (Byrd, 1975), Forties (Walmsley, 1975) and Brent (Bowen, 1975). In the Central and Northern North Sea, it is considered unlikely that any more giant fields such as these await discovery. The majority of the discoveries over the past decade have been of much smaller size, and their successful, economic development is reliant upon a spectrum of technologies, of which biostratigraphy is one. However, on the UK continental shelf, there remain significant underexplored areas, such as West of Shetland and some parts of the North Sea, and it may be that large accumulations lie undiscovered in these regions.

#### 4.3.3.1 Well evaluation

Once wells have been drilled, samples are selected for study over intervals of interest. This is followed by sample processing and identification of marker species, allowing the breakdown of the well succession into systems, stages, substages and biozones. Potential or actual reservoirs are studied closely, to establish their age and environment of deposition, and to attempt a biozonation. At this stage, amounts of missing section (unconformities) may also be identifiable. Once all these data are generated, it is possible to correlate the well stratigraphy with seismic markers and sequences. This seismic calibration is a very important step as it allows prediction of subsurface stratigraphy in undrilled areas by study of seismic lines.

It is also a pre-requisite for the well "post mortem" exercise during which the actual well results are compared with the pre-drilling well prognosis. The latter process provides an important means of evaluating well success or failure and of refining knowledge of the subsurface which can be applied to future wells.

This initial analysis work, which shortly follows well termination, comprises the well evaluation exercise. Once this has been undertaken on several wells in an area a correlation study may be undertaken involving a further more detailed study of each well. For particularly significant wells, many separate studies over a period of years may be performed on the same intervals, maybe utilizing new processing techniques or studying different fossil groups, in an attempt to extract as much biostratigraphic data as possible from the well.

Once knowledge is built up by well analysis and the biostratigraphic character of the lithostratigraphic units becomes known, it is possible to use this to interpret subsequent wells while drilling to aid operational decisions.

#### 4.3.3.2 Drilling problems

Successful well completion, that is drilling all the well objective horizons and reaching a pre-established terminal depth (T.D.), is an essential requirement for testing prospects seen in seismic data and in fulfilling government drilling licence obligations. In this process, there are a number of drilling activities which can be aided substantially by micropalaeontology. This can be undertaken at the well site if a rapid turnaround for results is required, or at an onshore base. Often, however, such work is undertaken on the drilling rig to allow efficient decision making and the saving of as much drilling time as possible. With the high cost of offshore drilling, the saving of drilling time or contribution to successful well completion can result in cost savings from thousands to millions of pounds Sterling. Nowadays, with the invention of non polluting self contained palynology processing units, all micropalaeontological subdisciplines can be performed offshore. This is having the effect of substantially increasing the impact of wellsite micropalaeontology in that a range of microfossil groups can be studied, thereby enhancing the problem solving potential.

*Setting casing*

The well bore is lined with metal casing to maintain the hole in good condition (see Figure 4.8).

Figure 4.8 Offshore exploration drilling rig (semisubmersible), surface and subsurface components.

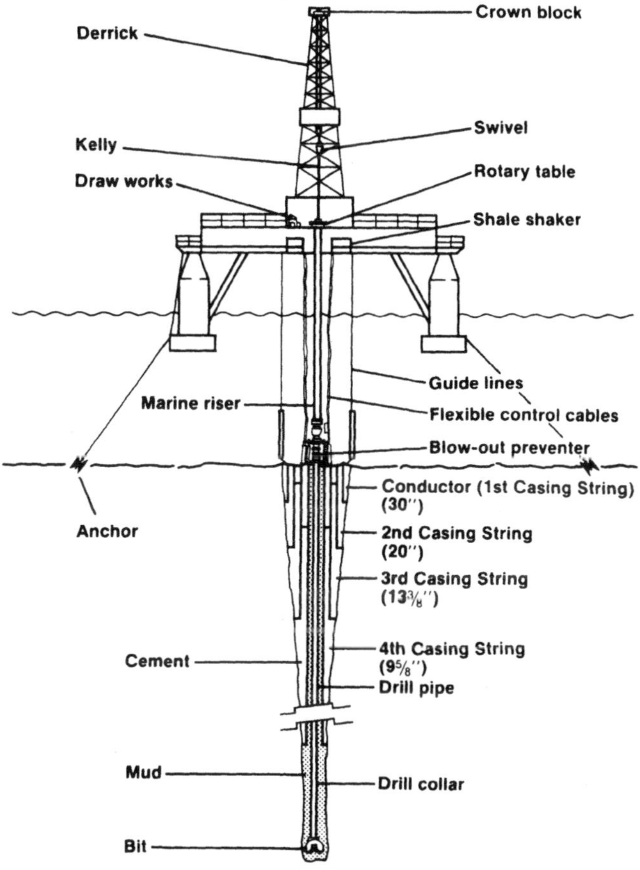


In open hole (i.e. uncased) conditions, weak rocks typically crumble off (or "washout") into the hole as cavings, which may cause problems for the drill bit or cause, at worst, a stuck drill pipe, due to the development of swelling clays. This could threaten well completion. Good hole condition can only be maintained for a relatively short time in the open hole and it becomes necessary to line or case the hole at particular levels in the geological succession. Such casing also allows high fluid pressures often encountered during deep drilling to be contained. Where abnormally high pressures (overpressuring) are expected at a particular stratigraphic level, it is easier to contain this pressure if the amount of open hole above the high pressure zone is at a minimum. Thus it becomes necessary to set the new casing just above the high pressure zone. If this pressure is not contained, a blow out will occur, with the risk of loss of life for rig personnel, and possible marine pollution.

Micropalaeontology can help by allowing the subsurface stratigraphy to be interpreted while drilling. This provides a check on the seismic prognosis which will have predicted the high pressure zone (usually a sandstone sequence) at a certain depth. Biostratigraphic formation fingerprinting will allow lithostratigraphic units to be identified in association with lithological examination of the drill cuttings. Study of nearby wells may allow the stratigraphy to be predicted ahead of the bit with knowledge of the local thickness of rock units and the depth between significant microfossil events ("tops" or flood abundances).

Such circumstances are typical of drilling in the axial zone of the North Sea Central Graben, where Jurassic sandstone objectives may occur at depths of up to 20,000 ft and contain overpressured fluids. In these wells it is common to set the 9 5/8" casing at or near the top of the Kimmeridge Clay Formation, which contains the sandstones. Wellsite micropalaeontology allows the various horizons of the Lower Cretaceous sequence to be identified while drilling, and predictions to be made of the depth of the top of the Kimmeridge Clay.

*Coring point selection*

A great deal of valuable information regarding a reservoir can be obtained if it is cored. Its quality (porosity and permeability) can be measured and studied and it can be examined for sedimentary structures indicative of reservoir type and environment of deposition. Coring is, however, expensive and it is generally only used to study particular reservoirs or where hydrocarbons are encountered. As mentioned previously, reservoirs can often be characterized and separated micropalaeontologically. Thus, at the well site, it is possible to use this technique to identify the particular sand to be cored. This is a valuable aid to sand identification where several sands of similar lithological character are developed in a well sequence, but where previous exploration experience indicates that not all have the same potential for hydrocarbon entrapment (due to the development of sealing lithologies between the sands acting as vertical barriers to hydrocarbon migration).

Furthermore, it may be important, for sedimentary facies analysis, to core the upper sand/shale contact of a reservoir. This is a more difficult task for the wellsite micropalaeontologist as it involves accurately predicting the depth of the top of the sand ahead of the drill bit. It can, however, be successfully performed if there are sufficient nearby wells which can be studied to allow the establishment of a database showing the depth range between bioevents and the top of the sand.

*T.D. (terminal depth) decisions*

Operators' drilling obligations to the government and pre-drill well prognoses usually require wells to be terminated at particular geological horizons. In the North Sea, this may be the top Cretaceous for a well testing the Palaeocene, the Jurassic for a well testing the Upper Cretaceous, or the Carboniferous for a well testing the Permo-Triassic. Often this T.D. (terminal depth) horizon can be predicted at an approximate depth prior to drilling by seismic interpretation, or is to be located at a particular lithostratigraphic boundary. However, a significant number of T.D. horizons do not correspond to a seismic reflector and thus need to be picked by other means. Also, a back up technique may be required to identify a lithostratigraphic unit. In these cases, micropalaeontology is utilized to help identify the T.D. horizon. For instance, the top Jurassic does not equate to a seismic marker in the North Sea (the "near base Cretaceous" reflector, equating to the top Kimmeridge Clay Formation, lies within the lowermost Cretaceous), but can be readily identified by microfossil events, particularly in the dinoflagellate cyst and radiolaria populations. In cases where the T.D. horizon does equate to a seismic marker and can therefore be depth predicted, micropalaeontological confirmation is nevertheless frequently required. This can be undertaken at the wellsite, but is frequently performed by rapid analysis at an onshore base on samples flown from the rig ("hot shot" samples). The decision whether or not to undertake the analysis at the well site is determined mainly by the required turnaround time for results, the distance of the rig from the onshore base and the frequency of helicopters.

*Stratigraphic monitoring while drilling*

In frontier (or "wildcat") areas, there is often a desire to obtain detailed information on the well stratigraphy as a well is being drilled. In other words, the routine biostratigraphic well evaluation process needs to be advanced and brought forward to the well site by the on rig study of well samples. This situation is frequently associated with uncertainty over the identity and age of seismic markers identified prior to drilling. Should these, on being drilled, prove to be much older than predicted, this could lead to an early well termination based on the well site micropalaeontology results. This would incur savings on drilling time and therefore also save costs.

### 4.3.4 Appraisal of discoveries

Once a hydrocarbon discovery has been made, subsequent wells drilled to test the extent of the accumulation are termed appraisal wells (or "step-out wells" in North American terminology). At this stage, the work of the micropalaeontologist becomes more detailed as attention focuses on the relatively short reservoir section of the discovery and appraisal wells.

#### 4.3.4.1 Well correlation

The most important task for the micropalaeontologist is to establish a reservoir correlation between the discovery and appraisal wells. Detailed micropalaeontological analysis, integrated with wireline log interpretation and depositional sequence analysis has proved to be the most reliable method of achieving this correlation. Integration with seismic data should also be undertaken, although, in the appraisal stage, correlations are often undertaken at scales which are beyond seismic resolution.

It is important that correlations should not be attempted by analysis of wireline log signatures alone, although this is a popular approach with petroleum geologists. Experience indicates that different stratigraphic units can have virtually identical log only signatures. Missing sections in a well can also make log correlation alone difficult and unreliable. Oil industry files contain many erroneous multi-well log only correlations, upon which past geological assessments have been made.

#### 4.3.4.2 Reservoir distribution and reserves estimates

Correlation between discovery and appraisal wells is fundamental in establishing the lateral and vertical extent of the reservoir. This allows calculation of the rock volume to be made, which, combined with a knowledge of the vertical distribution of hydrocarbons in each well, enables a volumetric estimate of the reserves to be made. This estimate of the amount of recoverable reserves is the single most important factor in the assessment of the economic viability of the hydrocarbon accumulation.

4.3.4.3 Trap evaluation

Results of the well correlation also aid assessments of the nature and style of the trap to the accumulation. Micropalaeontology is of particular value in demonstrating the existence of two types of trap:

Stratigraphic trap (shale out or pinch out): caused by the lateral facies change of, for example, sand to shale.

Unconformity trap: caused by the lateral or vertical disappearance of reservoir facies due to truncation beneath or onlap above an unconformity.

In both cases the trap is created by the development of a permeability barrier at the sand/shale contact. This presents a barrier to the vertical movement of hydrocarbons in the reservoir, allowing hydrocarbons to accumulate. It is important in the exploration/appraisal phase to determine which of these configurations is trapping the hydrocarbons because a sand shale out may be associated with a reduction in reservoir quality over a substantial area in the vicinity of the shale out, due to sand/shale interdigitation, particularly in the distal facies of submarine fans. On the other hand, an unconformity trap may preserve reservoir quality (porosity and permeability) in clastic reservoirs up to the actual reservoir termination. In carbonate reservoirs an unconformity, if associated with uplift and exposure to meteoric waters, can significantly enhance porosity and permeability by carbonate dissolution and this is very important to determine.

The evaluation of these stratigraphic relationships is fairly straightforward once the well correlation and distribution of bioevents between the wells has been worked out, as summarized in Figure 4.9. Obviously, recognition of unconformities in such situations relies on having seen relatively complete sequences of biozones elsewhere in the basin.

It is also important to test these relationships deduced from well correlations against seismic data. It can be difficult, for instance, to be certain whether missing section between wells is due to onlap or truncation at an unconformity. If seismic data quality is adequate, the actual relationship can sometimes be clarified.

The crucial role of biostratigraphy in the appraisal stage of exploration is illustrated by two North Sea examples (Figures 4.10 and 4.11). These are real examples from the North Sea, and the various correlations described are those that were actually made firstly without biostratigraphy and subsequently with biostratigraphy. Figure 4.10a shows discovery well A and its correlation with three appraisal wells B, C and D, based on log correlation alone. Note that well A discovered oil whereas the three appraisal wells B, C and D are dry holes (i.e. water was present but no hydrocarbons), and that the sands were interpreted as laterally continuous between wells A, C and D but terminated laterally between wells D and B by a syn-depositional fault. Figure 4.10b shows the significantly changed correlation once the micropalaeontological correlation was applied to the wells, using four regionally significant bioevents. This showed that two sands were present, with the upper oil-bearing sand less widespread than the lower water-bearing sand. The upper sand shales out between wells A and D, trapping the oil, and the lower sand shales out between wells D and B (all bioevents are present in well B suggesting a complete, but shale-dominated sequence). The effect of the

Figure 4.9 Demonstration of the development of a stratigraphic trap or an unconformity trap according to the relationships of correlative bioevents between two wells A (dry hole) and B (hydrocarbon discovery). Based on real North Sea examples from clastic settings (i.e. sandstone reservoirs sealed by shales). Bioevents i to v are numbered in depth (and age) order.

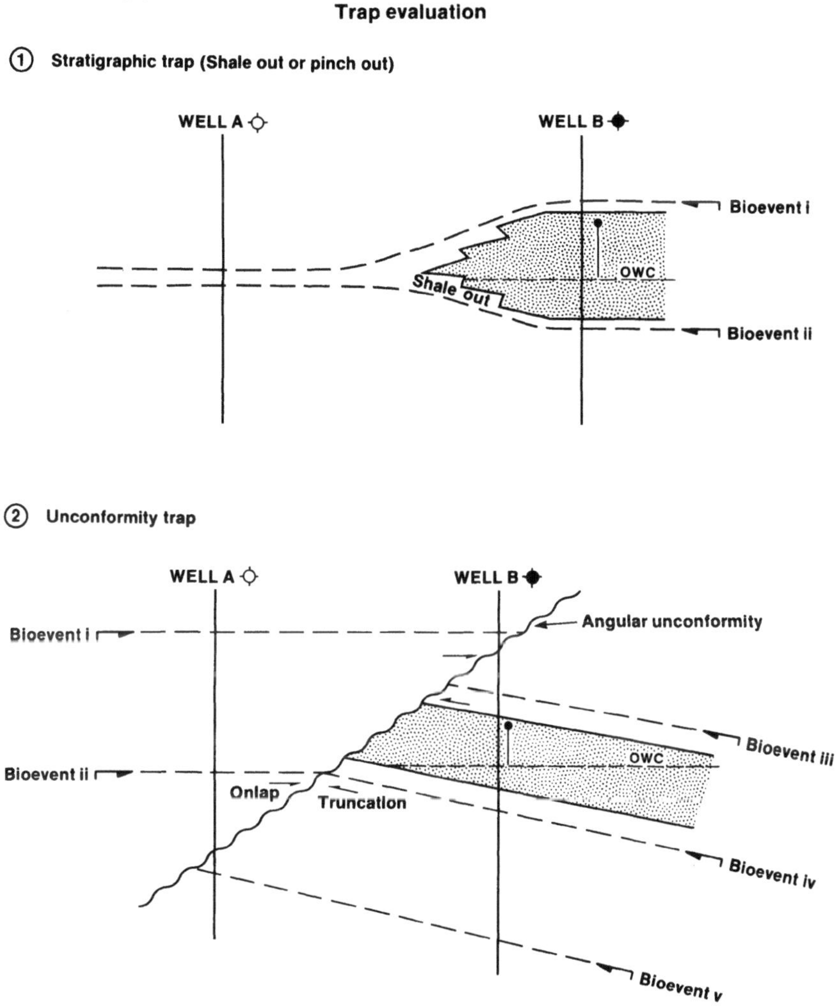


biostratigraphic correlation on reserves is substantially detrimental in establishing that the rock volume of the upper, hydrocarbon-bearing sand is markedly less than in the original interpretation. This effectively downgrades the reserves estimates of oil in place and shows the oil pool to be much smaller than was originally thought. In this case the biostratigraphy shows that embarkation upon a programme of platform construction and development could be less profitable than originally thought on the basis of the erroneous log correlation model.

Figure 4.10 North Sea appraisal case history no. 1. Correlation of oil discovery well (A) with three subsequent appraisal wells (B, C and D), initially, using lithostratigraphy (wireline logs) alone (Figure 4.10a) and, subsequently, using biostratigraphy (Figure 4.10b). Final interpretation downgrades reserves estimates.

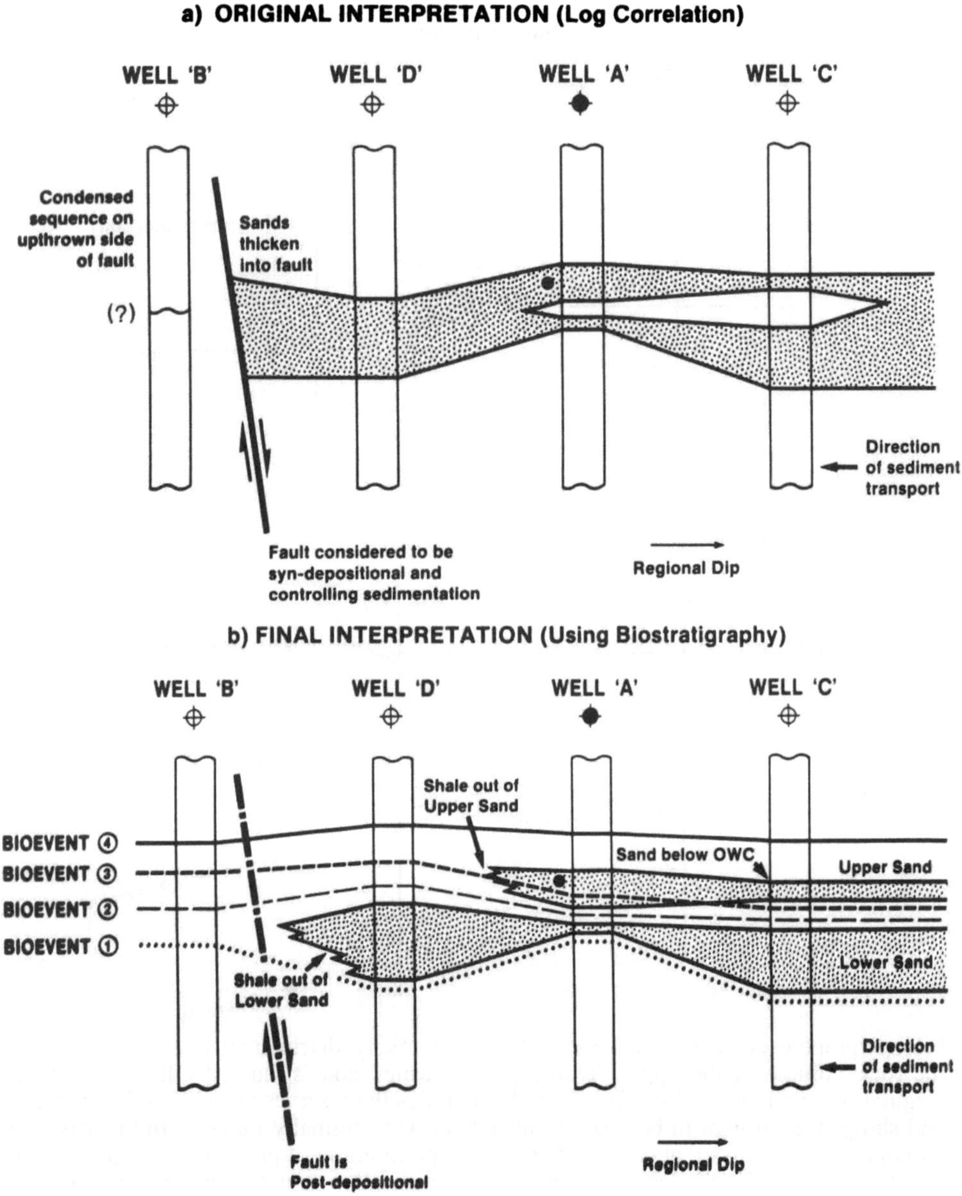

Figure 4.11a shows a similar, simplistic, log only correlation of a discovery well B with three nearby wells A, C and D. Each well contained two gross sand units separated by a thin shale. As these lithostratigraphic units showed similar wireline log character, they were correlated "layer cake" fashion as in Figure 4.11a. Using this correlation it was difficult to explain why the hydrocarbons at the top of the upper sand in well B were not seen in well A, thus no trapping mechanism could be proposed. This significantly increased the risk on the development of the hydrocarbon accumulation. Once a detailed micropalaeontological study had been undertaken, the well correlation changed significantly (see Figure 4.11b), with three, rather than two sands being identified. This explained the occurrence of hydrocarbons, with gas being confined to the upper sand unit and oil to the middle sand. Both these sands pinched out between wells B and A, the latter containing only the lower sand unit. A trapping mechanism was now established as an unconformity trap truncating the hydrocarbon-bearing sands beneath a post-upper sand unconformity. The lower sand also overlies an unconformity. Dating of the overlying unconformity (by the age of the oldest sediments overlying the unconformity, immediately pre-bioevent 2) indicates the geological age of trap creation. The effect of the micropalaeontological correlation has been to produce a reservoir zonation, to indicate a trap mechanism (and timing) and to enhance the prospect by allowing more accurate reservoir distribution and reserves assessments. It also suggests a location for the next well to be updip of well B at which point a maximum hydrocarbon column should be encountered. The precise location of this, and other, appraisal wells can only be pinpointed by interpretation of seismic data. Thus it is crucial to integrate the new correlations with seismic sequences by calibration in the well sections and seismic extrapolation into undrilled areas (see Figure 4.12). These examples are typical of appraisal micropalaeontological studies and show that while prospects may be either enhanced or downgraded, such work allows more accurate reservoir distribution and reserves estimates, with important impact on petroleum exploration decision making.

The main applications of micropalaeontology to exploration and appraisal are summarized in Figure 4.13.

### 4.3.5 Field development

Once a hydrocarbon accumulation has been discovered its distribution mapped and recoverable reserves calculated, its economic viability as a potential producing field is assessed. This assessment takes many factors into account including present and future oil (and gas) price, cost of development (including production method, proximity of existing pipelines, number of development wells required, tax), availability of markets for the hydrocarbons, and politics. Should the decision be made to produce the field and government approval is obtained, the development phase begins.

Development involves the drilling of several, and sometimes tens of development wells. While the primary role of these wells will be as producers or water injectors, the geological information derived from them has significant impact on development. As new wells are drilled, the additional reservoir information from each is used to reappraise the geological model upon which the development strategy is based and also to reevaluate the hydrocarbon reserves.

Figure 4.11 North Sea appraisal case history no.2. Correlation of four North Sea wells (A,B, C and D) initially using logs alone (Figure 4.11a) and, subsequently, using biostratigraphy (Figure 4.11b). Final interpretation enhances the prospect.

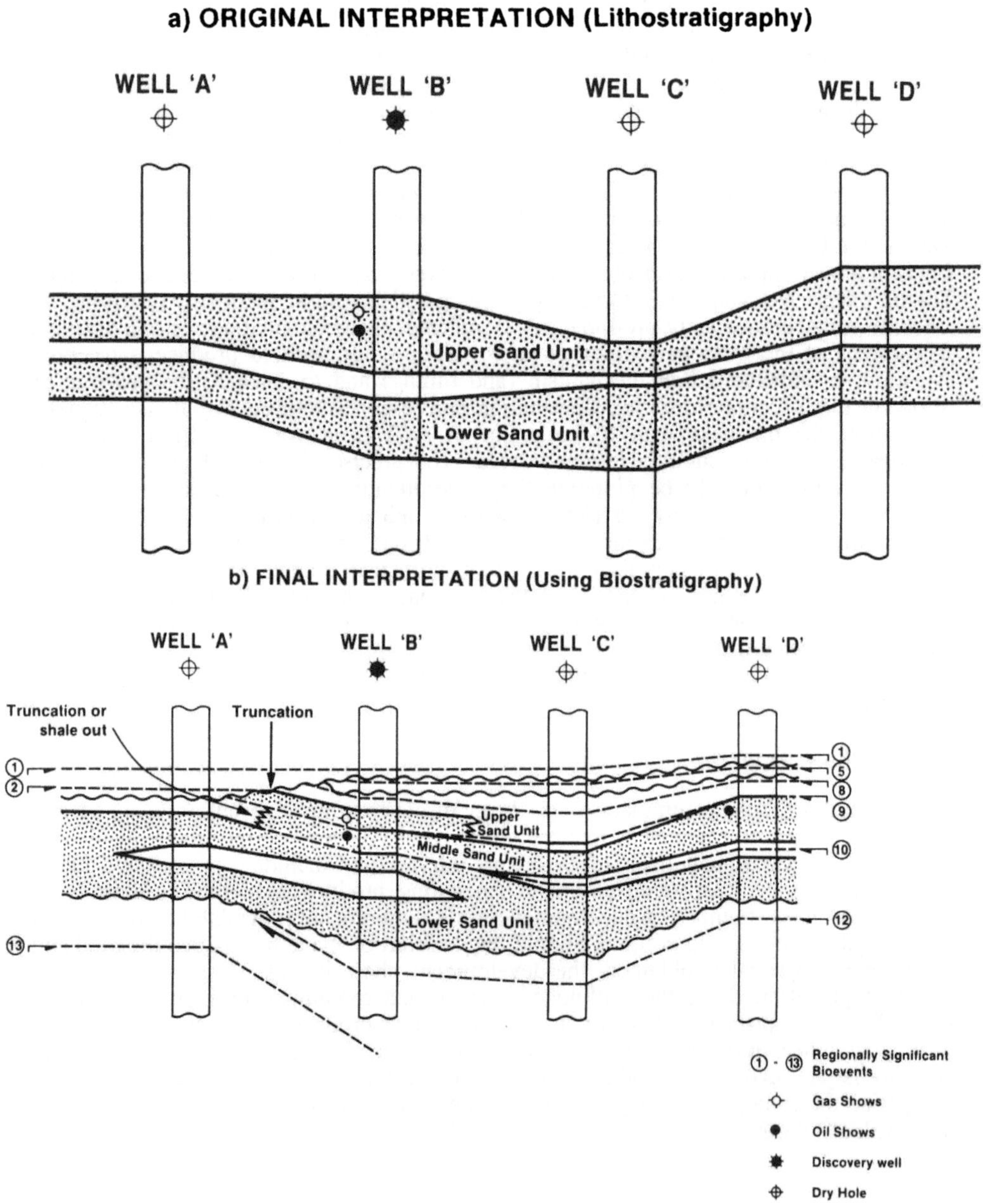

Figure 4.12 Crucial steps between a discovery and accurate appraisal by integration of well data with seismic interpretation.

**SUCCESSFUL EXPLORATION AND APPRAISAL DEPENDS ON**

**Accurate well evaluation**

⇩

**Detailed well to well correlation**

⇩

**Assessment of reservoir distribution**

⇩

**Interpretation of trap type**

⇩

**Pinpoint by seismic interpretation**

Figure 4.13 Summary of applications of micropalaeontology to exploration and appraisal activities.

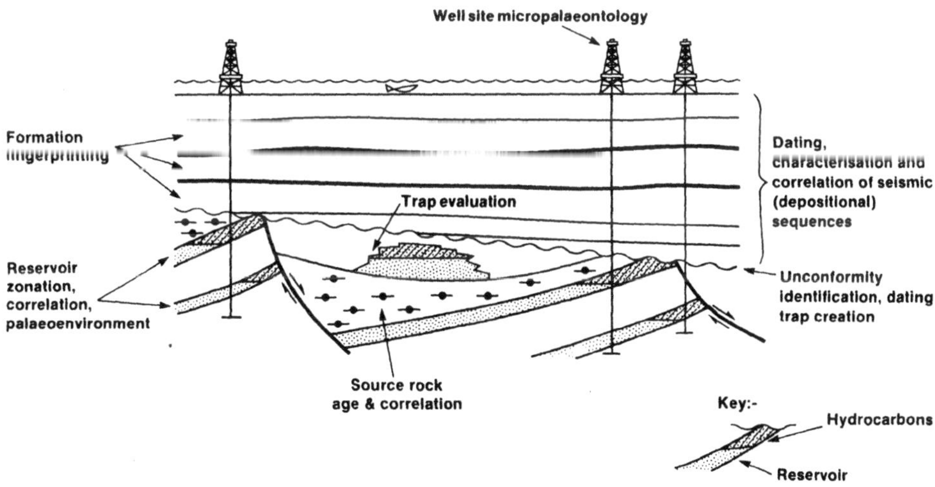

In the early phase of development drilling, detailed micropalaeontological studies are usually undertaken routinely on each development well. The principles and aims of this work are the same as for appraisal studies, namely to produce a detailed reservoir zonation and correlation of each well with its predecessors as soon after drilling as possible. One advantage at this stage is that substantial amounts of core are often available in the reservoir, this being the best quality sample type available; exploration and appraisal studies are often reliant on primarily ditch cuttings samples supplemented by variable numbers of sidewall cores. This allows biozonation previously established to be refined and enhanced by the recognition of local micropalaeontological changes within the reservoir of the field. It is often the case that a field reservoir may occupy only two or three regional biozones, thus local bioevents have to be erected to establish a refined biozonation.

The correlations thus established are used to assess the all important aspect of reservoir continuity. This has to be known as precisely as possible to ensure the success of the development plan. For instance, a widespread, uniform sand reservoir can be drained with fewer wells than can stacked, discontinuous sand reservoirs. Subsurface facies variations are invariably more complex than early geological models suggest and it is essential to reassess and, if necessary, modify a development well programme to incorporate new knowledge regarding reservoir distribution. It is important to integrate these biostratigraphic studies with sedimentological facies analysis, also undertaken on the cores, to place the reservoir facies models into a chronostratigraphic context. Several of the papers in Abbots (1991) and that on the Fulmar Field by Johnson et al. (1986) illustrate how geological models can evolve during field life as more wells are drilled. It is a universal fact that North Sea reservoir models have always become more complex as more understanding is gained of the subsurface geology. The application of biostratigraphic studies is an integral underpinning part of this increased understanding.

Allied to gaining knowledge of the lateral distribution of the reservoir, it is also important to assess the completeness of the reservoir sequence in each well. It is often found, for instance, that parts of deep water sand reservoirs pass laterally into shale between wells. This has been reported for example, in the Upper Jurassic Brae Formation reservoir sands of the South Brae field and its satellites (Turner et al., 1987) and in the Palaeocene-Eocene sands of the Gannet Complex (Armstrong, Ten Have and Johnson, 1987). Unconformities and faults can also be developed, cutting out parts of the reservoir between wells. Faulting is a particular problem in Brent Province fields such as Brent and Statfjord, where complex fault zones on the flanks of tilted fault blocks frequently lead to repetition of section in deviated (i.e. non vertical) development wells (see Bowen, 1975; Kirk, 1980). Interpretation of these complex repeated sequences is a difficult task for a micropalaeontologist, but is of great impact in field development studies. On the upthrown side of these Brent fault blocks, unconformities are developed cutting out varying amounts of reservoir; evaluation of the completeness of the section by recognition of present and absent biozones (in tandem with facies and log studies) allows the amounts of missing reservoir to be ascertained. The development of crestal unconformities is typical of many fields in addition to those of the Brent province, including the Upper Jurassic reservoir fields of Claymore (Maher and Harker, 1987) and Piper (Maher, 1981; Schmitt and Gordon, 1991).

In field development, reservoir correlation is aided by reservoir pressure data obtained from each development well. Reservoir continuity and fluid communication of individual sands can be indicated by similarity of pressure values between wells. It may be the case, however, that sands in pressure communication occur in different biozones, and therefore belong to successive phases of sand deposition. As long as interbedded shales do not form a fluid migration barrier (i.e. a seal), reservoir fluids will move across bedding planes. Conversely, pressure discontinuities between individual sands (the individuality of the sands being indicated by biostratigraphic correlation), would be evidence that interbedded shales had sealing properties. The latter situation is problematic for the production geologist as more wells may be required to drain discontinuous reservoirs. The occurrence of widespread shale intervals acting as communication barriers is typical of northern North Sea fields with the Middle Jurassic Brent Group coastal sand/shale succession as reservoir. In the Brent Field, this has created five separate stacked drainage units (Johnson and Stewart, 1985), and, in the Thistle Field, three units (Hallett, 1981). In the latter case the development plan was modified in line with the newly acquired knowledge. This was also the case in the Forties Field where the discovery of a separate reservoir, the Charlie Sand, necessitated an amendment of the development strategy (Carman and Young, 1981). No such pressure barriers appear to have been encountered in the stacked Palaeocene - Lower Eocene sands of the Frigg Field (Heritier, Lossel and Wathne, 1979). These inter-reservoir claystones often correlate with maximum flooding surface condensed sections.

It is important to establish good well correlations at an early stage in field life to enable accurate reserves reassessment and prediction of reservoir performance. This is the basis of cash flow forecasts which provide financial security for the field operator and partners. It is also important to understand reservoir distribution and connectivity prior to determining well locations for secondary recovery (e.g. water injection).

In cases where a field spans more than one licence and crosses a block boundary, it is necessary to determine the relative amounts of hydrocarbons falling in each licence area, and therefore how much belongs to the two licence groups (each comprises an operating company and usually one or more co-venture partners). This leads to an equity determination, the equity reflecting the percentage of hydrocarbons in each licence, which are thereby distributed between the operators and partners. The percentage equity is the basis of distribution of field production costs as well as profits between the operator and partners. Equity determinations, or redeterminations can occur more than once during the life of a field as more knowledge is acquired on reservoir and hydrocarbon distribution across the field.

A number of North Sea fields span more than one operated licence, such as Statfjord (Norwegian blocks 33/9 and 33/12, UK Blocks 211/24 and 211/25), Ninian (UK blocks 3/3 and 3/8), Frigg (Norwegian blocks 25/1 and 30/10, UK Blocks 10/1, 10/6, 9/10a and 9/5, spanning four licences) and Brae (UK blocks 16/7a and 16/7b). As a knowledge of reservoir distribution is crucial to these hydrocarbon volumetric assessments, detailed well correlations are usually carried out as an essential part of the geological evaluation, integrating micropalaeontological biozone events with pressure data and wireline log signatures. Such correlation work has had major financial impact in North Sea fields as even a small increase in one licence group's equity can result in a large increase in its reserves base and revenue.

Published case histories describing the use of micropalaeontology in the development of North Sea fields are relatively few, though there are many instances in which the application of biostratigraphy has provided essential underpinning of a reservoir model and stratigraphic correlation framework. Many of the published papers on North Sea fields (e.g. in the conference proceedings volumes edited by Illing and Hobson, 1981; Brooks and Glennie, 1987; Abbots, 1991; Hardman, 1992; Proceedings 4th Conference on Petroleum Geology of NW Europe, in press) contain references to the use of micropalaeontology in reservoir studies, though few provide details of the use and impact of biostratigraphy.

Figure 4.14 Cross sections of the Frigg Field showing the palynological zonation of the multiple sandstone reservoir sequence. Note the discontinuous nature of the sand developments and the numerous lateral sand pinch outs providing stratigraphic trapping (from Heritier, Lossel and Wathne, 1979).

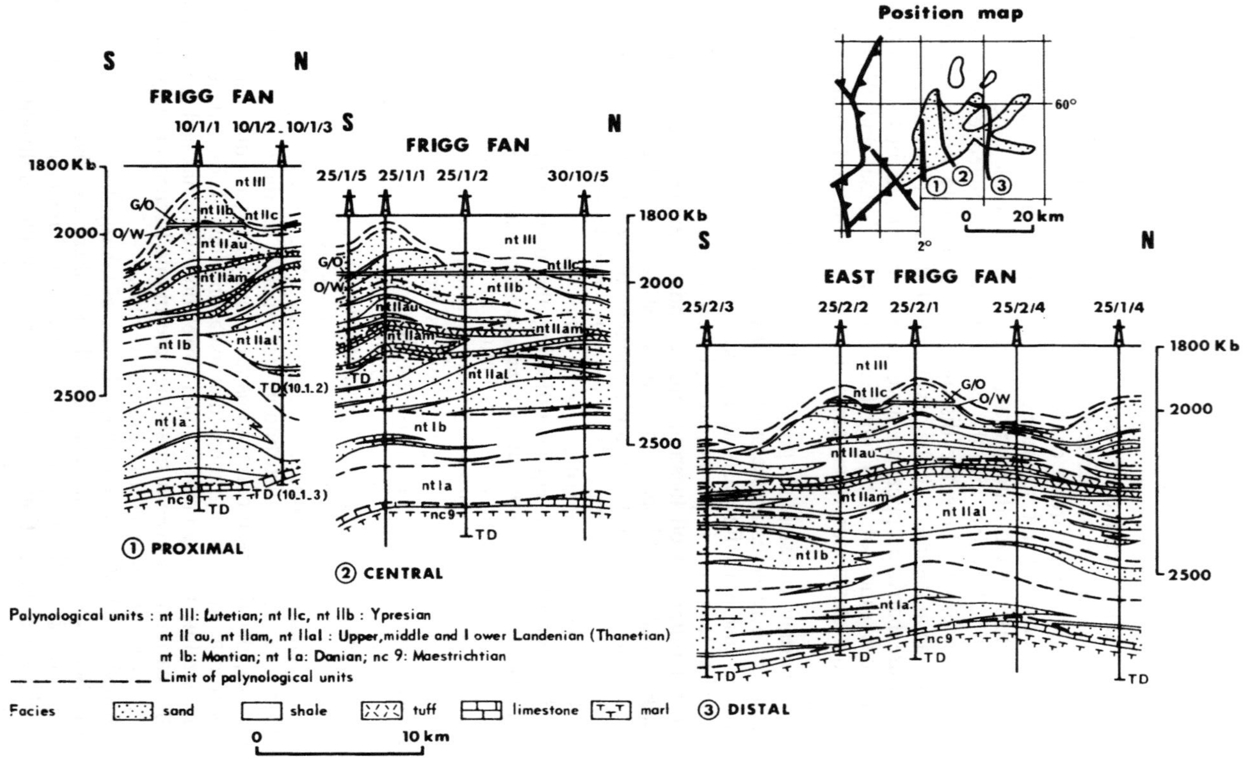

Figure 4.15 Cross section of the Tay Formation multiple reservoir sands in the Gannet Field Complex, showing the use of Shell's PT biozones for zonation and correlation. As with the Frigg Field, note the discontinuous development of the reservoir sands (from Armstrong, Ten Have and Johnson, 1987).

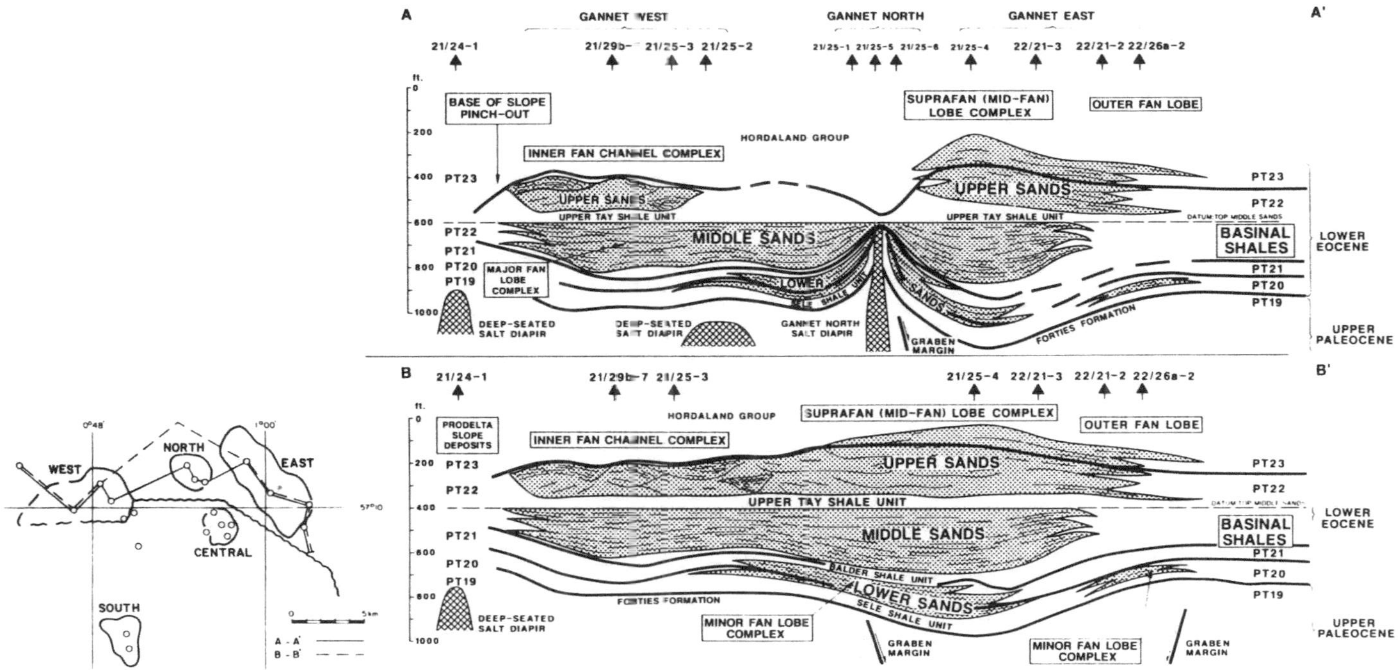

Some field papers, however, do provide more than just passing reference to biostratigraphic studies. In the Frigg Field, where the multiple sand reservoirs span the Lower Palaeocene - Lower Eocene interval, reservoir zonation and subdivision is based on a series of palynological and foraminiferal events (Heritier, Lossel & Wathne, 1979; Conort, 1986; Brewster, 1991) (see figure 4.14). In the Gannet Field complex, zonation and correlation of the Palaeocene Middle-Eocene Forties and Tay Formation sand reservoirs is reliant upon recognition of stratigraphically significant changes in populations of dinocysts, pollen, foraminifera, diatoms and radiolaria. The use of Shell's PT biozones to zone the reservoir is shown in figure 4.15 (from Armstrong, Ten Have and Johnson, 1987).

A good example of the applied use of micropalaeontology in a production context is that of Riley, Roberts and Connell (1989), who described the application of palynology (dinoflagellate cysts, miospores) to the correlation of Kimmeridgian - Volgian multiple sand reservoirs in the South Brae Field area. In this type of reservoir, correlation on log character alone is extremely difficult and of little reliability, as different sands and interbedded shales have very similar log signatures. Figure 4.16 shows the correlation of two wells in South Brae only 3950 ft (1204m) apart, based on logs. Figure 4.17 shows how this correlation changed fundamentally once the biozonal framework (JB Zones) was utilized in conjunction with the logs. The biozonation has proved to be the only reliable means of reservoir correlation in the field area and demonstrates sharp lateral facies changes from conglomerate to sand to shale in passing from proximal to distal fan positions. These facies changes provide a major control on hydrocarbon distribution within the Brae area and their understanding is fundamental to reservoir management. It is significant to observe that the major biostratigraphic events and shale breaks correspond with correlatable maximum flooding surface condensed sections in the Brae area. These can be correlated both within the greater Brae area between the South Brae, Central Brae, North Brae, East Brae (see data in Turner and Connell, 1991), as well as throughout the North Sea basin. The condensed sections are thus fundamental to reservoir correlation and zonation, and also often act as local seals within stacked multiple reservoir successions to form pressure barriers to fluid communication. A further similar example of the applied use of biostratigraphy to production geology is that of Riley, Harker and Green (1992), who described the application of a local palynology zonation to the reservoir zonation and correlation in the Lower Cretaceous of the Scapa Field (UK North Sea block 14/19). In this field, the use of a palynology zonation was essential to the establishment of a confident reservoir correlation and enabled the recognition of three reservoir sand units, with separate and mappable distributions. As in the Brae Upper Jurassic example, it is evident from an examination of the data presented in Riley, Harker and Green (1992), and its placement in a regional context, that the claystone units between the sands represent regional condensed sections probably deposited during times of maximum flooding, by comparison with known basinwide condensed sections. It is notable that only palynology has been used in the Scapa Field for zonation. In general, much more reliable results are usually obtained when other micropalaeontological subdisciplines are used in conjunction, and the Hauterivian-Ryazanian section, within which the Scapa sands are developed in the Outer Moray Firth, is known also to contain abundant foraminifera, radiolaria, ostracoda and calcareous nannofossils.

Figure 4.16 Correlation of two wells (H and F) in the South Brae Field on the basis of log signature (from Riley, Roberts and Connell, 1989).

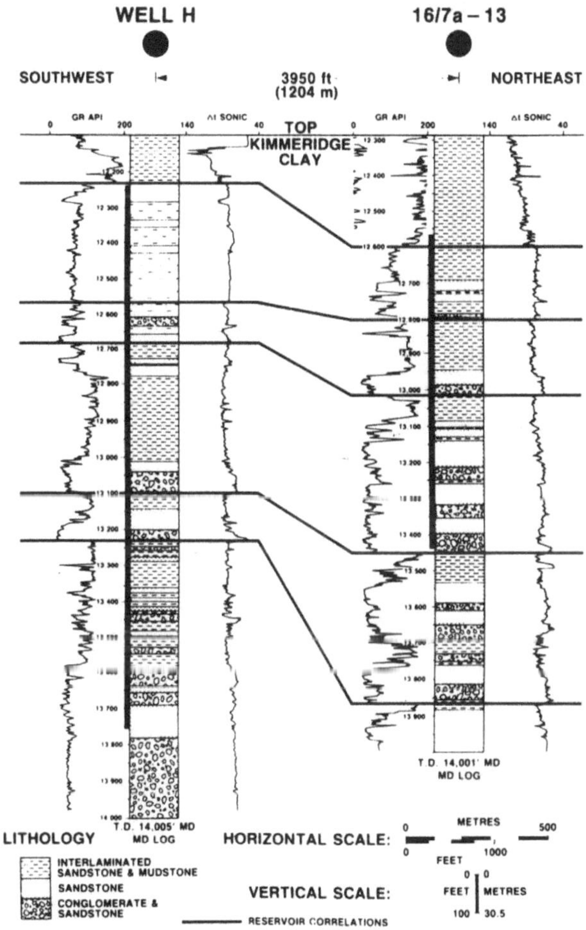

Figure 4.17 Revised correlation of wells H and F using biostratigraphy, logs and pressure data. JB zones are based on palynology. Note the variable lithology between the wells (only 3950 ft apart) due to rapid lateral facies changes in the submarine fan reservoir. It is impossible to correlate the wells on log character alone.

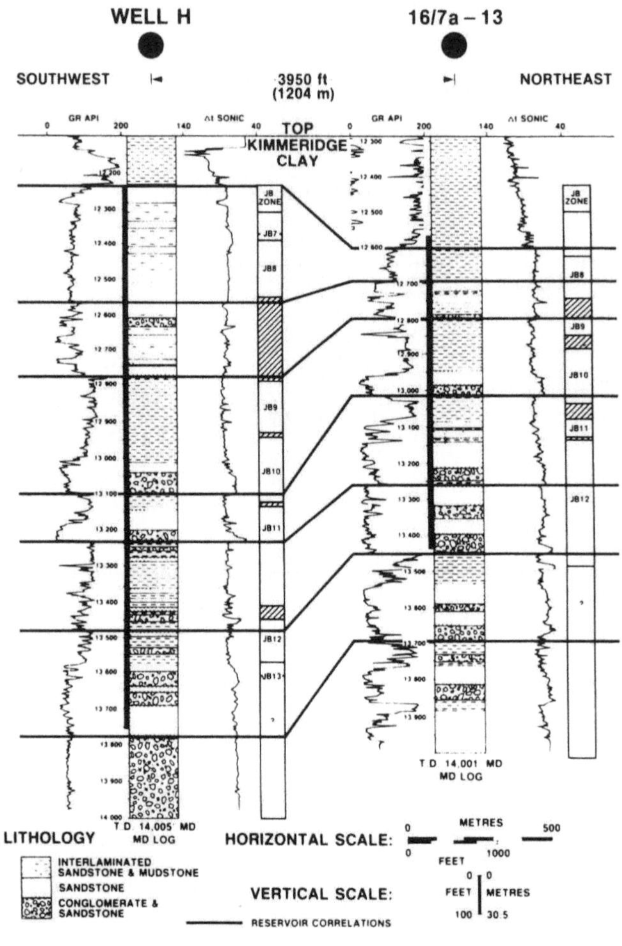

In the Draugen Field (Offshore Mid Norway, block 6407/9), Van der Zwan (1990) has published a detailed description of the application of palynology and palynofacies to the correlation and palaeogeographic and palaeoenvironmental evaluation of the Upper Jurassic (Middle Oxfordian) to Lower Cretaceous (Ryazanian) reservoir section. The Rogn Formation shallow marine sandstone reservoir, of Kimmeridgian to Early Volgian age, has been zoned and correlated by dinoflagellate cysts. The paper describes the zonation used in considerable detail, and also describes the application of palynofacies analysis to the reservoir section (see above).

Other papers referring to the applied use of micropalaeontology in a production context include Eide (1989), on the application of palynology to the correlation of the Triassic Lunde Formation in the Snorre Field area of offshore Norway (blocks 33/9, 34/4, 34/7 and 34/10) and Pedersen et al. (1989), on the integrated use of palynology in the correlation and palaeoenvironmental evaluation of the Heidrun Field reservoir (Triassic-Upper Jurassic; offshore Mid Norway, blocks 6507/7 and 6507/8).

In contrast to exploration and appraisal studies in production geology, integration of reservoir zonations and correlations with seismic activities, and extrapolation into undrilled areas of fields is difficult, although some success has been achieved especially as a result of improved techniques of seismic data acquisition and processing. This is because reservoir intervals are relatively thin compared to whole well sections, and the level of detailed evaluation required in field development work is often beyond seismic resolution.

## 4.4 North Sea hydrocarbon plays and micropalaeontology

A wide range of microfossil groups is used in the North Sea to evaluate a diversity of plays varying from Palaeozoic to Eocene. However, most of the known fields and undeveloped accumulations occur within the Triassic to Eocene interval.

### 4.4.1 Northern North Sea

Fields in the Northern North Sea region (Viking Graben) fall into two stratigraphic groupings, Triassic - Jurassic and Lower Tertiary. The stratigraphic distribution of the reservoirs and the microfossil groups used for zonation and correlation are shown in Figure 4.18.

Known fields with Jurassic reservoirs include Statfjord (Kirk, 1980), Brent (Bowen, 1975; Johnson and Stewart, 1986; Struijk and Green,1991), Alwyn (Johnson and Eyssautier, 1987), Ninian (Albright, Turner and Williamson, 1980; van Vessem and Gan, 1991), Thistle (Hallett, 1981; Williams and Milne, 1991), Heather (Gray and Barnes, 1981; Penny, 1991), Magnus (De'Ath and Schuyleman, 1981; Shepherd, 1991), Sleipner (Larsen and Jaarvik, 1981), Brae and satellites (Turner et al., 1987; Turner and Connell, 1991) and T-Block fields (Kessler and Moorhouse, 1984).

Fields with Lower Tertiary reservoir sands include Maureen (Cutts, 1991), Andrew, Cyrus (Mound, Robertson and Wallis, 1991), Heimdal (Conort, 1986), Balder (Sarg and Skjold, 1982) and Frigg (Heritier, Lossel and Wathne, 1979).

Figure 4.18 Stratigraphic distribution of the reservoirs and microfossil groups used for zonation and correlation in the Northern North Sea. Note that the Palaeocene/Eocene boundary would now be placed within the upper part of the Sele Formation (see Powell, 1988), rather than near the top of the Balder Formation.

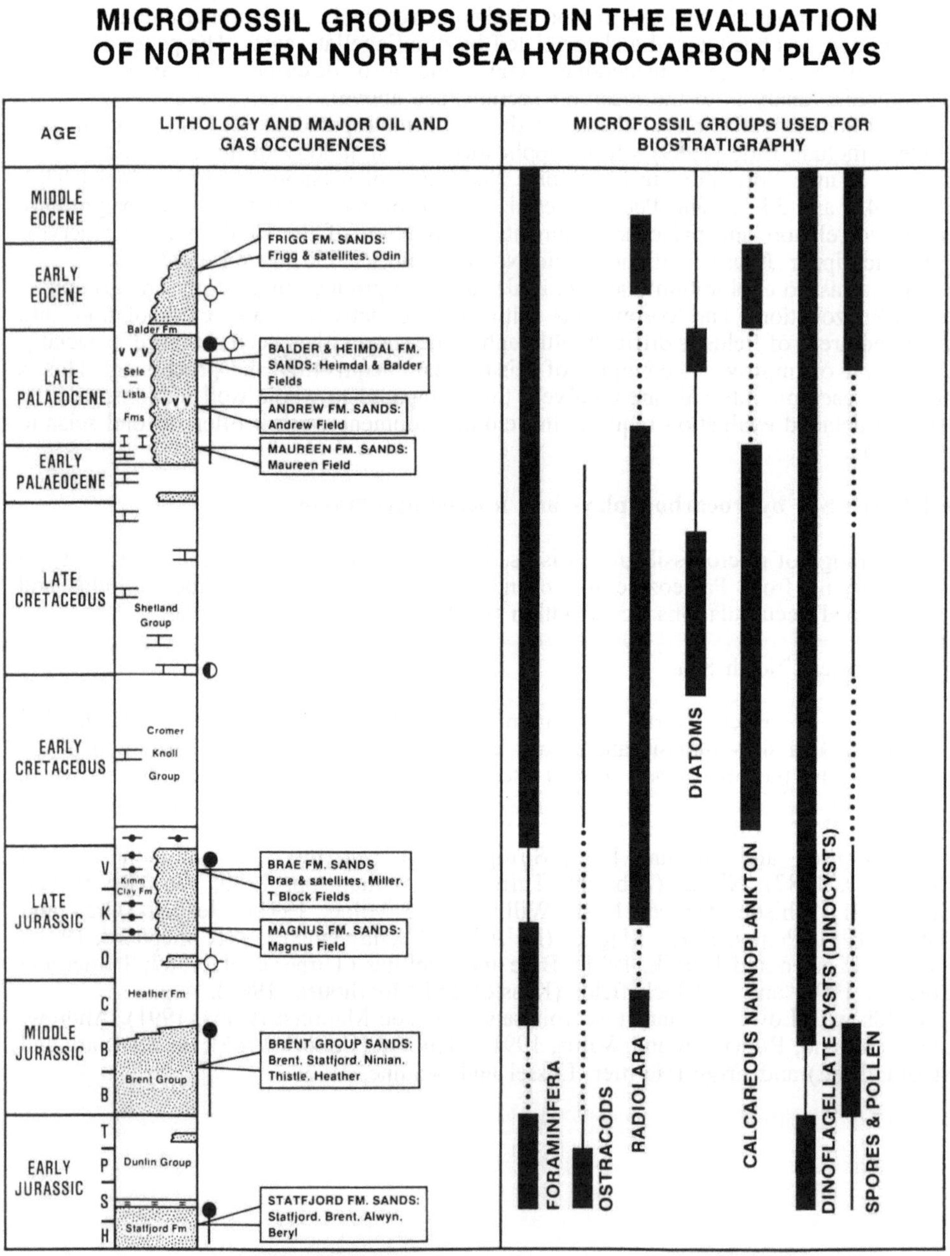

Figure 4.19 Stratigraphic distribution of the reservoirs and microfossil groups used for zonation and correlation in the Central North Sea. Note that the Palaeocene/Eocene boundary would now be placed within the upper part of the Sele Formation (see Powell, 1988), rather than near the top of the Balder Formation.

**MICROFOSSIL GROUPS USED IN THE EVALUATION OF CENTRAL NORTH SEA HYDROCARBON PLAYS**

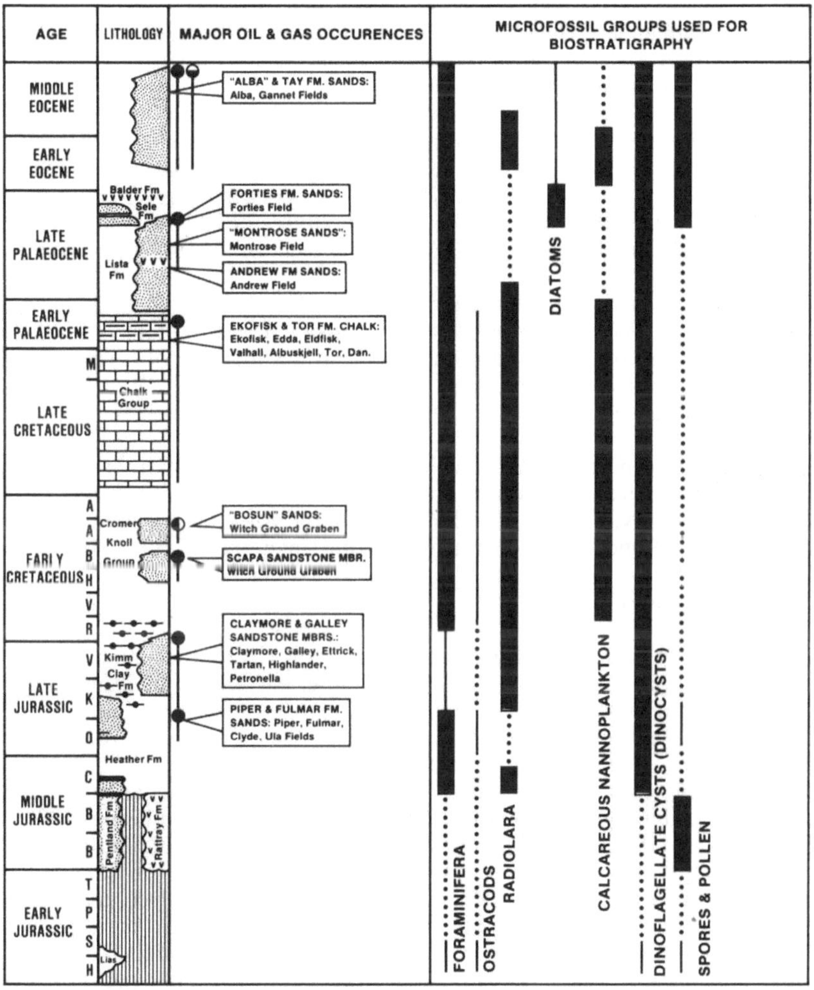

### 4.4.2 Central North Sea

Fields and known accumulations in the Central North Sea area are less stratigraphically polarized than in the Viking Graben. The stratigraphic distribution of the reservoirs and the fossil groups used for zonation and correlation are shown in Figure 4.19.

Known fields with Jurassic reservoirs in the Outer Moray Firth basin of the Central North Sea include Piper (Maher, 1981, Schmitt and Gordon, 1991), Chanter (Schmitt, 1991), Claymore (Maher and Harker, 1987; Harker et al., 1991), Petronella (Waddams and Clark, 1991), Rob Roy, Ivanhoe, Scott (Boldy and Brealey, 1990; Parker, 1991), Galley, Tartan, Highlander (Harker, Gustav and Riley, 1987) and Ross (blocks 13/28 and 13/29). The only field in the Inner Moray Firth basin is Beatrice (Linsley et al., 1980), which has a Jurassic reservoir.

Further south in the Central Graben are additional fields with Jurassic reservoirs, including Kittiwake (Glennie and Armstrong, 1991), Guillemot (Armstrong et al., 1987), Fulmar (Johnson, Mackay and Stewart, 1986; Stockbridge and Gray, 1991), Clyde (Smith, 1987; Stevens and Wallis, 1991), Ula (Spencer, Home and Wiik, 1986), Erskine (block 23/26), Puffin (blocks 29/4a, 29/5a, 29/9a, 29/10), Acorn (blocks 29/8a, 29/8b), Beechnut (29/9b), Jacqui (30/13), Argyll and Duncan (30/24) and Angus (31/26; Hall, 1992).

Lower Cretaceous sands form the reservoir in the Scapa Field (Maher and Harker, 1987) and southern Witch Ground Graben "Bosun Field" (blocks 15/30-16/26) (Bisewski, 1990). The south Central Graben is also the location of a major group of Chalk Group reservoir accumulations and fields, including Ekofisk (Van den Bark and Thomas, 1980) and its neighbours (e.g. Edda, Eldfisk, Valhall, Albuskjell, Tor and Dan, see D'heur, 1984).

Palaeocene and Eocene sands are also important in the Central Graben, comprising the reservoirs in the Andrew, Montrose (Fowler, 1975; Crawford, Littlefair and Affleck, 1991), Forties (Walmsley, 1975; Hill and Wood, 1980; Carman and Young, 1981), Alba (Harding et al., 1990) and Gannet (Armstrong, Ten Have and Johnson, 1987).

### 4.4.3 Micropalaeontology

Despite the relative wealth of published petroleum geological information on the North Sea and its fields, as cited above, relatively little has been published regarding micropalaeontological biostratigraphy and taxonomy from the North Sea.

As previously mentioned, several papers cite the use of biozones in field development studies, including Heritier et al. (1980, Frigg Field), Armstrong et al. (1987, Gannet Field complex) and Riley et al. (1989, Brae Field complex). Papers describing Jurassic micropalaeontology are rare but include Nagy (1980a, 1980b) and Nagy and Johansen (1991) describing Lower - Middle Jurassic foraminifera from the Dunlin and Brent Groups of the Statfjord area. With the aim of stimulating the application of radiolaria to subsurface North Sea correlation, Dyer and Copestake (1989) published a review of the use of radiolaria for zonation and correlation in the Upper Jurassic and lowermost Cretaceous of the northern and central North Sea. This remains the only paper to date on North Sea radiolaria. The only paper to date on Lower Cretaceous foraminifera is that of Burnhill and Ramsay (1981) from the Buchan Graben area of the Central North Sea. The latter paper described a zonation scheme, based on planktonic and benthic foraminifera, which can be applied to the correlation and subdivision of the Turonian-Albian interval. This work has been built upon by Crittenden, Cole and Harlow (1991), and applied to the accurate subdivision of the same interval into new lithostratigraphic units. From the Barremian-Aptian, Banner,

Copestake and White (1992) present a taxonomic revision of North Sea planktonic foraminifera, together with an informal zonation scheme which is of use for the correlation of the subsurface Lower Cretaceous.

Four useful papers have been published on North Sea nannofossil zonations, on the Upper Cretaceous (Mortimer, 1987), Lower Cretaceous (Jakubowski, 1987) and Lower Palaeocene (Varol, 1989a, 1989b). More published information is available from the Cenozoic. King (1983) has proposed a useful basic zonation for the Tertiary to Quaternary of the North Sea based on foraminifera, diatoms and radiolaria. Jacqué and Thouvenin (1975) also described some diatoms from the uppermost Palaeocene - lowermost Eocene volcanic sediments (Balder Formation) of the North Sea; a greater diversity of diatoms, together with silicoflagellates and dinocysts were later described from this formation by Malm et al. (1984). Knox, Morton and Harland (1981) have also utilized dinocysts for their regional stratigraphic studies of the Palaeocene. The study of Palaeocene-Oligocene deep water agglutinated foraminifera in the North Sea was also given a considerable boost by the publication of Gradstein and Berggren's paper in 1981. These authors have continued to research Tertiary deep water agglutinated foraminifera, as evidenced by the publications of Gradstein, Kaminski and Berggren (1988), Gradstein and Kaminski (1989) and Gradstein et al. (1992). The latter paper is important in containing an integrated biozonation scheme for the North Sea Tertiary to Quaternary, based on foraminifera, radiolaria and dinoflagellate cysts. This scheme, however, is far less refined than unpublished, proprietary biozonation schemes currently in use in the North Sea. This work on Cenozoic agglutinated foraminifera is supplemented by the comprehensive taxonomic review by Charnock and Jones (1990). This is an extremely useful publication for micropalaeontologists working in the area in that it includes much more information than is usually available publicly on North Sea microfaunas.

The second edition of A Stratigraphical Atlas of Fossil Foraminifera (Jenkins and Murray, eds., 1989) is an important new volume containing several chapters with significant North Sea data on Jurassic (Copestake and Johnson, 1989; Morris and Coleman, 1989; Shipp, 1989), Cretaceous (King et al., 1989) and Cenozoic (King, 1989) foraminifera. The latter paper updates King's (1983) earlier biozonation scheme. The emphasis in King's work is on the benthic foraminifera and a comprehensive research on the often diverse planktonic foraminiferal faunas present in the North Sea well successions has yet to be published. An interval particularly rich in planktonic foraminifera is the Lower Palaeocene (Maureen and Ekofisk Formations), which contains worldwide markers for standard planktonic foraminiferal biozones, as alluded to by Mudge and Copestake (1992a,b) and Copestake and Dyer (1981).

The proceedings of the 4th Conference on Petroleum Geology of North West Europe, held in April, 1992, are due to be published in 1993, and will contain several papers which aptly demonstrate the crucial part played by detailed micropalaeontological studies in supporting the search and extraction of North Sea hydrocarbons.

The most useful fossil groups in evaluation of North Sea reservoirs appear to be (see also Figures 4.18 and 4.19):

Middle Jurassic sands: spores and pollen, dinocysts.

Upper Jurassic sands: dinocysts, radiolaria, foraminifera.

Lower Cretaceous sands: dinocysts, radiolaria, foraminifera, calcareous nannofossils.

Upper Cretaceous - Lower Palaeocene chalks: calcareous nannofossils, foraminifera.

Palaeocene sands: dinocysts, radiolaria, foraminifera, diatoms.

Uppermost Palaeocene - lowermost Eocene sands: spores and pollen, dinocysts, diatoms.

Eocene Sands: dinocysts, foraminifera, radiolaria.

## 4.5 Pitfalls in biostratigraphic interpretations

No review of the application of micropalaeontological biostratigraphy is complete without mention of some of the commonest problems related to the application of the science to problem solving in the oil industry.

To some extent, the review of biostratigraphic techniques presented above is idealized, and the case study examples were chosen for which good biostratigraphic data was available. In many instances, however, the available data is less than ideal owing to many factors.

### 4.5.1 Factors affecting data quality

Various drilling techniques, such as the use of downhole motors ("turbodrilling") and diamond drill bits, are popular with drillers because rates of penetration are often increased. However, these tools are frequently (but not always) destructive to microfossils and can render the stratigraphic interpretation of a succession impossible.

In the deeply buried parts of the North Sea, microfossils can become destroyed due to the high temperatures and pressures. In particular, organic fossils are often completely destroyed at great depths due to the organic maturation process which converts organic matter to hydrocarbons. The depth at which the destruction takes place is usually greater than around 12,000 ft, but the present day burial depth is not a reliable guide to microfossil destruction owing to several probable factors, including heat flow variations (for instance related to the presence of salt pillows or diapirs), degree of structural inversion and overpressure.

### 4.5.2 Reworking and caving

Caving is the contamination of ditch cuttings samples from a particular depth by material from a shallower part of the borehole. The term caving refers to the falling of sediment from the sides of the hole to leave a washout. Washouts can be identified from the caliper log which measures the diameter of the well bore. Caving is a well known problem, which can render invalid the use of stratigraphic bases in wells. However, it is often the case that the preservation of microfossils in cavings is much better than in situ assemblages from the caved horizon. This is due to the fact that the cavings have not been affected or damaged by the drill bit, thus the judicious evaluation of caved assemblages can aid in the stratigraphic evaluation of the well section. Sidewall cores can often also contain contamination as part of the sidewall core represents the mud cake, which is made up of a mixture of sediment and drilling mud adhering to the well sides.

Reworking of fossils from older sediments into younger sediments is more common than is often realized, and is particularly a problem for the palynologist. This is due to the small size of the organic microfossils which renders them prone to reworking. With some fossil groups, such as foraminifera, it is sometimes relatively easy to recognize reworking, due to the presence of characteristic fossil preservation features, such as colour and general appearance which can highlight the origin of a fossil as being recycled from an older sediment. Reworking, if not correctly diagnosed, can result in zones and stratigraphic tops being interpreted erroneously high in the section. It is thus important to undertake more than one type of micropalaeontological discipline in tandem to provide a cross check on this phenomenon.

The correct identification of reworking and the recognition of the likely source of the reworked material can be of value in indicating the presence of a sequence boundary (above which reworking is more the rule than the exception) or a possible sediment provenance area.

### 4.5.3 Age interpretations

A very common pitfall relates to the age interpretation placed upon microfossil assemblages. It is a fact that different age interpretations can be placed on the same fossil assemblages by different biostratigraphers currently working in the North Sea oil industry. One reason for this is that interpretations of the age significance of microfossils in the subsurface offshore is usually based on the comparison with the range of the same fossils in onshore sections in which calibration against standard zones can be made. Unfortunately, however, some important North Sea species have different ranges offshore compared to onshore, rendering offshore age interpretations uncertain. Unfortunately, many geologists and biostratigraphers base many of their stratigraphic evaluations upon the recognition and correlation of interpreted age intervals. The basic framework for many operating and biostratigraphic consultancy company stratigraphic schemes is an age breakdown based on units such as Late Volgian, Middle Volgian, Late Palaeocene etc., and these age units are added to well composite logs (which summarize well lithology and stratigraphy against selected wireline logs). Thus a error pitfall is often made by geologists correlating the age units from composite logs and correlating wells on this basis.

### 4.5.4 Taxonomic nomenclature

Variable taxonomic nomenclature between workers and companies is a problem and correlatable microfossil assemblages may not appear to be so due to the application of different names by separate workers to the same fossils. This is exacerbated by the fact that a significant number of North Sea stratigraphically important marker species are as yet undescribed in the literature, thus each biostratigraphic company will typically apply a different in house nomenclature to these taxa. This problem is particularly acute for fossil groups which have not been adequately described taxonomically. Palaeocene-Eocene diatoms are a good example; at least six unpublished nomenclature schemes are currently in use.

4.5.5. Preparation techniques

Sample preparation techniques are an often neglected area of micropalaeontological activity. Owing to the pressure applied to the micropalaeontologist to obtain rapid results for the geologist, sample preparation is frequently rushed and routine methods are typically applied to bulk quantities of samples. The development of an effective preparation technique is a skill which is not consistently applied between workers and companies. This can and has made the difference between different biostratigraphers obtaining either good yields or sometimes no yields at all from the same sample set.

4.5.6 Microfossil and zonal identification

Once the palaeontological slides are presented to the biostratigrapher, pressure of work and identification skills come to play, and often the identifications between different workers, of the same basic assemblages, can be markedly disparate. An important factor affecting the identifications is the availability of detailed, well illustrated publications of North Sea microfossils. It is only fairly recently that any papers illustrating North Sea assemblages have become available.

Traditionally the recognition of biozones has been based on absolute stratigraphic ranges, with no consideration being paid to abundance changes. This reliance on "tops" and "bases" overlooks the fact that fossil species are often rare near their extinction points. In the industrial environment, with small samples of variable quality and time constraints on processing and identification, these rare tops may often be missed. The first downhole record of a species is more likely to relate to an increased abundance level, and the correlation of numerical increases of microfossil taxa is usually more reliable than the reliance on recognition of "tops" alone.

4.5.7 Palaeoenvironmental controls

The palaeoenvironmental control upon microfossil stratigraphic ranges can be acute within a basin. This fact has long been recognized from a theoretical point of view but not many industry biostratigraphers appear to pay heed to this. As Powell (1992) has mentioned, fossils can have significantly different times of appearance and disappearance from one well location to another, if the wells are situated in different palaeoenvironmental settings. For instance, in the North Sea Tertiary, in which coeval shelf and basin successions occur, some key marker taxa are known, from integrated studies, to range younger in the basin than on the shelf.

4.5.8 Practical correlation and biostratigraphy

Accurate correlation between wells is the major and most consistently required applied use of micropalaeontology. It is important to realize, however, that often it is not possible to correlate wells on biostratigraphic data alone. Miscorrelations can be made owing to various combinations of all the factors discussed above. It is often found that bioevents may occur lower than usual in a well section due to poor data quality, or occur higher, due to the overlooking of reworking. It is therefore essential that biostratigraphic interpretations of all types are made within a framework of an integrated study, which incorporates all available data. In an ideal subsurface study this will include the availability of good quality samples, the application of more than one micropalaeontological discipline if possible (palaeoenvironmental setting notwithstanding), availability of a suite of wireline logs, knowledge of drilling parameters, and information from seismic data regarding gross subsurface relationships. As a minimum, more than one micropalaeontological discipline should be applied to the same sample set and the resulting biostratigraphic data tied to wireline logs. The current popularity of sequence stratigraphy is timely as this provides the ideal integrated medium in which to conduct biostratigraphic work.

## 4.7 Development of micropalaeontology in North Sea exploration

As with other exploration sciences such as seismic acquisition and processing, well logging and hydrocarbon enhanced recovery techniques, the discipline of micropalaeontology has developed virtually beyond recognition since the 1960's when the first wells were drilled in the North Sea Basin.

In the 1960's, micropalaeontology was focused upon attempted age determination of offshore sequences using classic fossil groups such as foraminifera and ostracoda. It was found, however, that the North Sea basin contained rocks of different, generally deeper water facies than onshore north west Europe, with indigenous microfossil assemblages. Thus it proved difficult to correlate with standard onshore sequences and thereby produce refined age interpretations. Groups useful for zonation onshore, such as ostracoda, were found to be only sporadically developed offshore.

The discipline developed significantly in the 1970's due to the elucidation of a zonation for the Jurassic - Lower Cretaceous based on dinocysts (tabulated in Rawson and Riley, 1982, later expanded upon in Riley et al., 1989). This zonation (though it has still not been fully documented) was a fundamental milestone because it enabled detailed subdivision and correlation of Jurassic reservoir sands and provided a tie to onshore standard sequences in which the same, cosmopolitan dinocysts were discovered. It also provided an important age calibration of the indigenous microfossil assemblages which co-occur with the dinocysts. This represented a first step in the increasing use of palynology in North Sea exploration, a discipline still being used intensively today (e.g. Herngreen, Lissenberg and White, 1988).

More recent developments in palynology have been in Upper Cretaceous and Tertiary clastic sequences, mainly based on dinocysts, but also, latterly, spores and pollen. However, palynology does have its drawbacks: relatively high cost (due to long sample processing and identification time), propensity for reworking of palynomorphs, in good state of preservation, into younger sediments, palynomorph sparsity due to high temperature destruction in deep wells; and reduced abundance in deep water and calcareous sediments.

From the mid 1970s to mid 1980s these drawbacks were not generally realized and palynology was often undertaken to the exclusion of all other micropalaeontological disciplines, as is illustrated by published papers from this period. However, the 1980s and 1990s have seen more attention being paid to other fossil groups. In the Jurassic, foraminifera, and particularly radiolaria (Dyer and Copestake, 1989) are known to be of great value for correlation and zonation, either as an additional refinement to the palynology, or as a replacement in deep wells where palynomorphs either become destroyed due to the high thermal maturity or are absent due to palaeoenvironmental constraints. In the Cretaceous - Lower Palaeocene, much work has been undertaken on calcareous nannofossils (Jakubowski, 1987; Mortimer, 1987; Van Heck and Prins, 1987; Varol, 1989a,b), especially in Chalk Group reservoirs of the Greater Ekofisk area. One drawback here, however, is that diagenetic processes can hamper nannofossil preservation and identification, as is also the case with associated foraminifera. Nannofossils are poorly represented or absent from most of the preceding Jurassic and overlying Tertiary, which are mostly developed in deep water, non calcareous facies.

In recent years, Tertiary biostratigraphy has developed rapidly due to the discovery of significant amounts of hydrocarbons in the Palaeocene - Eocene. A refined dinocyst zonation has been developed (Costa and Manum, 1988; Powell, 1988) and parts of the Tertiary section are being intensively studied for pollen and spores. In addition, again in tandem with hydrocarbon discoveries, diatom, foraminifera and radiolaria zonations have been developed (e.g. King, 1983) and are being continuously updated and refined (e.g. King, 1989). It has been found repeatedly that micropalaeontology has its greatest impact in exploration when more than one subdiscipline is used. Errors, often significant, are inherent in the use of one fossil group, and the most refined correlation, which can solve the most problems, results when several groups are used, for instance dinocysts, radiolaria, spores, pollen, foraminifera and diatoms in the Palaeogene.

### 4.6 Concluding remarks - how can micropalaeontology help to find oil?

A crucial question for an exploration manager looking for hydrocarbons and wishing to cut costs, is how can disciplines such as micropalaeontology help him find petroleum in the subsurface? Without reiterating the wide range of applications of the science to petroleum finding, a simple answer is that it can help indirectly, by impacting on the evaluation of the four vital elements of reservoir (fingerprinting, biozonation, extent, continuity, completeness, correlation, rock volume), trap (nature of seal and timing), trap type (whether unconformity, structural or stratigraphic) and source (dating, zonation, fingerprinting, correlation). In addition, by supporting sequence stratigraphic analysis, it can identify and calibrate seismically mappable features, and can also, in conjunction with palaeoenvironmental evaluations, be used predictively to suggest areas of enhanced reservoir development.

At present micropalaeontology is used extensively and intensively to support exploration, appraisal and development programmes. As the North Sea hydrocarbon province moves into its mature phase, it is crucial, from an economic point of view, that all available technology, including micropalaeontology, is used to maximize the location and extraction of hydrocarbons from ever smaller accumulations. The discipline thus has a vital role to play in prolonging the life of the province. The emphasis in the future must be on integrated technology, bringing together such disciplines as seismic, biostratigraphy, log interpretation, sedimentology and reservoir engineering.

In the final analysis, the successful application of micropalaeontology will rely on the awareness of biostratigraphers of the value of their science and how it can be used in an applied way to support the industry. Only if this is effectively achieved will the industry continue to support the discipline, by providing research funds and employment. The key to effective application is for micropalaeontologists to think about their subject.This will involve an honest appreciation of its pitfalls, as well as strengths, and communication of all these aspects successfully to the industry.

**Acknowledgements**

The author is grateful to BP Exploration Company Limited for provision of facilities and permission to publish. Diagrams and information regarding the application of palynology in the Brae area were kindly provided by Dr L A Riley. Nigel Hooker is thanked for provision of advice and information with regard to palynological interpretations. Graham Booth and Tony Welsh are thanked for their constructive comments on the manuscript and for their support in obtaining approval to publish.

**References**


Abbots, I.L. (Ed). 1991. United Kingdom Oil and Gas Fields 25 Years Commemorative Volume. *Geological Society Memoir No 14*, 507pp.

Albright, W.A., Turner, W.L. and Williamson, K.R. 1980. Ninian Field, U.K. Sector, North Sea. *Giant Oil and Gas Fields of the Decade, Amer. Assoc. Petr. Geol. Mem., 30.* 173-193.

Anderson, R.O. 1983. *Radiolaria*. Springer-Verlag, 355pp.

Anderton, R. In press. Sedimentation and basin evolution in the Palaeogene of the Northern North Sea and Faeroe-Shetland Basins. In: Parker, J.R. (Ed) *Petroleum Geology of Northwest Europe: Proceedings of the 4th Conference. Geological Society, London.*

Andrews, I.J. and Brown, S. 1987. Stratigraphic evolution of the Jurassic, Moray Firth. In: Brooks, J and Glennie, K.W. (Eds), 785-795

Armentrout, J.M., Malecek, S.J., Fearn, L.B., Sheppard, C.E., Naylor, P.H., Miles, A.W., Desmarais, R.J. and Dunay, R.E. In press. Log-motoif analysis of Palaeogene depositional systems tracts in the North Sea. In: Parker, J.R. (Ed) *Petroleum Geology of Northwest Europe: Proceedings of the 4th Conference. Geological Society, London.*

Armstrong, L., Ten Have, A. and Johnson, H.D. 1987. The geology of the Gannet Fields, Central North Sea, U.K. sector. In: Brooks, J. and Glennie, K.W. (Eds), 533-548.

Banner, F.T., Copestake, P. and White, M.R. In press. Barremian-Aptian Praehedbergellidae of the North Sea area: a reconnaissance. *Bull. British Museum Nat. Hist.*

Berggren, W.A. and Miller, K.G. 1988. Paleogene tropical planktonic foraminiferal biostratigraphy and magnetobiochronology. *Micropaleontology, 34.* 362-380.

Bisewski, H. 1990. Occurrence and depositional environment of the Lower Cretaceous sands in the southern witch Ground Graben. In: Hardman, R.F.P. and Brooks, J. (Eds), 325-338.

Blow, W.H. 1969. Late Middle Eocene to Recent planktonic foraminiferal biostratigraphy. *Proceedings First International Conference on Planktonic Microfossils. Geneva. 1967.* 1, 199-422. Elsevier.

Blow, W.H. 1979. *The Cainozoic Globigerinida.* E.J. Brill, Leiden, 1413pp.

Boldy, S.A.R. and Brealey, S. 1990. Timing, nature and sedimentary result of Jurassic tectonism in the Outer Moray Firth. In: Hardman, R.F.P. and Brooks, J. (Eds), 259-279.

Bolli, H.M, Saunders, J.B. and Perch-Neilsen, K. (Eds). 1985. *Plankton Stratigraphy.* Cambridge University Press, 1006pp.

Boulter, M.C. and Riddick, A. 1986. Classification and analysis of palynodebris from the Palaeocene sediments of the Forties Field. *Sedimentology, 33*, 871-886.

Bowen, J.M. 1975. The Brent Oil - Field. In: Woodland, A.W. (Ed), 353-362.

Brennand, T.P., Van Hoorn, B. and James, K.H. 1990. Chapter 1. Historical Review of North Sea Exploration. In: Glennie, K.W. (Ed). *Introduction to the Petroleum Geology of the North Sea.* 3rd Edition. Blackwell Scientific Publications, 402pp.

Brewster, J. 1991. The Frigg Field. In: Abbots, I.L. (Ed), 117-126.

Brooks, J. and Glennie, K.W. (Eds) 1987. Petroleum Geology of N.W.Europe. *Proceedings 3rd Conference on Petroleum Geology of North West Europe.* Graham and Trotman. London, 1219pp.

Burckle, L.H. 1978. Marine Diatoms. In : Haq, B.U. and Boersma, A., 245-266.

Burnhill, T.J. and Ramsay, W.V. 1981. Mid-Cretaceous palaeontology and stratigraphy, central North Sea. In : Illing, L.V. and Hobson, G.D. (Eds), 245 - 254.

Byrd, W.D. 1975. Geology of the Ekofisk Field, Offshore Norway . In: Woodland, A.W. (Ed), 439-446.

Carman, G.J. and Young, R. 1981. Reservoir geology of the Forties oilfield. In: Illing, L.V. and Hobson, G.D. (Eds), 371-379.

Charnock, M.A. and Jones, R.W. 1990. Agglutinated foraminifera from the Palaeogene of the North Sea. In: Hembleben, C. et al. (Eds) *Paleoecology, Biostratigraphy and Taxonomy of Agglutinated Foraminifera*, Kluwer Academic Publishers, The Netherlands, 139-244.

Cockings, J.H., Kessler, L.G., Mazza, T.A. and Riley, L.A. 1992. Bathonian to mid-Oxfordian sequence stratigraphy of the South Viking Graben, North Sea. In: Hardeman, R.F.P. (Ed), 65-105.

Conort, A. 1986. Habitat of Tertiary Hydrocarbons, South Viking Graben. *Habitat of Hydrocarbons on the Norwegian Continental Shelf. Norwegian Petroleum Society.* Graham and Trotman, 159-170.

Copestake, P. and Dyer, R. 1981. Early Palaeocene (Danian) planktonic foraminifera from the North Sea. British Micropalaeontological Society Demonstration Meeting Booklet, 1-33.

Copestake, P. and Johnson, B. 1989. The Hettangian to Toarcian (Lower Jurassic). In : Jenkins, D.G. and Murray, J.W. (Eds), 129-188.

Costa, L.I. and Manum, S.B. 1988. The description of the interregional zonation of the Palaeogene (D1-D15) and the Miocene (D16-D20). In: Vinken, R. (Ed) *The Northwest European Tertiary Basin. Results of the International Geological Correlation Programme, Project No 124. Geologisches Jahrbuch, Reihe A, Heft 100*, 321-330.

Costa, L.I., Manum, S.B. and Meyer, K.-J. 1988. The regional distribution of dinoflagellates; correlation of the interregional zonation with the local zones and with the regional lithostratigraphy. Great Britain-Norway, the Viking Graben. In: Vinken, R. (Ed) *The Northwest European Tertiary Basin. Results of the International Geological Correlation Programme, Project No 124. Geologisches Jahrbuch, Reihe A, Heft 100*, 330-332.

Crawford, R., Littlefair, R.W. and Affleck, L. G. 1991. The Arbroath and Montrose Fields, Blocks 22/17, 18, UK North Sea. In: Abbots, I.L. (Ed), 211-217.

Crittenden, S. 1986. Planktonic foraminifera and the biostratigraphy of the Early Tertiary strata of the North Sea basin; a brief discussion. *Newsl. Stratigr. 15,* 163-171.

Crittenden, S., Cole, J. and Harlow, C. 1991. The Early to "Middle" Cretaceous lithostratigraphy of the Central North Sea (UK Sector). *Journal of Petroleum Geology, 14,* 387-416.

Cutts, P.L. 1991. The Maureen Field, Block 16/29a, UK North Sea. In: Abbots, I.L. (Ed), 347-352.

De'ath, N.G. and Schuyleman, S.F. 1981. The geology of the Magnus oilfield. In: Illing, L.V. and Hobson, G.D. (Eds), 342-351.

Deegan, C.E. and Scull, B.J. (compilers). 1977. A standard lithostratigraphic nomenclature for the central and northern North Sea. *Rep. Inst-Geol. Sci.* No 77/25.

Den Hartog Jager, D.G., Giles, M.R. and Griffiths, G.R. In press. Evolution of Palaeogene fans of the North Sea in space and time. In: Parker, J.R. (Ed) *Petroleum Geology of Northwest Europe: Proceedings of the 4th Conference. Geological Society, London.*

D'heur, M. 1984. Porosity and hydrocarbon distribution in North Sea chalk reservoirs. *Marine and Petroleum Geology 1*, 211-238.

Donovan, A.D., Djakic, A.W., Garfield, T.R., Ioannides, N.S. and Joanes, C.R. In press. Sequence stratigraphic control on Middle and Upper Jurassic reservoir distribution within the UK Central North Sea. In: Parker, J.R. (Ed) *Petroleum Geology of Northwest Europe: Proceedings of the 4th Conference. Geological Society, London.*

Duval, B., Cramez, C. and Vail, P.R. In press. Types and hierarchy of stratigraphic cycles. *International Symposium on Mesozoic and Cenozoic sequence stratigraphy of European basins, Dijon,* May, 1992.

Dyer, R. and Copestake, P. 1989. A review of Late Jurassic to earliest Cretaceous radiolaria and their biostratigraphic potential to petroleum exploration in the North Sea. In: Batten, D.J. and Keen, M.C. (Eds). *Northwest European Micropalaeontology and Palynology*, 214-235, Ellis Horwood Ltd., Chichester.

Eide, F. 1989. Biostratigraphic correlation within the Triassic Lunde Formation in the Snorre area. In: Collinson, J.D. (Ed) *Correlation in Hydrocarbon Exploration*, 327-338. Graham and Trotman for the Norwegian Petroleum Society.

Fowler, C. 1975. The Geology of the Montrose Field. In: Woodland, A.W. (Ed), 467-476.

Galloway, W.E. 1989. Genetic stratigraphic sequences in basin analysis 1: architecture and genesis of flooding-surface bounded depositional units. *Bull.Amer. Assoc.Petrol.Geol. 73*, 125-142.

Galloway, W.E., Garber, J.L., Sloan, B.J. and Lui, X. In press. Sequence stratigraphic framework of the Cenozoic fill, Central and Northern North Sea Basin. In: Parker, J.R. (Ed) *Petroleum Geology of Northwest Europe: Proceedings of the 4th Conference. Geological Society, London.*

Glennie, K.W. and Armstrong, L.A. 1991. The Kittiwake Field, Block 21/18, UK North Sea. In: Abbots, I.L. (Ed), 339-346.

Gradstein, F.M. and Berggren, W.A. 1981. Flysch-type agglutinated foraminifera and the Maestrichtian to Paleogene history of the Labrador and North Seas. *Marine Micropaleontology 6*, 211-268.

Gradstein, F.M., Kaminski, M.A. and Berggren, W.A. 1988. Cenozoic foraminiferal biostratigraphy, Central North Sea. In: Rögl, F. and Gradstein, F.M. (Eds) Second Workshop on Agglutinated Foraminifera, Vienna 1986, Proceedings. *Abhandlungen Geologischen Bundesanstalt, 41*, 97-108.

Gradstein, F.M. and Kaminski, M.A. 1989. Taxonomy and biostratigraphy of new and emended species of Cenozoic deep-water agglutinated foraminifera from the Labrador and North Seas. *Micropaleontology, 35(1)*, 72-92.

Gradstein, F.M., Kristiansen, I.L., Loemo, L. and Kaminski, M.A. 1992. Cenozoic foraminiferal and dinoflagellate biostratigraphy of the central North Sea. *Micropaleontology, 38*, 101-137

Gray, W.D.T. and Barnes, G. 1981. The Heather oil field. In : Illing, L.V. and Hobson, G.D. (Eds), 335-341.

Hall, S.A. 1992. The Angus Field, a subtle trap. In: Hardman, R.F.P. (Ed), 151-186.

Hallett, D. 1981. Refinement of the geological model of the Thistle field. In: Illing, L.V. and Hobson, G.D. (Eds), 315-325.

Haq, B.U. 1978. Silicoflagellates and Ebridians. In: Haq, B.U. and Boersma, A. (Eds), 267-276.

Haq, B.U. and Boersma, A. (Eds). 1978. *Marine Micropaleontology*. Elsevier, 376pp.

Haq, B.U., Hardenbol, J. and Vail, P.R. 1987. Chronology of fluctuating sea-levels since the Triassic. *Science 235*, 1156-1167.

Harding, A.W., Humphrey, T.J., Latham, A., Lunsford, M. K. And Strider, M.H. 1990. Controls on Eocene submarine fan deposition in the Witch Ground Graben. In : Hardman, R.F.P and Brooks J. (Eds), 353-367.

Hardman, R.F.P. and Brooks, J. (Eds). 1990. Tectonic Events Responsible for Britain's Oil and Gas Reserves. *Geological Society Special Publication No. 55.*

Hardman, R.F.P. (Ed) 1993. Exploration Britain: Geological insights for the next decade. *Geological Society Special Publication No. 67.*

Harker, S.D., Gustav, S.H. and Riley, L.A. 1987. Triassic to Cenomanian stratigraphy of the Witch Ground Graben. In: Brooks, J. and Glennie, K.W. (Eds), 809-818.

Harker, S.D., Mantel, K., Morton, D. and Riley, L.A. 1991. Oxfordian-Kimmeridgian (Late Jurassic) reservoir sandstones in the Witch Ground Graben area, UK North Sea. *Abstract American Association of Petroleum Geologists Bulletin, 75.*

Harland, W.B., Cox, A.V., Llewellyn, P.G., Pickton, C.A.G., Smith, A.G. and Walters, R. 1982. *A geologic time scale.* Cambridge University Press.

Harland, W.B., Armstrong, R.L., Cox, A.V., Craig, L.E., Smith, A.G. and Smith, D.G. 1990. *A geologic time scale.* Cambrigdge University Press, 263 pp.

Hedberg, H.D. 1976 *International Stratigraphic Guide: A Guide to Stratigraphic Classification. Terminology and Procedure.* John Wiley, New York, 1-200.

Heilmann-Clausen, C., Laursen, G.V. Thomsen, E. and Michelsen, O. In press. Biostratigraphy and age of Cenozoic sequences in the Eastern North Sea. *International Symposium on Mesozoic and Cenozoic sequence stratigraphy of European basins, Dijon,* May, 1992.

Heritier, F.E., Lossel, P. and Wathne, E. 1979. Frigg field - Large Submarine - Fan Trap in Lower Eocene Rocks of the North Sea Viking Graben. *Bull. Amer. Assoc. Petr. Geol. 63*, 1999-2020.

Herngreen, G.F.W., Lissenberg, Th. and White, L.J. 1988. Dinoflagellate, sporomorph and micropalaeontological zonation of Callovian to Ryazanian strata in the Central North Sea Graben, The Netherlands. *Proceedings 2nd International Symposium on Jurassic Stratigraphy.* Lisbon, 1987. Centro de Geociencias da Universidade de Coimbra (INIC), Portugal. 745-762.

Hesjedal, A. and Hamar, G.P. 1983. Lower Cretaceous stratigraphy and tectonics of the south-southeastern Norwegian offshore. In : Kaaschieter, J.P.H. and Reijers, T.J.A. (Eds), *Petroleum Geology of the southeastern North Sea and the adjacent onshore areas. Geol. Mijnbouw 62,* 135-144.

Heusser, L. 1978. Spores and Pollen in the marine realm. In: Haq, B.U. and Boersma, A. (Eds), 341-357.

Higgins, A.C. and Austin, R.L. 1985. *A Stratigraphical Index of Conodonts*, Ellis Horwood Ltd, Chichester, 263pp.

Hill, P.J. and Wood, G.V. 1980 Geology of the Forties field, U.K. Continental Shelf, North Sea. *Giant Oil and Gas Fields of the Decade. Mem. Amer. Assoc. Petrol. Geol.* 30, 81-93.

Illing, L.V. and Hobson, G.D. (Eds) 1981. *Petroleum Geology of the Continental shelf of N.W. Europe. Proceedings of the Second Conference on Petroleum Geology of the Continental Shelf of NorthWest Europe.* Heyden and Son Ltd, 521 pp.

Jacque, M. and Thouvenin, J. 1975 Lower Tertiary Tuffs and Volcanic Activity in the North Sea. In: Woodland, A.W. (Ed), 455-466.

Jakubowski, M. 1987 A proposed Lower Cretaceous nannofossil zonation scheme for the Moray Firth area of the North Sea. *Abh. Geol. B.-A. 39*, 99-119.

Jansonius, J. and Jenkins, W.A.M. 1978. Chitinozoa. In : Haq, B.U. and Boersma, A. (Eds), 341-357.

Jenkins, D.G and Murray, J.W. (Eds) 1989. *Stratigraphical Atlas of Fossil Foraminifera*, 2nd Edition, Ellis Horwood Ltd., Chicester, 593pp.

Johnson, A. and Eyssautier, M. 1987. Alwyn North Field and its regional geological context. In : Brooks, J. and Glennie, K.W. (Eds), 963-978.

Johnson, H.D. and Stewart, D.J. 1985. Role of clastic sedimentology in the exploration and production of oil and gas in the North Sea. In: Brenchley, P.J. and Williams, B.P.J. (Eds), *Sedimentology, recent developments and applied aspects. Geol. Soc. London Special Publication No.18*, 249-310.

Johnson, H.D., Mackay, T.A. and Stewart, D.J. 1986. The Fulmar Oil-field (Central North Sea): geological aspects of its discovery, appraisal and development. *Marine and Petroleum Geology 3*, 99-125.
Johnson, R.W. 1984. The development of the Beatrice Field. *Institute of Petroleum. Annual Conference 1984. Offshore Europe*, 1-12.
Kaufmann, E. G. 1988. Concepts and methods of high-resolution event stratigraphy. *Ann. Rev. Earth Planet. Sci. 16*, 605-654.
Kent, D.V. and Gradstein, F.M. 1985. A Cretaceous and Jurassic geochronology. *Geol. Soc. America Bulletin 96*, 1419-1427.
Kessler, L.G. and Moorhouse, K. 1984 Depositional processes and fluid mechanics of Upper Jurassic conglomerate accumulations, British North Sea. In : Koster, E.H. and Steel, R.J. (Eds) *Sedimentology of Gravels and Conglomerates. Canad. Soc. Petrol. Geol. Mem. 10.* 383-397.
King, C. 1983. Cainozoic micropalaeontological biostratigraphy of the North Sea. *Rep. Inst. Geol. Sci. 82/7*, 1-40.
King, C. 1989. Cenozoic of the North Sea. In : Jenkins, D.G. and Murray, J.W. (Eds) , 418-489.
King, C., Bailey, H.W., Burton, C. and King, A.D. 1989. Cretaceous of the North Sea. In: Jenkins, D.G. and Murray, J.W. (Eds), 372-417.
Kiorboe, L. In press. Tertiary seismic sequence stratigraphy in the Faeroe Basin and the Faeroe Shetland Channel. *International Symposium on Mesozoic and Cenozoic sequence stratigraphy of European basins, Dijon*, May, 1992.
Kirk, R.H. 1980. Statfjord Field - A North Sea Giant. *Giant Oil and Gas Fields of the Decade. Amer. Assoc. Petr. Geol. Mem. 30*, 95-116.
Kling, S.A. 1978. Radiolaria. In: Haq, B.U. and Boersma, A. (Eds).
Knoll, A.H. and Swett, K. 1985. Micropalaeontology of the Late Proterozoic Veteranen Group, Spitsbergen. *Palaeontology 28*, 451-473.
Knox, R.W.O'B, Morton, A.C. and Harland, R. 1981. Stratigraphical relationships of Palaeocene sands in the UK sector of the central North Sea. In: Illing, L.V. and Hobson, G.D. (Eds), 267-281.
Larsen, R.M. and Jaarvik, L.J. 1981. The geology of the Sleipner Field Complex. *Norwegian Symposium on Exploration (Norsk Petroleumsforening). Article 15*, 1-33.
Linsley, P.N., Potter, H.C., McNab, G. and Racher, D. 1980 The Beatrice Field, Inner Moray Firth, U.K. North Sea. *Giant Oil and Gas Fields of the Decade. Mem. Amer. Assoc. Petrol. Geol. 30*, 117-129.
Maclennan, A.M. and Trewin, N. 1989. Palaeoenvironments of the late Bathonian - mid Callovian in the Inner Moray Firth. In : Batten, D.J. and Keen, M.C. (Eds), *Northwest European Micropalaeontology and Palynology*, 92-117, Ellis Horwood Ltd, Chichester.
Maher, C.E. 1981. The Piper Oilfield In: Illing, L.V. and Hobson, G.D. (Eds), 358-370.
Maher, C.E. and Harker, S.D. 1987 Claymore Oil Field In : Brooks, J. and Glennie, K.W. (Eds), 835-846.
Malm, O.A., Christensen, O.B., Furnes , H., Lovlie, R., Ruselatten, H. and Ostby, K.L. 1984. The Lower Tertiary Balder Formation: an organogenic and tuffaceous deposit in the North Sea region. In : Spencer, A.M. et al (Eds). *Petroleum Geology of the North European Margin*, 149-170, Graham and Trotman.
Martini, E. 1971. Standard Tertiary and Quaternary calcareous nannoplankton zonation In : Farinacci, A. (Ed) *Proceedings of the 2nd Planktonic Conference. Roma. 1970*, 739-785. Elsevier.
Miall, A.D. 1986. Eustatic Sea Level changes Interpreted from Seismic Stratigraphy. A Critique of the Methodology with particular reference to the North Sea Jurassic Record. *Amer. Assoc. Petr. Geol. Bull. 70*, 131-137.
Miall, A.D. 1992. Exxon global cycle chart: An event for every occasion? *Geology, 20*, 787-790.
Michelsen, O, Danielsen, M., Heilmann-Clausen, C., Jordt, H., Larsen, G. and Thomsen, E. In press. Cenozoic sequence stratigraphy in the Eastern North Sea. *International Symposium on Mesozoic and Cenozoic sequence stratigraphy of European basins, Dijon*, May, 1992.
Mitchell, S.M., Beamish, G.W.J., Wood, M.V., Malacek, S.J., Armentrout, J.A., Damuth, J.E. and Olson, H.C. In press. Palaeogene sequence stratigraphic framework of the Faeroe Basin. In: Parker, J.R. (Ed) *Petroleum Geology of Northwest Europe: Proceedings of the 4th Conference. Geological Society, London.*
Mitchener, B.C., Lawrence, D.A., Partington, M.A., Bowman, M.B.J. and Gluyas, J. 1992. Brent Group: sequence stratigraphy and regional implications. In: Morton, A.C., Haszeldine, R.S., Giles, M.R. and Brown, S. (Eds). Geology of the Brent Group. *Geological Society Special Publication No 61*, 506 pp.

Mitchum, R.M., Vail, P.R. and Thompson, S. 1977. Seismic Stratigraphy and Global Changes of Sea Level, Part 2: The Depositional Sequence as a Basic Unit for Stratigraphic Analysis. *Amer. Assoc. Petrol. Geol. Mem.26*, 53-62.

Morris, P.R. and Coleman, B.E. 1989. The Aalenian to Callovian (Middle Jurassic). In : Jenkins, D.G. and Murray, J.W. (Eds), 189-236.

Morrison, D., Bennet, G. G. and Byat, M. G. 1991. The Don Field, Blocks 211/13a, 211/14, 211/18a, 211/19a, UK North Sea. In: Abbots, I.L. (Ed), 89-94.

Mortimer, C.P. 1987. Upper Cretaceous calcareous nannofossil biostratigraphy of the Southern Norwegian and Danish North Sea area. *Abh. Geol.B.-A. 39*, 143-175.

Morton, A.C., Hallsworth, C.R. and Wilkinson, G.C. In press. Stratigraphic evolution of sand provenance during Palaeocene deposition in the northern North Sea area. In: Parker, J.R. (Ed) *Petroleum Geology of Northwest Europe: Proceedings of the 4th Conference. Geological Society, London.*

Mound, D. G, Robertson, I.D. and Wallis, R.J. 1991. The Cyrus Field, Block 16/28, UK North Sea. In: Abbots, I.L. (Ed), 295-300.

Mudge, D.C. and Copestake, P. 1992a. A revised Lower Palaeogene lithostratigraphy for the Outer Moray Firth, North Sea. *Marine and Petroleum Geol., 9*, 53-69.

Mudge, D.C. and Copstake, P. 1992b. Lower Palaeogene stratigraphy of the Northern North Sea. *Marine and Petroleum Geol.*, 9, 287-301.

Muller, K.L. 1978. Conodonts and other phosphatic microfossils. In : Haq, B.U. and Boersma, A. (Eds), 277-292.

Nagy, J. 1980a. Jurassic foraminiferal facies in the Statfjord Area, Northern North Sea - 1. *Journal of Petroleum Geology 8*, 273-295.

Nagy, J. 1980b. Jurassic foraminiferal facies in the Statfjord Area, Northern North Sea - II. *Journal of Petroleum Geology 8*, 389-404.

Nagy, J. and Johansen, H. O. 1991. Delta-influenced foraminiferal assemblages from the Jurassic (Toarcian-Bajocian) of the Northern North Sea. *Micropaleontology, 37*, 1-40.

Neal, J. E. In press. Palaeogene sequence stratigraphy of Central North Sea subsurface and selected outcrops of Northwest Europe. *International Symposium on Mesozoic and Cenozoic sequence stratigraphy of European basins, Dijon,* May, 1992.

Neale, J.W. 1988. Ostracods and palaeosalinity reconstruction. In: De Deckker, P., Colin, J-P., Peypouquet, J-P. (Eds) *Ostracoda in the Earth Sciences*, 125-156, Elsevier.

Nummendal, D. and Swift, D.J.P. 1987. Transgressive stratigraphy at sequence bounding unconformities: Some principles determined from Holocene and Cretaceous examples. *Canad. Soc. Econ. Palaeontol. and Min.*, 241-260.

Parker, R.H. 1991. The Ivanhoe and Rob Roy Fields. In: Abbots, I.L. (Ed), 331-338.

Parsley, A.J. 1990. North Sea Hydrocarbon Plays. In: Glennie, K.W. (Ed), *Introduction to the Petroleum Geology of the North Sea,* 3rd Edition, 362-388. Blackwell Scientific Publications.

Partington, M.A., Copestake, P. and Mitchener, B. In press. Biostratigraphic calibration of North Sea Jurassic genetic stratigraphic sequences. In: Parker, J.R. (Ed) *Petroleum Geology of Northwest Europe: Proceedings of the 4th Conference. Geologicial Society, London.*

Partington, M.A., Mitchener, B.C., Milton, N.J. and Fraser, A.J. In press. A genetic stratigraphic sequence stratigraphy for the North Sea Late Jurassic and Early Cretaceous: Stratigraphic distribution and prediction of Kimmeridgian - Late Ryazanian reservoirs in the Viking Graben and adjacent areas. In: Parker, J.R. (Ed) *Petroleum Geology of Northwest Europe: Proceedings of the 4th Conference. Geologicial Society, London.*

Pedersen, T., Harms, J.C., Harris, N.B., Mitchell, R.W. and Tooby, K.M. 1989. The role of correlation in generating the Heidrun Field geological model. In: Collinson, J.D. (Ed) *Correlation in Hydrocarbon Exploration* , 327-338. Graham and Trotman for the Norwegian Petroleum Society.

Penney, B. 1991. The Heather Field, Block 2/5, UK North Sea. In: Abbots, I.L. (Ed), 127-134.

Perch-Neilsen, K. 1985a. Silicoflagellates. In : Bolli, H.M., Saunders, J.B. and Perch-Neilsen, K. (Eds), 811-846.

Perch-Neilsen, K. 1985b. Mesozoic Calcareous Nannofossils; Cenozoic Calcareous Nannofossils. In : Bolli, H.M., Saunders, J.B. and Perch-Neilsen, K. (Eds), 329-554.

Peypouquet, J-P. (Ed) *Ostracoda in the Earth Sciences*, Elsevier, 125-156.

Pokorny, V. 1978. Ostracodes In : Haq, B.U. and Boersma, A. (Eds), 109 - 150.

Posamentier, H. W. and Vail, P.R. 1988. Eustatic controls on clastic deposition II - sequence and systems tracts models. In: Wilgus, C. K. et al. (Eds). *Sea-level changes : an integrated approach. Society of Economic Paleontologists and Mineralogists Special Publication 42,* 125-154.

Powell, A.J. 1988. A modified dinoflagellate cyst biozonation for latest Palaeocene and earliest Eocene sediments from the central North Sea. *Review of Palaeobotany and Palynology, 56* (3/4), 327-344.
Powell, A.J. 1992. Applying palynology. *American Association of Stratigraphic Palynologists Newsletter, 25 (2),* 18-23.
Rattey, P A and Hayward, A.B. In press. Sequence stratigraphy of a failed rift system. The Middle Jurassic to Early Cretaceous basin evolution of the Central and Northern North Sea. *Proceedings Tectonics and Seismic Sequence Stratigraphy. Geological Society Petroleum Group Meeting, September, 1991.*
Rawson, P.F. and Riley, L.A. 1982. Latest Jurassic - Early Cretaceous Events and the "Late Cimmerian Unconformity" in North Sea Area. *Amer. Assoc. Petr. Geol. Bull. 66*, 2628-2648.
Remane, J. 1985. Calpionellids. In : Bolli, H.M., Saunders, J.B. and Perch-Neilsen, K. (Eds), 555-572.
Riley, L.A, Harker, S.D. and Green, S.C.H. 1992. Lower Cretaceous palynology and sandstone distribution in the Scapa Field, UK North Sea. *Journal of Petroleum Geology, 15 (1)*, 97-110.
Riley, L.A., Roberts, M.J. and Connell, E.R. 1989. The Application of Palynology in the Interpretation of Brae Formation Stratigraphy and Reservoir Geology. In: Collinson, J.D. (Ed) *Correlation in Hydrocarbon Exploration*, 339-356. Graham and Trotman for the Norwegian Petroleum Society.
Rioult, M., Dugue, O., Jan du Chene, R., Ponsot, C., Fily, G., Moron, J.-M. and Vail, P.R. 1991. Outcrop sequence stratigraphy of the Anglo Paris Basin, Middle to Upper Jurassic (Normandy, Maine, Dorset). *Bull. Centres Rech. Explor.-Prod. Elf Aquitaine, 15*, 101-194.
Ross, C.A. and Haman, D. 1989. Suprageneric ranges of foraminiferida. *Jour. Foramin. Res. 19*, 72-83
Sarg, J.F. and Skjold, L.J. 1982. Stratigraphic Traps in Palaeocene Sands in the Balder Area, North Sea. In: Halbouty, M.T. (Ed). *The Deliberate Search for the Stratigraphic Trap. Amer. Assoc. Petrol. Geol. Mem. 32*,197-206.
Schmitt, H.R.H. 1991. The Chanter Field, Block 15/17, UK North Sea. In: Abbots, I.L. (Ed), 261-268.
Schmitt, H.R.H. and Gordon, A.F. 1991. The Piper Field, Block 15/17, UK North Sea. In: Abbots, I.L. (Ed), 361-368.
Shepherd, M. 1991. The Magnus Field, Blocks 211/11, 211/12, UK North Sea. In: Abbots, I.L. (Ed), 153-158.
Shipp, D.J. 1989. The Oxfordian to Portlandian (Upper Jurassic). In : Jenkins, D.G. and Murray, J.W. (Eds) 237-272.
Sloss, L.L. 1963. Sequences in the cratonic interior of North America. *Geol. Soc. America Bull. 74,* 93-114.
Smith, R.L. 1987. The structural development of the Clyde Field. In : Brooks, J. and Glennie, K.W. (Eds), 523-532.
Snelling, N.J. (Ed) 1985. The Chronology of the Geological Record. *Geol. Soc. London Memoir No 10.*
Spencer, A.M., Home, P.C. and Wiik, V. 1986. Habitat of hydrocarbons in the Jurassic Ula Trend, Central Graben, Norway. *Habitat of Hydrocarbons on the Norwegian Continental Shelf*, 111-127, Graham and Trotman.
Stephen, K.J., Underhill, J.R., Partington, M.A. and Hedley, R.J. In press. The sequence stratigraphy of the Hettangian-Oxfordian succession, Inner Moray Firth. *International Symposium on Mesozoic and Cenozoic sequence stratigraphy of European basins, Dijon,* May, 1992.
Stephen, K.J., Underhill, J.R., Partington, M.A. and Hedley, R.J. In press. The genetic sequence stratigraphy of the Hettangian to Oxfordian succession, Inner Moray Firth. In: Parker, J.R. (Ed) *Petroleum Geology of Northwest Europe: Proceedings of the 4ty Conference. Geologicial Society, London.*
Stevens, D.A. and Wallis, R.J. 1991. The Clyde Field, Block 30/17b, UK North Sea. In: Abbots, I.L. (Ed) 279-286.
Stevens, V. 1991. The Beatrice Field, Block 11/30a. In: Abbots, I.L. (Ed), 245-252.
Stockbridge, C.P. and Gray, D.I. 1991. The Fulmar Field. In: Abbots, I.L. (Ed), 309-316.
Struijk, A.P. and Green, R.T. 1991. The Brent Field, Block 211/29, UK North Sea. In: Abbots, I.L. (Ed), 63-72.
Summerhayes, C.P. 1986. Sea level curves based on seismic stratigraphy: their chronostratigraphic significance. *Paleogeog., Paleoclim., Paleoecol. 57,* 27-42.
Turner, C.C. and Connell, E.R. 1991. Stratigraphic relationships between Upper Jurassic submarine fan sequences in the Brae area, UK North Sea: the implications for reservoir distribution. *Offshore Technology Conference, OTC 6508*, Houston, Texas, 83-91.
Turner, C.C., Cohen, J.M., Connell, E.R. and Cooper, D.M. 1987. A depositional model for the South Brae Oilfield. In : Brooks, J. and Glennie, K.W. (Eds), 853-864.
Underhill, J.R. 1991. Implications of Mesozoic - Recent basin development in the Western Inner Moray

Underhill, J.R. 1991. Implications of Mesozoic - Recent basin development in the Western Inner Moray Firth, UK. *Marine and Petroleum Geology, 8*, 359-369.
Underhill, J.R. and Partington, M.A. In press. Implications of a regional tectonic control on the Intra-Aalenian ("Mid Cimmerian") sequence boundary for North Sea basin development, reservoir prospectivity and Exxon's sea level chart. *Bull. Amer. Assoc. Petrol. Geol.*
Vail, P.R. In press. Summary Mesozoic and Cenozoic sequence stratigraphy of European basins. Opening address. *International Symposium on Mesozoic and Cenozoic sequence stratigraphy of European basins, Dijon,* May, 1992.
Vail, P.R., Hardenbol, J. and Todd, R.G. 1984. Jurassic unconformities, chronostratigraphy and sea level changes from seismic stratigraphy and biostratigraphy. *Interregional unconformities and hydrocarbon accumulation. Amer. Assoc. Petrol. Geol. Mem. 36*, 129-144.
Vail, P.R., Mitchum, R.M. and Thompson, S. 1977. Seismic stratigraphy and Global Changes of Sea Level, Parts 3 and 4: Global Cycles of Relative Change of Sea Level. *Amer. Assoc. Petrol. Geol. Mem. 26*, 63-97.
Vail, P.R. and Todd, R.G. 1981. Northern North Sea Jurassic unconformities, chronostratigraphy and sea-level changes from seismic stratigraphy. In: Illing, L.V. and Hobson, G.D. (Eds), 216-235.
Vail, P.R. and Wornardt, W.W. 1990. Well log-seismic sequence stratigraphy: an integrated tool for the 90's. *GCSSEPM Foundation 11th Annual Research Conference Program and Abstracts, December 2, 1990,* 379-388.
Van Den Bark, E. and Thomas, O.D. 1980. Ekofisk: First of the Giant Oil Fields of Western Europe. *Giant Oil and Gas fields of the Decade. Amer. Assoc. Petrol. Geol. Mem 30*, 195-224.
Van Der Zwan, C.J. 1990. Palynostratigraphy and palynofacies reconstruction of the Upper Jurassic to lowermost Cretaceous of the Draugen Field, offshore Mid Norway. *Review of Palaeobotany and Palynology, 62* , 157-186.
Van Heck, S. E. and Prins, B. 1987. A refined nannoplankton zonation for the Danian of the Central North Sea. *Abh. Geol. B. -A. , 39*, 285-303.
Van Morkhoven, F.P.C.M., Berggren, W.A. and Edwards, A.S. 1986. Cenozoic cosmopolitan deep-water benthic foraminifera. *Bull. Centers Rech. Explor. - Prod. Elf - Aquitaine. Mem. 11*, 1-421.
Van Vessem, E.J. and Gan, T.L. 1991. The Ninian Field, Blocks 3/3 and 3/8, UK North Sea. In: Abbots, I.L. (Eds), 175-182.
Van Wagoner, J.C., Mitchum, R.M., Posamentier, H.W. and Vail, P.R. 1989. Seismic stratigraphy interpretation using sequence stratigraphy, part 2, key definitions of sequence stratigraphy. In: A.W. Bally, (Ed) *Atlas of seismic stratigraphy, vol.1. AAPG Studies in Geology 27*. 11-14.
Van Wagoner, J C , Mitchum, R.M., Campion, K.M. and Rahmanian, V.D. 1990. Siliciclastic sequence stratigraphy in well logs, cores and outcrops: concepts for high-resolution correlation of time and facies. *AAPG Methods in Exploration Series No 7.*
Varol, O. 1989a. Palaeocene calcareous nannofossil biostratigraphy In : Crux, J.A. and Van Heck, S.E. (Eds) *Nannofossils and their applications*, Ellis Horwood Ltd, Chichester, 267-310.
Varol, O. 1989b. Quantitative analysis of the *Arkangelskiella cymbiformis* Group and Biostratigraphic usefulness in the North Sea area. *Journal of Micropalaeontology, 8*, 131-134.
Vining, B.A., Ioannides, N.S. and Pickering, K.T. In press. Stratigraphic relationships of some Tertiary lowstand depositional systems in the Central North Sea. In: Parker, J.R. (Ed) *Petroleum Geology of Northwest Europe: Proceedings of the 4th Conference. Geologicial Society, London.*
Vollset, J. and Doré, A. G. 1984. A revised Triassic and Jurassic lithostratigraphic nomenclature for the Norwegian North Sea. *Norwegian Petroleum Directorate Bulletin No 3*, 1-53.
Waddams, P. and Clark, N.M. 1991. The Petronella Field, Block 14/20b, UK North Sea. In: Abbots, I.L. (Ed), 353-360.
Walmsley, P.J. 1975. The Forties Field. In: Woodland, A.W. (Ed) 477-486.
Whatley, R.C. 1988. Ostracoda and palaeogeography. In : De Deckker, P., Colin, J-P., and Peypouquet, J-P. (Eds) *Ostracoda in the Earth Sciences,* 103-124, Elsevier.
Whitakers, M.F. 1984. The usage of Palynostratigraphy and Palynofacies in defintion of Troll Field geology. In: *Offshore Northern Seas - Reduction of Uncertainties by Innovative Reservoir Geomodelling.* Norwegian Petroleum Society, Article G6.
Whittaker, M.F. , Giles , M.R. and Cannon, S.J. 1992. Palynological review of the Brent Group, UK sector, North Sea. In: Morton, A.C., Haszeldine, R.S., Giles, M.R. and Brown, S. (Eds). Geology of the Brent Group. *Geological Society Special Publication No 61*,169-202.
Williams, G. 1992. Palynology as a palaeoenvironmental indicator in the Brent Group, northern North Sea. In: Morton, A.C., Haszeldine, R.S., Giles, M.R. and Brown, S. (Eds). Geology of the Brent Group. *Geological Society Special Publication No 61*, 203-212.

Williams, G.L. 1978. Dinoflagellates, acritarchs and tasmanitids. In: Haq, B.U. and Boersma, A. (Eds), 293-326.
Williams, G.L. and Bujak, J.P. 1985. Mesozoic and Cenozoic dinoflagellates. In: Bolli, H.M., Saunders, J.B. and Perch-Neilsen, K. (Eds), 847-964.
Williams, R.R. and Milne, A.D. 1991. The Thistle Field, Blocks 211/18aand 211/19, UK North Sea. In: Abbots, I.L. (Ed), 199-210.
Woodland, A.W. (Ed) 1975. *Petroleum and the Continental Shelf of North-West Europe.* Applied Science Publishers Ltd., Institute of Petroleum, Great Britain.
Wornardt, W. W. In press. Abundance and diversity histograms- the key to interpreting systems tracts, condensed sections and sequence boundaries. *International Symposium on Mesozoic and Cenozoic sequence stratigraphy of European basins, Dijon,* May, 1992.
Wray, J.L. 1978. Calcareous Algae. In: Haq, B.U. and Boersma, A. (Eds), 171-188.

Philip Copestake
Integrated Exploration and Development Services Ltd
Enterprise House
Ilsom, Tetbury
Gloucestershire
GL8 8RX, UK.

# 5 PALYNOFACIES ANALYSIS

Richard V. Tyson

## 5.1 Introduction

As early as 1959, Jan Muller appreciated that the distribution patterns of palynomorphs and other particulate organic matter (POM) could be used for facies recognition and palaeogeographic reconstruction (Muller 1959). Such applications, and the increasing use of palynological methods for assessing hydrocarbon source rock potential led to the development of the palynofacies concept. The term palynofacies was first introduced by Combaz in 1964 to describe the quantitative and qualitative palynological study of the total particulate organic matter assemblage. Palynofacies analysis involves the identification of individual palynomorph, plant debris, and amorphous components, their absolute and relative proportions, size spectra, and preservation states (Combaz 1964, 1980). Palynomorph colour estimation of the level of organic maturation was also included in the original concept, although this is now usually regarded as a somewhat separate and distinct field of study.

The key to the success of palynofacies work is its inter-disciplinary nature and the fact that it forms a natural interface between palynology, sedimentology, and organic geochemistry.

Because of their objective and quantitative nature, geochemical methods currently form the basis of most routine industrial assessment of hydrocarbon source rock potential. Optical work is used either as a "screening technique", or to provide a qualitative to semi-quantitative control on the interpretation of geochemical bulk rock parameters. This combination of techniques has led to the development of the organic facies concept (Rogers 1980, Jones 1987), the characterisation of a rock unit based solely on the character of its organic constituents. The definition has no specific reference to methodology, but in practice organic facies work is usually based on elemental analysis or pyrolysis methods (Jones 1987). Palynofacies represents that aspect of the organic facies which may be determined by transmitted light microscopy.

For the sake of brevity it will be assumed that the reader has a basic familiarity with palynology and the nature and sources of other kinds of particulate organic matter. Those wishing more background information are referred to Tschudy & Scott (1969), Tappan (1980), Pocock (1982), Stach et al. (1982), Tissot & Welte (1984), Evitt (1985), Robert (1988), and Traverse (1988).



*D. G. Jenkins (ed.), Applied Micropaleontology*, 153–191.
© 1993 *Kluwer Academic Publishers.*

Figure 5.1: Schematic of factors controlling the palynofacies characteristics of marine sediments. Note that climate is the main controlling factor, because of its influence on terrestrial floras, freshwater runoff, and the hydrographic stability of marine waters (which in turn is a major determinant of primary productivity and the character of marine plankton assemblages).

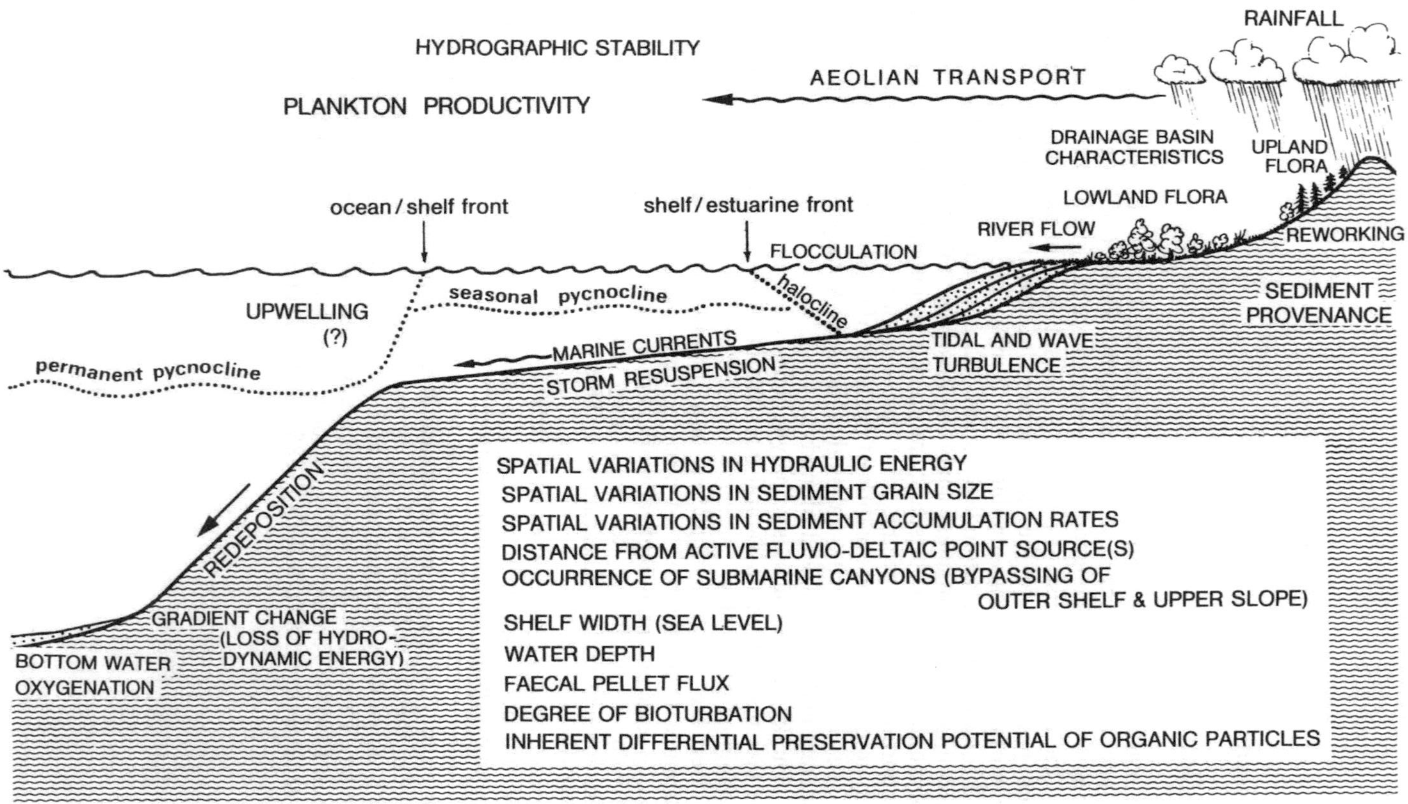

## 5.2 The uses of palynofacies

Despite the large number and complexity of variables that determine the palynofacies characteristics of sediments (see Figure 5.1; Cross 1964), adequate sampling, proper quantification, and the integration with sedimentological and geochemical data allow the palynofacies technique to be used for:

1) Determining the magnitude and location of terrigenous inputs (i.e. provenance and proximal-distal relationships with respect to clastic sediment source).
2) Determining depositional polarity (i.e. onshore-offshore direction).
3) Identification of regressive-transgressive trends in stratigraphic sequences (and thus depositional sequence boundaries).
4) Characterisation of the depositional environment in terms of: salinity (normal or saline lake waters, brackish "estuarine", or marine), oxygenation and redox conditions (strongly or moderately oxidising oxic conditions, and strongly or moderately reducing dysoxic to anoxic conditions), productivity (normal or upwelling), and water column stability (permanently stratified, seasonally stratified, or continuously mixed).
5) Characterisation and empirical subdivision of sedimentologically "uniform" facies, especially shales and other fine-grained sediments.
6) Deriving correlations at levels below biostratigraphic resolution.
7) Preliminary qualitative or semi-quantitative determination of hydrocarbon source rock potential, and qualification of bulk rock geochemical parameters.
8) Producing sophisticated and detailed organic facies models.

## 5.3 Kerogen - particulate organic matter in sedimentary rocks

The most commonly used term to describe the particulate organic matter (POM) contained in sedimentary rocks is kerogen. Various definitions of this term have been published (e.g. Durand 1980, Brooks 1981). Here it is used in a purely palynological sense to describe the dispersed particulate organic matter of sedimentary rocks that is insoluble in hydrochloric (HCl) and hydrofluoric acids (HF).

Three main groups of morphological constituents can be recognised within kerogen assemblages: *palynomorphs, phytoclasts* (both structured), and *amorphous organic matter* (structureless). The term "*palynomorph*" refers to all HCl and HF resistant organic-walled microfossils. The term "*phytoclast*" was introduced by Bostick (1971) to describe all dispersed clay- to fine sand- sized particles of plant-derived kerogen. This definition is too broad to be very helpful and here it is redefined to include only those particles that are literally "clasts" (Greek *klastos* = "broken in pieces"), by excluding palynomorphs and structureless (amorphous) materials.

The published kerogen classifications are characterised by a great deal of duplication of effort that has resulted in much superfluous jargon. A generally acceptable terminology for transmitted light work has proved elusive. A correlation of some existing terms is given in Table 5.1. The much more standardised and systematic maceral terminology used by organic petrologists (in reflected light studies ) should *never* be used in transmitted light work. Macerals can only be properly defined on reflected light characteristics; any other usage can produce pseudo-accuracy, unnecessary confusion, and futile controversy. A much-abbreviated list of macerals is given in Table 5.1 (for comparative purposes only).

Table 5.1: A correlation of some published transmitted light kerogen classifications (modified and extended from Tyson 1987). Approximately equivalent maceral group and maceral terminology are given only for comparative purposes (much simplified). The kerogen type (last column) is that which would correspond to an artificial kerogen assemblage dominated by the appropriate component in a well-preserved and unoxidised state (see Tissot & Welte 1984). DOM in the second column refers to dissolved organic matter.

| CATEGORY | | SOURCE | CONSTITUENT | MACERAL GROUP | MACERAL | Staplin 1969 | Correia 1971 |
|---|---|---|---|---|---|---|---|
| STRUCTURED | Zooclasts | Zooplankton & Zoobenthos | Graptolite debris<br>Arthropod debris | Faunal relics | | ? | ? |
| | Palynomorphs | Zoomorphs | Scolecodonts<br>Tectin foraminiferal linings<br>Chitinozoa | | | | |
| | | Phytoplankton (including meroplankton) | Prasinophyte phycomata | Exinite or Liptinite | Telalginite | Marine Phyto-plankton | Vegetale (MOV) |
| | | | Chlorococcales cyanobacteria | | | | |
| | | | Chlorococcales: Botryococcales<br>Hydrodictyales | | | Planktonic freshwater algae | |
| | | | Dinocysts<br>Acritarchs<br>Rhodophyte spores | | Lamalginite (pars) | Marine Phyto-plankton | |
| | | Sporomorphs | Miospores: microspores<br>pollen<br>Megaspores | | Sporinite | Spores and Pollen | |
| | Phytoclasts | Macrophyte plant debris | Cuticle | | Cutinite | Cuticles | |
| | | | Cortex | Vitrinite or Huminite | Telinite | Lignified Wood | Tracheides (MOT) |
| | | | Secondary xylem (wood) | | | | |
| | | | Charcoal<br>Biochemically oxidised wood | Inertinite | Pyrofusinite<br>Degradofusinite | Charcoal | Ligneuse (MOL) |
| | | Fungal debris | Hyphae | | Sclerotinite | Fungi | ? |
| STRUCTURELESS | Amorphous Organic Matter | Higher plant secretions | Intra-/extra-cellular resins | Exinite or Liptinite | Resinite | Resins | ? |
| | | Flocculated DOM | Organic aggregates | | Bituminite<br>Fluoramorphinite<br>Liptodetrinite | Sapropelic Amorphous | Colloidale (MOC) |
| | | Phytoplankton | Faecal pellets | | | | |
| | | Bacteria | Cyanobacteria/thiobacteria | | Lamalginite | | |
| | | Higher plant decomposition products | Humic cell-filling precipitates<br>Humic extracellular precipitates | Vitrinite or Huminite | Collinite<br>Hebamorphinite | ? | |

Table 5.1: *continued*

| Burgess 1974 | Bujak et al 1977 | Combaz 1980 | Claret et al 1981 | Parry et al 1981 | Massran & Pocock '81 Pocock 1982 | Whitaker '84 Bryant et al '88 | PURE KEROGEN TYPE |
|---|---|---|---|---|---|---|---|
| ? | ? | Chitinobiontes | ? | ? | ? | ? | III? |
| | | | ? | Foraminiferal linings | ? | Microforam test linings | |
| Algal | Phyrogen | Leiosphaer. | Microfossils | Marine Palynomorphs | Aqueous: Algal | Marine algae | I highly oil prone I |
| | | Algues: Botryococc. | | Freshwater algae | | Freshwater algae | |
| | | Dinoflagelles Acritarches | | Marine palynomorphs | Dinoflagellates Acritarchs "Round bodies" | Dinocysts Acritarchs | II oil prone II |
| Herbaceous | | Spores et Pollen | | Terrestrial Palynomorphs | Spores and pollen | Saccate, non-saccate sporomorphs | |
| | | Cuticules | Cuticles | Cuticle | Terrestrial: Leaf | Palynomaceral 3 | |
| Woody | Hylogen | Bois Ecorces Vaisseaux | Clear/opaque ligneous | Cortex | Root | Palynomaceral 1 & 2 | III gas prone |
| | | | | Brown wood | Stem | | |
| Coaly | Melanogen | Carbons | Coals | Black wood | Charcoal: Pyro- Degrado- | Palynomaceral 4 | IV inert |
| ? | ? | Champignons | ? | ? | Fungi | ? | |
| ? | ? | Resineux | ? | Resin | Resins | ? | I |
| Amorphous | Amophogen | Amorpho.: grumeleuse granuleuse pelliculaire | Amorphous | ? | Grey Amorphous | Structureless organic matter | II |
| | | | | | | | I/II |
| | | gelifée | | | Yellow-amber Amorphous | Palynomaceral 2 ? | III |

## 5.4 Simple classifications for rapid assessment of hydrocarbon potential

Classifications aimed at routine source rock "kerogen typing" are relatively simple and have a limited number of categories (usually four to six). They are almost solely concerned with identifying the relative proportions of inert, gas-prone, oil-prone, and very oil-prone material within the total kerogen assemblage.

There are three commonly cited classifications of this type (Correia 1971, Burgess 1974, and Bujak et al. 1977). Similar subdivisions of kerogen have also been used by Elf Aquitaine (Claret et al. 1981, Bustin 1988). All of these terminologies have arguable weaknesses with regard to their specific choice of terms or adequacy of descriptive criteria; in particular none of them were specifically designed for use with palynological microscopes equipped with fluorescence. This leads to inherent difficulty in assessing the proportion of highly oil-prone constituents and the preservation state (and hence hydrocarbon potential) of palynomorphs and amorphous material.

The essential categories that must be identified in order to provide the appropriate level of information on source rock potential are:

1) Inert material: non-fluorescent, opaque to semi-opaque, black, or very dark brown particles (even at the grain edge), representing oxidised or carbonized phytoclasts (including charcoal ); also fungal and tectin/chitinous materials.
2) Gas-prone material: non-fluorescent, generally orange or brown, translucent, phytoclasts, but also translucent, non-fluorescent structureless materials. Woody fragments are typical.
3) Oil-prone material: the volumetrically most important constituent is fluorescent amorphous organic matter, but fluorescent (non-oxidised) sporopollenin and "sporopolleninoid" palynomorphs, cuticle, and non-cellular membraneous debris are also included.
4) Highly oil-prone material: this includes very strongly fluorescent alginitic material derived from chlorococcale algae, prasinophyte algae, cyanobacteria, and some thiobacteria. The exact source of this fluorescent material is often not morphologically distinguishable. Resins are the only significant terrestrially-derived component belonging to this group.

It has been suggested that there is a poor correlation between the results of optical kerogen typing and geochemical measures of source rock potential (Powell et al. 1982, Waples 1985). However, while caution is always advisable, reasonable correlations do often exist between the optically determined relative volumetric abundance of oil-prone fluorescent kerogens and bulk-rock pyrolysis hydrocarbon yields (Mukhopadhyay et al. 1985, Stein et al. 1986, Senftle et al. 1987, Jones 1987, Tyson 1989, Nøhr-Hansen 1989, Gregory et al. 1990, Tyson 1990).

The combination of optically determined percentage data with Total Organic Carbon (TOC) values also permits the analysis of absolute and relative changes (although TOC is partly dependent on sedimentation rates). The actual amount of organic matter present, as well as the relative distribution of its constituents, must always be taken into account when interpreting kerogen data.

## 5.5 More detailed classifications for palaeoenvironmental analysis

Simple classifications are, by and large, adequate to describe hydrocarbon source rock potential, but cannot provide sufficient information to construct detailed organic facies models. For detailed palynofacies work a kerogen classification scheme should have at least ten categories and take into account the following variables wherever possible or practicable:

a) the biological provenance of the particles (if discernible),
b) any ecologically significant groupings that may be reflected by the particle types (e.g. marine versus freshwater aquatic palynomorphs, ecologically diagnostic indicator species where these are known, terrestrial versus marine phytoclasts),
c) the preservation states of the various particle types (partly a reflection of the distance or duration of transport and the redox status of the depositional environment),
d) any consistent or significant variation in size, morphology, or probable differences in density likely to be reflected in (or indicated by) the hydrodynamic behaviour of the particles,
e) any components with predictable differences in their geochemical character (hydrocarbon generating potential),
f) any recognisable differences in the geologic age or degree of organic maturation of any particle species within the assemblage.

Examples of more-detailed published classifications include those of Staplin (1969), Combaz (1980), Masran & Pocock (1981) and Pocock (1982), Tyson (1984, 1989), Whitaker (1984; cf. Bryant et al. 1988, Van der Zwaan 1990), and Hart (1986). None of these can be recommended wholeheartedly. They differ by the degree of emphasis placed on different aspects of the kerogen assemblage, providing more detailed subdivisions of the palynomorph, phytoclast, and/or amorphous components by greater attention to botanical source, morphology, and/or preservation states. However, it cannot be over-emphasised that the meaningful subdivision of amorphous materials and the proper assessment of preservation states demands the routine parallel use of blue light or ultra violet fluorescence observations.

Rather than reviewing the detailed classifications, this account will concentrate on a discussion and description of the various parameters most used in palynofacies analysis. This provides an indirect indication of the categories that should be utilised in kerogen classification. Particular emphasis is placed on those parameters that have been found to be most useful in the analysis of Mesozoic-Cenozoic marine facies (see Table 5.2).

Table 5.2: Summary of main palynofacies parameters with their generalised relative response to some palaeoenvironmental factors in marine sediments. "Redeposition" refers to characteristics of redeposited sediments compared with *in situ* basinal deposits. Note how the other factors can completely modify normal proximal-distal (onshore-offshore) trends.

**SUMMARY OF SOME GENERALISED MARINE PALYNOFACIES TRENDS**

| Environmental factor | Proximal→Distal trend | Increasing % sand | Distal anoxic facies | Redeposition | Upwelling (with arid hinterland) |
|---|---|---|---|---|---|
| ***Parameter responses:*** | | | | | |
| % phytoclasts of kerogen | high→low | increases | decreases | increases | decreases |
| % AOM of kerogen | low→high | decreases | increases | decreases | increases |
| % palynomorphs of kerogen | low→high | decreases | decreases | may increase | may increase |
| AOM fluoresence | variable | decreases | increases | decreases | may increase |
| opaque: translucent phytoclasts | low→high | increases | increases | decreases | may increase |
| % cuticle of phytoclasts | high→low | may increase | negligible | increases | decreases |
| phytoclast coarse fraction | high→low | increases | decreases | increases | decreases |
| % sporomorphs of palynomorphs | high→low→high | increases | increases | may increase | decreases |
| % microspores of miospores | high→low | increases | decreases | may increase | decreases |
| % thick/ornamented microspores (of total microspores) | high→low | increases | very low | may increase | decreases |
| % bisaccate pollen of miospores | low→high | decreases | increases | decreases | increases |
| % small round pollen of miospores | low→high | decreases | increases | may decrease | increases |
| frequency of megaspores | high→low | may increase | negligible | may decrease | negligible |
| frequency of tetrads | high→low | may increase | may increase | decreases | decreases |
| absolute sporomorph abundance | high→low | decreases | decreases | increases | decreases |
| % plankton of palynomorphs | low→high→low | decreases | decreases | may increase | increases |
| % chlorococcales of plankton | high→low | decreases | usually low | may increase | decreases |
| % acritarchs of marine plankton | high→low | may increase | usually low | may increase | decreases |
| % dinocysts of marine plankton | low→high→low | may decrease | decreases | may increase | increases |
| % prasinophytes of marine plankton | low→high | no data | increases | decreases | decreases |
| long: short-spined acritarchs | low→high | decreases | may increase | decreases | no data |
| % chorates of dinocysts | low→high | decreases | increases | may decrease | decreases? |
| peridinoid: gonyaulacoid dinocysts | high→low | may increase | may increase | may decrease | increases |
| dinocyst species diversity | low→high | decreases | decreases | may increase | decreases |
| dinocyst species dominance | high→low | may increase | increases | may decrease | increases |
| absolute dinocyst abundance | low→high→low | decreases | decreases | may increase | increases |
| frequency of foraminiferal linings | high→low | decreases | decreases | increases | increases |
| % reworked palynomorphs | high→low | may increase | decreases | increases | decreases |

## 5.6 Palynofacies parameters for palaeoenvironmental analysis

### 5.6.1 Bulk kerogen parameters

#### 1. Percentage of phytoclasts

The great majority of phytoclasts are derived from the terrestrial flora. The percentage of phytoclasts is high in three main situations, reflecting either their high supply, their preferential preservation, or preferential sedimentation. The detailed significance of this parameter depends upon the nature of the phytoclast assemblage.

(a) High supply

Sufficient supply to dilute all other components is characteristic of proximal situations where deposition takes place close to the parent flora. There has generally been insufficient transport for significant sorting of the phytoclasts, and such proximal assemblages therefore reveal a more mixed composition (i.e. including tissues not normally preserved) and a greater variety of particle sizes.

Nearly all terrestrial organic matter that reaches the sea does so via rivers but is at least partly and often almost completely removed in the estuarine zone (Wollast 1983). In many estuarine systems the bulk of the suspended load, including much of the POM, appears to sediment out before the salinity reaches 3-10 $^{o}/_{oo}$ (Wollast 1983). Most terrestrial organic matter is deposited near the mouths of rivers, and remains trapped on the inner part of the shelf; significant amounts reach the outer shelf only where the discharge of adjacent rivers is high or the shelf is narrow (Muller 1959, Hedges & Parker 1976, Pocklington & Leonard 1979). Even in the case of the Amazon, there is a 60% decrease in suspended terrestrial organic matter in the first 120km offshore of the river mouth (Cai et al. 1988). This organic matter is sedimented mainly on the shelf above the 100m isobath and within 250km offshore of the mouth; TOC values rarely exceed 1.0% due to the high dilution by sediment (Showers & Angle 1986).

Degens and Mopper (1976) suggest that only sediments deposited in estuaries or close to shorelines reveal a strong terrigenous influence, regardless of whether transgressive or regressive conditions prevail. However, it is clear that global peaks in oceanward redeposition of organic (and inorganic) terrestrial material coincide with major sea level lows when the shelves are narrow and cut by canyons, and there is a higher frequency of turbidity currents (Habib 1982, Summerhayes 1987). In the deep-sea the highest phytoclast abundances are strongly related to the occurrence and frequency of turbidites (Tyson 1984). Rupke and Stanley (1974) found plant debris to be concentrated in the upper d Bouma division of sand-silt turbidites and the lower parts of turbiditic mud divisions, but "virtually absent" in the normal hemipelagic muds. Similar patterns have been observed in prealpine flysch deposits (Weidmann 1967).

(b) High relative preservation

The percentage of phytoclasts is also often high in oxidising situations, where the highly resistant (refractory) nature of lignin leads to woody debris being the only organic material to survive. Two situations are typical: fluvial and delta-top settings with strongly fluctuating water tables (where there is strong post-depositional oxidation of phytoclast material in sandy sediments), and distal oxidising environments dominated by small woody fragments that have survived a long distance or duration of transport (see below). The phytoclasts are often opaque to semi-opaque, and the absolute amounts of material and TOC values are typically low, especially in the latter case.

(c) Hydrodynamic equivalence

The percentage of phytoclasts is also influenced by hydrodynamic equivalence effects. Being relatively large and dense organic particles, woody debris is often concentrated in sediments rich in coarse silt and/or very fine sand. The total organic carbon content of clastic sediments is generally observed to correlate with the silt and clay content rather than the median grain size (Buchanan 1958). However, this partly depends on the predominant nature of the organic matter (i.e. whether it is discrete particles such as phytoclasts and palynomorphs, or disseminated amorphous material), and also the particle size (phytoclasts being generally larger than palynomorphs). For example, Hennessee et al. (1986) observe that TOC is correlated with silt content in the upper part of Chesapeake Bay where terrestrial organic matter is dominant, but with clay content in the middle part of the Bay where the organic matter is primarily plankton-derived. The percentage of phytoclasts is often correlated with coarse silt or (VF-F) sand content (Reyre 1973).

2. Percentage of fluorescent amorphous organic matter

High percentages of fluorescent amorphous organic matter (AOM) reflect enhanced preservation under reducing conditions and, to a lesser extent, sedimentation removed from active sources of terrestrial organic matter.

Most AOM is produced as organic aggregates derived from dissolved organic matter (Riley 1970) or as faecal pellet material (Porter & Robbins 1981). Carbon isotopic evidence indicates that all typical marine AOM is ultimately phytoplankton-derived (Lewan 1986). In volumetric terms, nearly all the marine organic matter in sediments is represented by AOM; this is because resistant structural support tissues are unnecessary in most aquatic environments, and because the aquatic biomass is dominated by unicellular algae and bacteria. Only a very small proportion of unicellular phytoplankton produce fossilising structures. The amount of amorphous organic aggregate material generally exceeds that of the living phytoplankton, and represents the largest reservoir of marine particulate organic matter (Riley 1970). Amorphous material is also produced by benthic filamentous cyanobacteria of well-lit shallow waters, and by the benthic and pelagic sulphur bacteria of oxygen deficient environments (Williams 1984, Glikson & Taylor 1986).

The relative and absolute abundance of AOM is strongly correlated with areas of low (dysoxic) bottom water oxygen values, especially (but not necessarily) those underlying areas of high primary productivity (Davey & Rogers 1975, Tissot & Pelet 1981, Summerhayes 1983). AOM consists of easily degraded material that survives only where the duration and extent of aerobic degradation is limited (Tyson 1987, 1989). It frequently dominates the kerogen assemblages in which it is found. This is because of the large reservoir of organic aggregate material that is present in aquatic environments, which if preserved under sufficiently reducing conditions, usually swamps all other components. In oxygen deficient basins with high AOM preservation, allochthonous terrestrial material is only dominant in the immediate vicinity of fluvio-deltaic sources, or within turbidites (Tyson 1984, 1987). In carbonate facies it may be the only organic matter available for preservation.

3. Percentage of palynomorphs

Being the least abundant of the three main morphological groups of particulate organic matter, the relative abundance of palynomorphs is primarily controlled by the extent of AOM or phytoclast dilution. The percentage of palynomorphs in the kerogen assemblage is therefore highest in moderately oxidising settings with low AOM preservation, and in situations removed from the immediate vicinity of active fluvio-deltaic sources of phytoclasts. They are also associated with situations characterised by initial low phytoclast production or transport (as in arid climates associated with little vegetation and low runoff).

The detailed interpretation is dependent upon the composition of the palynomorph assemblage. Where the assemblage is dominated by sporomorphs, high palynomorph percentages usually occur in low energy, distal, moderately oxidising situations where buoyant or wind-blown pollen (especially bisaccates) have become preferentially concentrated, though usually in low to moderate absolute numbers. The coarse phytoclast fraction, if initially present, has usually been deposited before such areas have been reached, and does not therefore dilute the palynomorphs. If the palynomorph assemblage is dominated by dinocysts, high palynomorph percentages (and high absolute abundances) may also be related to areas of high primary productivity. Areas of coastal upwelling are typically associated with arid conditions on the adjacent land areas, and are therefore usually poor in phytoclasts. As most palynomorphs are hydrodynamically equivalent to medium to fine silt grade clastic material (Muller 1959, Wall et al. 1977) they may become locally concentrated by sorting effects (e.g. within silty turbidites).

4. Ratio of opaque to translucent woody phytoclasts

Opaque phytoclast material is mainly derived from the oxidation of translucent woody material during prolonged transport or post-depositional alteration. It is also produced as charcoal during natural wildfires.

Fisher (1980), Denison and Fowler (1980), Parry et al. (1981), Batten (1982), Whitaker (1984), Boulter and Riddick (1986), and Bustin (1988) show a clear association between high ratios of opaque phytoclasts and relatively coarse-grained, high energy, organic-poor facies. This includes distributary channel sands, point bars, levees, proximal crevasse splay deposits, shoreface, offshore, and submarine channel sands. This correlation is often attributed to the minimal buoyancy of opaque phytoclast material and its consequent hydrodynamic equivalence to sand-sized clastics (Fisher 1980, Denison & Fowler 1980, Parry et al. 1981). Opaque (inertinite) material is certainly more dense than translucent (vitrinite) material, but these are diagenetically modified products. Although laboratory experiments suggest charcoal particles become quickly waterlogged (Davis 1967), their porous structure results in a low effective density, making the particles initially buoyant and capable of flotation (Patterson et al. 1987, Sander & Gee 1990). Whitaker (1984) considers this material to be the most buoyant type of phytoclast.

High ratios of opaque phytoclasts also often result from post-depositional oxidation of woody material. Fisher (1980), Denison and Fowler (1980), Hancock and Fisher (1981), and Fisher and Hancock (1985) indicate an abundance of non-charcoal opaque phytoclasts in deltaic facies. This material probably represents the in situ post-depositional bio-oxidation of normal wood particles during seasonal fluctuations in water table conditions. Such oxidation also takes place in littoral sediments with tidally fluctuating water tables (Pocock 1982). The correlation with sandy sediments is at least partially a reflection of their higher permeability and the increased potential for in situ oxidation.

Habib (1982) shows that the kerogen assemblages of oceanic sediments are typically dominated by small (<20 μm), equidimensional, opaque to semi-opaque particles, especially during transgressive periods of low terrestrial organic matter supply. Such oxidised woody material is also characteristic of continental slope facies (Masran 1984). It represents a background input of charcoal or woody material that has been comminuted and oxidised during prolonged transport within the basin, or during reworking on the shelf. Wind-blown charcoal may constitute ≤50% of the TOC in oxidising pelagic sediments (Herring 1985), but the particles are usually no more than several microns in diameter (Smith et al. 1973) and thus easily underestimated by palynologists.

In marine sediments there is often an offshore increase in the ratio of opaque to total phytoclasts (Summerhayes 1987). This is also expressed in a general offshore increase in the ratio of opaque to well-preserved translucent woody material (Tyson 1989). Such trends reflect fractionation during transport, producing a relative offshore increase in the fine-grained refractory opaque material, but an inevitable absolute decrease as phytoclast TOC declines offshore. In proximal settings the opaque material is diluted by the overall greater supply of "fresh" phytoclasts.

5. Percentage cuticle debris

The great majority of cuticle debris is derived from leaves. Whole leaves cannot generally be transported very far before they are destroyed; they are therefore deposited fairly close to their source (Rich 1989). However, fine leaf debris can easily be transported by even the slowest currents (Dance 1981). Fisher (1980) considers that cuticle debris is the most buoyant variety of structured terrestrial organic matter, and is thus especially characteristic of facies resulting from the settling out of flotation and suspension loads under low energy conditions.

Cuticle debris is abundant in modern sediments of the Orinoco Delta but decreases rapidly offshore (Muller 1959). Like all other phytoclast debris, it is concentrated in belts opposite the larger Orinoco distributaries. Cross et al. (1966) observe identical trends in Recent sediments of the Gulf of California. Data in Batten (1973) suggest a strong association between the distribution of cuticle and those fluvio-deltaic and lacustrine facies rich in macroscopic plant fragments. Relatively large pieces of cuticle, as well as entire leaves, are especially characteristic of prodelta facies (Batten 1974). Parry et al. (1981) and Nagy et al. (1984) also find that high percentages of cuticle (15-40%) are characteristic of delta top embayment, prodelta, and distributary facies.

Significant amounts of cuticle debris only appear to reach deeper waters when they are funnelled down submarine canyons and deposited on submarine fans (Cross et al. 1966, Habib 1982). Boulter and Riddick (1986) observe that cuticle is relatively more abundant in the higher energy parts of submarine fan systems, especially the channel sandstones. They suggest that most of this "buoyant" material would normally by-pass the fan, but is sometimes trapped by rapid sedimentation during gravity flow sand deposition. Stanley (1986) also finds that coarse phytoclast material, much of it cuticle-rich leaf debris, is commonly concentrated in sandy overbank deposits in proximal base-of-slope and upper and mid- (or supra-) fan valley channels.

6. Phytoclast particle size trends

Phytoclast particle sizes are rarely recorded in a systematic fashion because of time constraints. The use of automated image analysis techniques now makes this a practical possibility (e.g. Lorente 1990), and qualitative observations suggest that it may be a useful parameter for assessing proximal-distal trends (Caratini et al. 1983). The particle sizes of tracheid, charcoal and cuticle phytoclasts have all been observed to decrease in an offshore direction (Muller 1959, Cross et al. 1966, Batten 1974, Habib 1982, Herring 1985, Patterson et al. 1987). Redeposition will, of course, locally modify such trends. The particle size distribution of phytoclasts is likely to be strongly affected by the granulometric composition of the sediment, proximity to source, and by the presence or absence of macroscopic debris that may break down into smaller particles during sample maceration.

7. Fluorescence intensity of AOM matrix

This parameter is not only essential for determining source rock potential (Tyson 1990, and references therein), it is also useful for assessing redox-controlled levels of preservation in amorphous organic components and palynomorphs. In most aquatic sediments the AOM is plankton- derived, and its preservation is controlled by the extent of aerobic degradation. Oxygen deficient (dysoxic to anoxic) conditions preserve the labile, hydrogen-rich components which fluoresce when exposed to ultra violet or blue light excitation. The fluorescence of the matrix of heterogeneous amorphous particles is most significant, as the matrix is the part most easily oxidised, and the total fluorescence response (including inclusions) is also dependent on the nature of the original plankton (e.g. the abundance of the highly fluorescent alginitic prasinophyte or chlorococcale algae, and cyanobacteria).

5.6.2 Palynomorph parameters

1. Percentage sporomorphs (of total palynomorphs)

Sporomorphs is a general collective term used to describe all the palynomorphs produced by terrestrial macrophytes. The relative abundance of sporomorphs in relation to other palynomorphs (especially phytoplankton) is mainly a reflection of proximity to terrestrial sources and the productivity of fossilising organic-walled microplankton. Dense, thick-walled, strongly ornamented, or large spores may also become preferentially concentrated in silty or sandy facies because there are few hydrodynamically equivalent palynomorphs produced by the phytoplankton.

Spores, and especially the pollen of anemophilous plants, are produced in large numbers, and depositional sites close to the parent flora (e.g. in delta top facies) typically show high absolute abundances and high percentages of sporomorphs. The sporomorph assemblages are characterised by a marked over-representation of the local flora, and by extreme variability (Muller 1959, Darrell & Hart 1970). The high percentage of sporomorphs is also partly because of the low representation of fossilising phytoplankton in such environments.

As the main vector of sporomorph dispersal is river water (Muller 1959), sporomorph concentrations capable of diluting all other palynomorphs are generally restricted to the vicinity of active fluvio-deltaic sources, where lowered salinities also suppress the productivity of fossilising phytoplankton (Muller 1959, de Jekhowsky 1963, Stanley 1965, Cross et al. 1966, Williams & Sarjeant 1967, Davey 1970, Reyre 1973, Davey & Rogers 1975, Heusser & Balsam 1977, Mudie 1982). Such prodelta facies show higher sporomorph percentages and assemblages characterised by high absolute abundances and moderate diversities (i.e. higher than on the delta top; Batten 1974).

Further offshore, under normal shelf conditions, the declining absolute input of sporomorphs and the increasing representation of fossilising organic-walled microplankton leads to a decrease in the percentage of sporomorphs. However, dinocyst abundance declines with increasing depth and distance offshore, often reaching a maximum on the continental slope (Wall et al. 1977). Oceanic or other permanently stratified basins are therefore often dominated by high percentages (but low absolute abundances) of sporomorphs, mainly small, thin-walled, or saccate pollen (Heusser & Balsam 1977, Habib 1982, Melia 1984). The same pattern is often also seen in the central parts of marine black shale basins, where permanent stratification and high stability of the basin, along with possible natural eutrophication effects, inhibits the in situ production of dinocysts (Tyson 1984, 1987, 1989).

Simple sporomorph plankton ratios have not been included as a major parameter in this chapter. Although they are frequently cited in the palynological literature, their interpretation depends critically upon the nature of both the sporomorphs and plankton involved.

2. Percentage of organic-walled microplankton (of total palynomorphs)

The relative abundance of organic-walled microplankton is of course inversely related to that of sporomorphs. It is consequently highest in those temperate to subtropical shelf areas removed from active fluvio-deltaic sources (i.e. outer shelf areas, and areas between river mouths), in areas that are poorly vegetated (e.g. due to arid climates), and in areas of enhanced primary productivity. Areas of coastal upwelling often meet both the latter two conditions, as the wind patterns that produce the upwelling often result in arid climates onshore (Davey 1970, Davey & Rogers 1975, Melia 1984, Hooghiemstra et al. 1986). As described above, the percentage of plankton in pelagic sediments is often low, but it may be increased in turbidites due to the redeposition of dinocysts from the continental shelf and slope (Tyson 1984). Percentages are also high in pelagic sediments deposited during periods of high sea level, when all terrestrial material (including sporomorphs) tends to be trapped on the wider continental shelves, but in this case the absolute abundances are very low (see Habib 1983).

3. Percentage microspores of miospore sporomorphs

The primary control on the abundance of microspores is the occurrence of the humid conditions necessary to support significant pteridophyte growth (at least locally). Spores are mainly produced in low lying swampy deltaic areas, and are therefore typically well represented in many coals and associated deltaic facies (Chaloner & Muir 1968). Their general abundance in the regional palynoflora can therefore be used as an indicator of humid climatic conditions provided the sediments were not deposited too far offshore (Courtinat 1989). The relative abundance of spores over pollen is generally a useful indicator of proximity to fluvio-deltaic source areas, since they are produced in lower numbers than anemophilous pollen and tend to be transported less efficiently (Hughes & Moody-Stuart 1967, Tschudy 1969, Heusser & Balsam 1977, Mudie 1982, Habib 1982, Mutterlose & Harding 1987, Tyson 1989).

4. Percentage of dense, large, or thick-walled microspores of total microspores

During transport, the total spore population is fractionated according to hydrodynamic equivalence, such that the proportion of the large, dense, thick-walled, or strongly ornamented spores (which are less easily transported) decreases away from their source. Examples of this trend are reported by Hughes and Moody-Stuart (1967), Batten (1974), Parry et al. (1981), Habib (1982), and Lund and Pedersen (1985). This pattern is strongly influenced by changes in sediment granulometry (Hughes & Moody-Stuart 1967).

5. Percentage of saccate pollen of total sporomorphs

Saccate pollen grains, especially bisaccate conifer pollen, are probably the most buoyant and most easily transported of all sporomorphs (Hopkins 1950, Traverse 1988). Although their absolute concentration declines, their percentage contribution to the total sporomorph population typically increases in an offshore direction (Heusser & Balsam 1977, Heusser 1983, 1988), and is therefore useful as an indicator of relative distance from active fluvio-deltaic sources (Chaloner & Muir 1968, Habib 1982, Mutterlose & Harding 1987, Tyson 1989).

A few notes of caution should be sounded. Firstly, once water-logged the bisaccate pollen behave like other, denser, pollen (Brush & Brush 1972). Secondly, all bisaccate pollen should not be treated equally. Mudie (1982) shows that there is a relative size sorting of bisaccate pollen, which become smaller with increasing distance offshore, and that some large and denser bisaccate pollen are never carried very far from river mouths before being sedimented. Thirdly, if the distribution of the parent plant is widespread, the whole basin may be "saturated" with saccate pollen, with little meaningful variation being apparent (Hughes & Moody-Stuart 1967).

Bisaccate pollen, being suited to long distance dispersal by wind, may dominate any depositional site where sporomorph supply is predominantly by aeolian transport, as in arid areas (Courtinat 1989) or distal offshore settings (Melia 1984). This is especially the case where there are persistent wind patterns to transport the pollen from upland coniferous source areas (Melia 1984, Hooghiemstra et al. 1986).

6. Percentage of small, thin-walled, sphaeromorph pollen of total sporomorphs

Like saccate pollen, the percentage of small simple spherical pollen tends to increase in an offshore direction, as shown by the distribution patterns of both modern (Muller 1959, Tschudy 1969), and ancient pollen (Hughes & Moody-Stuart 1967, Habib 1979). It can be similarly used as an indicator of relative proximity to fluvio-deltaic source areas.

7. Frequency of megaspores

The controls on megaspore distribution are primarily the same as those of spores in general, and of the large, dense, thick-walled, or strongly ornamented spores. However, they are much less easily transported. Abundances of over 500 per gram of sediment are generally limited to fluvial, marsh, swamp, and lagoon facies (Speelman & Hills 1980). With increasing transport from the parent flora their abundance decreases sharply and the assemblage composition becomes more uniform (Speelman & Hills 1980). Because of their hydrodynamic equivalence they are frequently most abundant in medium to coarse silts rather than clays (Batten 1974).

8. Frequency of tetrads and sporomorph masses

Tetrads and sporomorph masses are often most abundant near to the parent flora, and therefore tend to increase in a proximal direction (Hughes & Moody-Stuart 1967). They are particularly common in prodelta facies (Batten 1974). This is not simply because their cumulative size leads them to be sedimented quickly; if the individual sporomorphs are of low density they should be capable of long distance flotation, and therefore other factors which control the probability of the tetrad or mass retaining its integrity are of equal importance. The mode of transport is especially important; the lower the energy of the transporting and depositing mechanism, the more likely the tetrad or mass will stay intact. Low energy distal hemipelagic or pelagic facies may therefore exhibit a surprisingly high frequency of tetrads and masses of buoyant pollen. By comparison, interbedded turbidites may often exhibit lower frequencies, presumably because of disaggregation during the higher energy transport (Tyson 1984).

9. Absolute sporomorph abundance (numbers per gram of sediment)

The absolute abundance of spores and pollen in marine sediments generally shows a more or less logarithmic (exponential) offshore decrease away from river mouths (Hoffmeister 1954, Woods 1955, Mudie 1982). However, this is strongly dependent on the size and character of the drainage basin and the magnitude and variability of river discharge (Mudie 1982, Heusser 1983, 1988). Very high abundances (15,000 to 100,000 per gram) are found in deltaic areas, but these decrease rapidly offshore (Muller 1959, Darrell & Hart 1970). Concentrations on the adjacent shelves are strongly influenced by local sediment grain size trends, as the sporomorphs are sorted and become concentrated in silty clays and winnowed out of sands (Davey 1971, Heusser & Balsam 1977). Hoffmeister (1960) suggested values of greater than 7,500-8,000 per gram can be used as indicators of proximity to land, or at least to river mouths (Stanley 1965). Normal values on the continental slope, away from river influence, are generally only a few thousand per gram, and values in continental rise and abyssal sediments are in the region of 10-500 grains per gram (Heusser 1983).

Very few geological studies have reported absolute sporomorph concentrations. However, it is clear that at times of low sea level there is a much greater supply of sporomorphs into deep water, with values of several tens of thousands of grains per gram being encountered even in relatively distal parts of the central Atlantic during the Early Cretaceous (Habib 1979, 1982, 1983).

Although preparation methods have to be more exacting, the use of "absolute" abundance variations (for all palynomorph groups) needs to be much more widely exploited by palynofacies workers as the information provided is much more definitive than relative percentage data. Combined use greatly increases the interpretative capability of the palynofacies technique (Muller 1959, de Jekhowsky 1963, Cross et al. 1966, Reyre 1973). Abundances expressed as numbers per gram are, of course, not truly absolute as they are affected by sediment accumulation rates (dilution by clastic or biogenic mineralic material). However, the concentration of sporomorphs in deltaic and submarine fan facies generally remains high despite this effect (Cross 1975).

10. Percentage chlorococcale algae of total plankton

The chlorococcale colonial green algae are exclusively freshwater. The two main fossilising genera are Botryococcus (?Carboniferous-Recent) and Pediastrum (early-Cretaceous-Recent). Both are well known from ancient lacustrine sediments where they are often responsible for the formation of excellent petroleum source rocks (Cane 1976, Hutton 1987, 1988). In comparison with Botryococcus, Pediastrum appears to be less tolerant of elevated salinities, and to flourish in deeper, more permanent, better mixed, harder water, and more nutrient-rich lake waters (Wake & Hillen 1980, Reynolds 1984, Talbot & Livingstone 1989). Where they do occur together, the much greater growth rate of Pediastrum generally results in its dominance, and thus high Botryococcus:Pediastrum ratios generally reflect suppression of the latter (Tyson, in press). Most fossil records of Botryococcus are from lacustrine, fluvial, lagoonal, and deltaic facies (Piasecki 1986, Batten & Lister 1988), but it is also recorded from facies that are at least temporarily hypersaline (Hunt 1987).

The oily composition of senescent Botryococcus colonies makes them highly buoyant (Burns 1982), and they are quite commonly flushed out of deltaic areas to become a minor but conspicuous component of adjacent shelf sediments (Loh et al. 1986, Robert 1988); similar occurrences are noted for Pediastrum (Evitt 1963, Liengjarern et al. 1980, Hutton 1988). Both genera are also sometimes redeposited into deepwater facies (McLachlan & Pieterse 1987). In marine sediments they can be used as indicators of relative proximity to fluvio-deltaic source areas, or as indicators of redeposition from the same.

11. Percentage of acritarchs of total plankton

In Mesozoic and Cenozoic sediments, acritarchs are generally only significant components of the phytoplankton assemblage in shallow water marginal marine facies. They are usually abundant only in areas where dinocysts are inhibited by brackish (and perhaps hypersaline) conditions (Wall 1965, Downie et al. 1971, Davey 1971, Riley 1974, Burger 1980, Schrank 1984). However, small acritarchs are also very abundant in Callovian black shales from the central Atlantic (Tyson 1984).

Being the most tolerant kind of marine phytoplankton, acritarchs can be useful for recognising saline influences within Mesozoic-Cenozoic delta top facies (Hancock & Fisher 1981). It should be noted that the small size of acritarchs means that they are easily overlooked, especially if sieving has been used during sample preparation. If present in AOM they may be completely undetectable unless fluorescent light observations are made.

During their Palaeozoic acme (Ordovician to Devonian), the significance of the acritarchs was probably much like that of the dinocysts in the Mesozoic to Recent. It is probable that many were similarly meroplanktonic (Dufka 1990).

12. Percentage of dinocysts of the total plankton

The cyst-forming habit of modern meroplanktonic dinoflagellates represents an adaptation to the unstable hydrographic regimes of temperate to subtropical shelf waters (Wall et al. 1977). Very few oceanic or tropical dinoflagellates produce fossilising organic-walled cysts (Wall et al. 1977, Dale 1986). In most marine situations dinocysts are the predominant form of fossilising phytoplankton, and they normally form a very high percentage of the total organic-walled microplankton assemblage. This is particularly the case in areas of seasonal coastal upwelling where the production of dinocysts may be very high (Davey 1971, Davey & Rogers 1975, Melia 1984). In brackish and hypersaline environments they may be at least partially replaced by acritarchs (see below). The percentage therefore generally increases offshore, but then declines in basinal pelagic or hemipelagic facies, except within interbedded turbidites (Tyson 1984).

13. Percentage prasinophyte algae of the total plankton

The phycomata (fossilising structures) of prasinophyte algae are often conspicuous components of both shelf and oceanic organic-rich sediments deposited under dysoxic to anoxic conditions (Tyson 1984, 1987, 1989). As the kerogen assemblages of black shale facies are usually dominated by AOM, and the percentage of palynomorphs is generally low, the abundance of prasinophyte algae is probably often exaggerated by palynologists. The number of individuals can reach "literally thousands per sample" (Guy-Ohlson 1988). This is not exceptional for fossil microplankton, but being often large and thick-walled, they are far more conspicuous than most other marine palynomorphs.

The prasinophyte signal of organic-rich shales is at least partly a relative effect produced by the decline of in situ dinocyst production in permanently stratified basins (see earlier). The ratio of prasinophyte algae to dinocysts therefore represents an index of hydrographic stability, as it increases in more "pelagic" stratified basin facies (Tyson 1984, 1987). A similar interpretation may apply to prasinophyte:acritarch ratios in the Early Palaeozoic (Dufka 1990). The prasinophyte algae are holoplankton, and the phycomata are only sedimented after the contents have been released; their distribution pattern is typically pelagic, while that for dinocysts is not. The abundance of prasinophytes has also been regarded by some as an indicator of brackish surface water conditions. Although this could explain the observed association with anoxic facies if runoff resulted in a permanent halocline, prasinophyte-dominated assemblages often lack other palynological and palaeontological characteristics indicative of major freshwater inputs. The modern phycomata-producing forms are almost exclusively marine.

Local dominance of prasinophyte algae has also been recorded from restricted (hypersaline?), and partly anoxic, lagoon and shallow water carbonate facies in the Late Jurassic (Bernier & Courtinat 1979, Hunt 1987). These occurrences also probably reflect an inhibition of dinocyst production and thus a relative increase in prasinophytes. However, they could also simply reflect the fact that the floating phycomata reach such areas in larger amounts than resuspended dinocysts. Abundances will also be enhanced by low sedimentation rates, and it is noteworthy that they are often highest in the most organic-rich, most radioactive, relatively condensed, anoxic facies deposited under "starved basin" conditions.

14. Ratio of long- to short-spined acritarchs

The length of the processes ("spines") of acanthomorph acritarchs appears to be an indicator of the energy of the depositional environment. Relatively distal assemblages tend to show a dominance of longer and more fragile processes, while those from higher energy marginal facies typically have shorter and stubbier ones (Wall 1965, Riley 1974).

15. Percentage chorate dinocysts of total dinocysts

Vozzhennikova (1965) was the first to suggest that offshore facies were characterised by a dominance of thin-walled delicate dinocysts with long processes (i.e. chorate morphotypes whose processes are >30% of the shortest diameter of the central body). Experimental model studies demonstrate that the presence of such processes does lead to measurable decreases in settling rates, but that the effect is modest and less important than size variations (Anderson et al. 1985, Sarjeant et al. 1987). The distribution patterns of modern dinoflagellates generally give little support to the notion of functional morphology (Dale 1983). However, the relative increase of chorate dinocysts in offshore and lower energy distal facies of Callovian to Tertiary age has been documented by many studies (e.g. Scull et al. 1966, Davey 1970, Riley 1974, Scott & Kidson 1977, Davies & Norris 1980, Sarjeant et al. 1987, Tyson 1989). This appears to be a purely passive sorting effect, but it may be advantageous by virtue of enhancing cyst dispersal. It can be a useful parameter for assessing proximal-distal and transgressive-regressive trends, especially where sporomorphs are infrequent or non-diagnostic.

16. Ratio of peridinioid to gonyaulacoid dinocysts

Downie et al. (1971) interpreted Palaeogene peridinioid-dominated assemblages as being indicative of nearshore lagoonal, estuarine, or brackish water environments, and gonyaulacoid-dominated assemblages as being characteristic of open marine facies. Harland (1973) has suggested the use of the "gonyaulacacean ratio" (gonyaulacoid cyst species : peridinioid cyst species) as a possible guide to palaeosalinity variations and proximity to shoreline. A number of mid-Cretaceous to Tertiary studies have confirmed the association of peridinioid dinocysts with nearshore facies (e.g. Davey 1970, Scott & Kidson 1977, Martinez-Hernandez et al. 1980, Mutterlose & Harding 1987, Lister & Batten 1988).

The gonyaulacacean ratio should not be interpreted only as an expression of salinity variations or onshore-offshore trends. Wall et al. (1977) and Bujak (1984) have observed that cyst assemblages from upwelling areas are also enriched in peridinioid forms. In addition, the gonyaulacacean ratio is also dependent on preservational factors, as peridinioid cysts are apparently grossly under-represented in Recent marine sediments, and are more susceptible to destruction during palynological processing (Dale 1976). Evolutionary factors must also be taken into account; peridinioid dinocysts underwent a major phase of diversification during the mid Cretaceous to Palaeogene so the significance of this parameter probably varies markedly through time.

17. Dinocyst diversity

Wall et al. (1977) observe that the maximum number of dinocyst species in Recent Atlantic samples is only about 20 and reaches its peak in warm water shelf facies. The diversity of fossil cyst assemblages is often much higher, especially in fine grained, normal marine shelf facies, where assemblages may contain 40-70 or more species (Mutterlose & Harding 1987). This is particularly true during Mesozoic and Palaeogene periods of high global sea level (see Bujak & Williams 1979, Goodman 1987).

Although in general terms diversity shows an irregular offshore increase, variability in diversity increases onshore and is greatest in estuarine assemblages, which frequently show high dominance (Wall et al. 1977, Morzadec-Kerfourn 1977, Tyler et al. 1982). Low diversities are typically associated with stressful environments, especially where these have good "trophic resources" that can be exploited by the few forms that are able to tolerate the conditions. The low diversity, high density cyst assemblages that occur in such environments are usually dominated by eurytopic cosmopolitan species (Wall et al. 1977).

Relative change in dinocyst diversity is a useful parameter for characterising normal open marine environments from more nearshore salinity-stressed conditions, and can therefore be used to identify transgressive-regressive trends. Examples of nearshore facies showing low diversity, high dominance, and high density assemblages have been documented by Goodman (1987), Liengjareren et al. (1980), Piasecki (1986), and Hunt (1987). In some instances low diversity, high dominance and high density can also be used to infer eutrophic, high productivity conditions (Davey & Rogers 1975, Schrank 1984, Honigstein et al. 1989).

18. Absolute dinocyst abundance (numbers per gram of sediment)

Cyst abundance is not simply related to dinoflagellate primary productivity (Wall et al. 1977). This is partly because cyst-formers are a small and inconsistent proportion of both the dinoflagellate and the total phytoplankton population, and partly because their distribution is strongly modified by sedimentological factors. Dinocysts behave as other fine sedimentary particles; they are generally coarse to medium silt-sized, but are hydrodynamically equivalent to fine silt or clay (Davey 1971, Dale 1976). Dinocyst concentration therefore tends to increase with the percentage of silt and clay, but stabilises at 3,000-10,000 per gram once this reaches 50-60% of the sediment (Wall et al. 1977). Grain size patterns strongly modify the observed onshore-offshore trends (Davey 1971).

In temperate latitudes the mean abundance of dinocysts increases by a factor of five from estuaries to the continental slope and rise, where values often exceed over 3,000-5,000 per gram in fine grained sediments (Wall et al. 1977). Cyst abundance subsequently declines with increasing depth and distance offshore.

High cyst densities (≤10,000 per gram) have been reported in association with estuarine frontal systems (Tyler et al. 1982), and it is quite clear that high cyst concentrations should not necessarily be equated with offshore "open shelf" conditions. The combined use of other parameters is required for reliable interpretation. However, in large data sets there is a statistical association between concentrations less than 100 per gram and water depths less than 70m (Wall et al. 1977); this probably reflects sediment dilution.

High to moderate abundances are found in areas of coastal upwelling (Davey 1971, Davey & Rogers 1975, Fauconnier & Slansky 1980, Melia 1984). This reflects both the high productivity and the typically low terrigenous sedimentation resulting from arid onshore climatic conditions.

19. Frequency of foraminiferal test linings

The distribution of foraminiferal linings is primarily controlled by that of the foraminifera which produce them. Foraminifera are larger and heavier than most palynomorphs, are not transported in the same fashion, and are generally much less abundant in terms of numbers of individuals per volume of sample. Furthermore, as only part of the foraminiferal population present in a sample will produce linings, it is not suprising that they are usually an insignificant component of the kerogen (usually less than 1%, and rarely more than 5% on a percentage particle abundance basis).

Their presence appears indicative of normal marine conditions (Muller 1959, Hoffmeister 1960, Tschudy 1969, Martinez-Hernandez et al. 1980, Courtinat 1989). Normal shelf facies may have abundances in the range of 50 to 500 per gram (Muller 1959, Traverse & Ginsburg 1966). Much higher abundances (1,000- 3,000 per gram) occur in the Gulf of California where their distribution is strongly correlated with the pattern of upwelling and reflects the high foraminiferal biomass (Cross et al. 1966). Moderately high abundances (<800 per gram) are also associated with areas of upwelling on the NW African margin (Melia 1984).

The fact that the abundance of linings decreases with increasing water depth (Melia 1984) supports the theory that they are derived from benthic rather than pelagic foraminifera (but productivity decreases offshore over deeper water). In ancient pelagic sediments significant abundances appear to be strongly correlated with redeposition (Tyson 1984, and unpublished data). Anoxic facies generally have low abundances, but benthic foraminifera can be abundant under what are, for metazoans, "anaerobic" conditions, and may even increase due to lack of grazing pressure (Douglas 1981).

20. Percentage of reworked palynomorphs (of total palynomorphs)

The proportion of identifiably reworked palynomorphs is partly a function of the ages and lithologies of the areas undergoing active erosion, the climate (weathering patterns, productivity of contemporary palynomorphs, and volume of runoff), and the distance and duration of transport. The ability of palynomorphs to withstand erosion and re-transportation is dependent on their level of maturation, their composition, and their morphology (size, structure, and wall-thickness). Generally the reworked forms are always strongly diluted by the contemporary palynomorphs unless the former are selectively concentrated because of differences in durability or hydrodynamic equivalence. Examples of the latter effect probably include the concentration of reworked palynomorphs in the levees of the Orinoco Delta (Muller 1959), and of large, pyrite-filled, sporomorphs in NW Atlantic contourites (Heusser & Balsam 1985).

The proportion of reworked palynomorphs normally declines away from their source and/or the point where they enter the basin (Williams & Sarjeant 1967, Hughes & Moody-Stuart 1967, Eshet et al. 1988). Their supply is clearly related to general clastic sediment supply, and is therefore highest during periods of low sea level and high runoff (Stanley 1969, Traverse 1988, Eshet et al. 1988). The percentage of reworked elements is a useful parameter for determining provenance and transgressive-regressive trends.

## 5.7 Comments on data interpretation

Subtle but significant variations in palynofacies character can only be assessed by proper counting and the presentation and analysis of numerical data. The most common kind of data used in palynofacies studies are percentages. Unfortunately, the literature contains many examples of percentage data having been interpreted as if they were mutually independent absolute abundances. Percentages are a property of the population, not of any individual component within that population. They only establish effect, not cause; changing any component of the population automatically alters the values for all the other components, and the same percentage value can therefore be produced in a variety of different ways.

Percentage data contain inherent correlations; percentages of one component are always correlated with that of the other components because they are calculated using the same sum (i.e. that which corresponds to 100%). For example, if the percentage of phytoclasts is high, the percentage of AOM has to be low, and causal relationships can only be established by cross correlation with other kinds of data (e.g. absolute abundances, AOM preservation). Such inherent arithmetic phenomena are relatively obvious when the number of components is small. However, when it is large (e.g. the percentages of many different palynomorph species within a total assemblage), they can easily mislead.

It is often advisable to break palynofacies data into logical subsets and calculate the percentages for each separately (e.g. different plankton groups as a percentage of only the marine palynomorphs, opaque phytoclasts only as a percentage of the total phytoclasts). The percentage data for each subset can then be compared with values from other subset(s) in order to evaluate more genuine correlations that may exist within the data. However, it is necessary to ensure that there is a statistically valid number of counts within each subset as well as within the data set as a whole.

As much of the interest in the data concerns the relative proportions of different components, it is also often more effective to plot specific ratios rather than, or as well as, some or all of the actual percentage values. Ratios are often a far more effective means of illustrating stratigraphic trends than are the crude data. One of their major advantages is that they are unaffected by the often significant proportion of unidentified or undifferentiated counts that may result in large uncertainties within percentage data (especially in palynomorph data).

All palynofacies workers are strongly advised to load their data (and derived percentages and significant ratios) into a good computer software package that permits easy data manipulation and rapid production of various graphic plots. The latter allows meaningful relationships to be identified very quickly and with very little effort. The data base can then be analyzed by statistical software packages to evaluate the significance of these relationships. The data should be organised in a way which allows it to be distinguished (and separately analyzed or grouped) according to vertical interval, section, lithology, lithofacies, and biostratigraphic age.

## 5.8 Diagrammatic representation of palynofacies data

Two common graphic means of summarising percentage data should be avoided, namely "pie-charts" (where 360° = 100%) and "strip logs" (where a uniform full column width indicates 100%). This is especially true if one is dealing with more than three or four relatively evenly distributed components. These diagrams are undesirable because it is very difficult to read off actual values or directly assess the relative ratios of different components, and the potentially significant minor components cannot be represented effectively. It is much better to show individual components in separate columns; values can be easily read off, the scales can be adjusted to emphasise potentially significant relative differences in the frequency of minor components, and error bars can also be shown.

In order to reduce the "signal to noise ratio", data from different lithologies and lithofacies should either be plotted separately or clearly distinguished (e.g. Tyson 1984), and also segregated prior to statistical analysis. Failure to do this can often produce such a large degree of scatter that meaningful trends can be obscured and therefore potentially overlooked.

Like other compositional data in geology, one of the most useful diagrammatic representations for displaying variations in assemblages is the use of ternary (triangular) diagrams. These are a very effective means of presenting percentage data for real populations or artificial groupings with three components. However, if these three components are not fairly evenly represented, or do not show a reasonable amount of variation, such diagrams may have little advantage over normal "binary" plots. The main advantage of ternary diagrams is that the data plots with a spatial separation that is useful for grouping samples into empirically-defined associations or assemblages.

One must always ensure that the data plotted is based on a sufficient number of counts. Note that although it is possible to expand the apparent resolution of the plot (i.e. if all the data plots in one corner, to enlarge this corner to the total size of the whole normal diagram), the added "precision" is often illusory, and does no more than artificially amplify small and statistically insignificant variations. Much of the data probably only has an accuracy of ±5-10%.

### 5.8.1 Total kerogen plots

Tyson (1989) uses an AOM-phytoclast-palynomorph plot of percentage particle abundances (i.e. relative numerical particle frequency) to characterise kerogen assemblages into eleven (independently sedimentologically-defined) marine shale facies (see Figures 5.2 and 5.3, and Table 5.3). The plot picks out the differences in relative proximity to terrestrial organic matter sources, and the redox status of the depositional subenvironments that control AOM preservation. As the emphasis is on palaeoenvironment rather than source rock potential, cuticle is incorporated within the phytoclast group, not the palynomorph group. Note the good discriminating power possible even with such very simple data.

In order to demonstrate correlations between different data sets, the percentage of various palynomorph categories, or geochemical values, can be plotted on to this diagram at the point defined by the corresponding kerogen composition.

### 5.8.2 Total palynomorph plots

Federova (1977) and Duringer and Doubinger (1985) use a microplankton-spore-pollen plot to indicate onshore-offshore depositional environments and transgressive-regressive trends (Figure 5.4).

Burger (1980) uses a sporomorph-acritarch-dinocyst plot to characterise shallow marine (neritic), brackish marine, or nonmarine-brackish depositional environments (Figure 5.4). Any plot using acritarchs may be difficult to use because of their low abundances in many facies, and because of the way in which small acritarchs are frequently overlooked, masked by AOM, or lost by sieving. Because the ecological significance of the acritarchs has changed with time, the interpretation shown for the plot really applies only to post-Triassic sediments. The significance of the percentage of dinocysts is also somewhat unreliable prior to their evolutionary diversification in the Callovian (late Middle Jurassic).

Table 5.3: Key to marine palynofacies fields defined on upper part of Figure 5.2 (simplified from Tyson 1989).

| Palynofacies field and Environment | | Comments | Spores: Bisaccates | Microplankton | Kerogen type |
|---|---|---|---|---|---|
| I | Highly proximal shelf or basin | High phytoclast supply dilutes all other components. | usually high | very low | III, gas prone |
| II | Marginal dysoxic-anoxic basin | AOM diluted by high phytoclast input, but AOM preservation moderate to good. Amount of marine TOC dependent on basin redox state. | high | very low | III, gas prone |
| III | Heterolithic oxic shelf ("proximal shelf") | Generally low AOM preservation; absolute phytoclast abundance dependent on actual proximity to fluvio-deltaic source. Oxidation and reworking common. | high | common to abundant dinocysts dominant | III or IV, gas prone |
| IV | Shelf to basin transition | Passage from shelf to basin in time (eg increased subsidence/water depth) or space (eg basin slope). Absolute phytoclast abundance depends on proximity to source and degree of redeposition. Amount of marine TOC depends on basin redox state. IVa dysoxic-suboxic, IVb suboxic-anoxic. | moderate to high | very low-low | III or II, mainly gas prone |
| V | Mud- dominated oxic shelf ("distal shelf") | Low to moderate AOM (usually degraded). Palynomorphs abundant. light coloured bioturbated, calcareous mudstones are typical. | usually low | common to abundant dinocysts dominant | III > IV, gas prone |
| VI | Proximal suboxic-anoxic shelf | High AOM preservation due to reducing basin conditions. Absolute phytoclast content may be moderate to high due to turbiditic input and/or general proximity to source. | variable, low to moderate | low to common dinocysts dominant | II, oil prone |
| VII | Distal dysoxic-anoxic "shelf" | Moderate to good AOM preservation, low to moderate palynomorphs. Dark-coloured slightly bioturbated mudstones are typical. | low | moderate to common dinocysts dominant | II, oil prone |
| VIII | Distal dysoxic-anoxic shelf | AOM-dominated assemblages, excellent AOM preservation. Low to moderate palynomorphs (partly due to masking). Typical of organic-rich shales deposited under stratified shelf sea conditions. | low | low to moderate dinocysts dominant, % prasinophytes increasing | II >> I, oil prone |
| IX | Distal suboxic-anoxic basin | AOM-dominated assemblages. Low abundance of palynomorphs partly due to masking. Frequently alginite-rich. Deep basin or stratified shelf sea deposits, especially sediment starved basins. | low | generally low prasinophytes often dominant | II ≥ I, highly oil prone |

Figure 5.2: AOM-Phytoclast-palynomorph plot (after Tyson 1989). Upper: Stylised diagnostic fields for Late Jurassic marine shale palynofacies (see Table 5.3). Lower: Actual data fields for sedimentologically-defined facies: 1. Oxidising shelf facies, 2. Basin slope facies, 3. Highly proximal shelf or basin facies, 4. Proximal submarine fault scarp fan facies, 5. Distal submarine fault scarp facies, 6. Distal graben facies. 7. Offshore basinal facies.

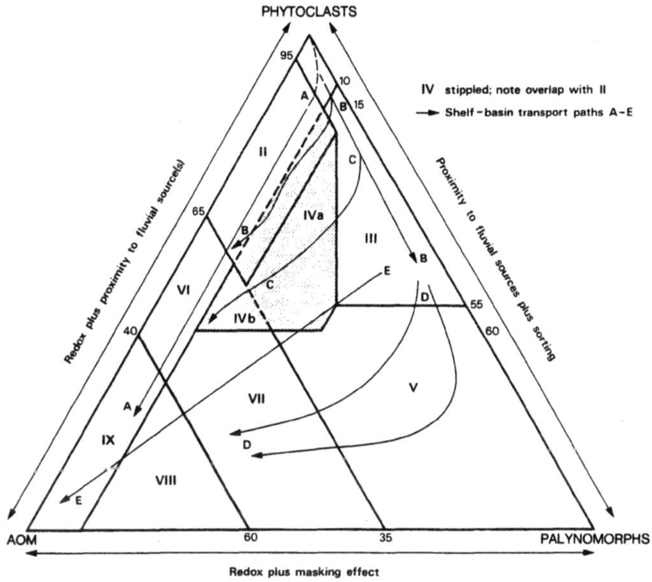


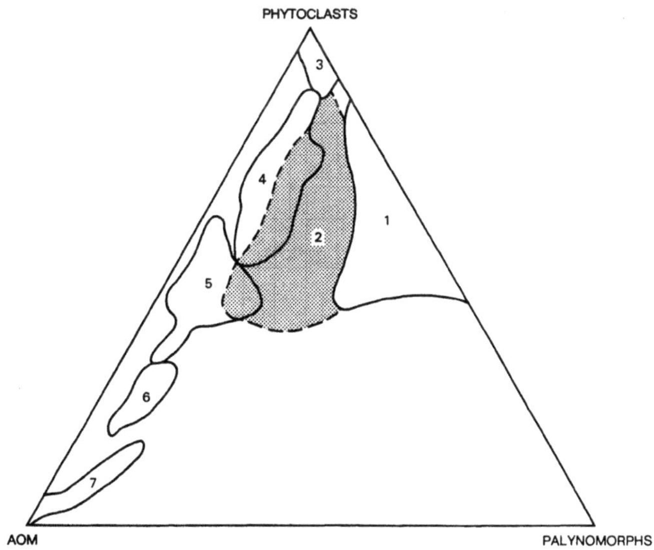

Figure 5.3: AOM-Phytoclast-palynomorph plots for a variety of other facies. Interpretation is consistent with that shown in Figure 5.2 (Upper), and supported by independent sedimentological and palaeoecological evidence.
**A**. Regressive Lower Jurassic sequence from the Mochras borehole, NW Wales (P.C. Jeffcock, unpublished). 1. Hettangian-Lower Sinemurian, 2. Sinemurian, 3. Lower Pliensbachian, 4. Toarcian.
**B**. Middle Jurassic deltaic facies (Brent equivalent), well 25/4-1 Norwegian North Sea (A. Wylde, unpublished).
**C**. Callovian Oxford Clay, Clapgate Farm borehole, Peterborough (P. Smith, unpublished). 1. Lower Oxford Clay (upper *calloviense*-lower *athleta* zones), 2. Middle Oxford Clay (upper *athleta* zone).

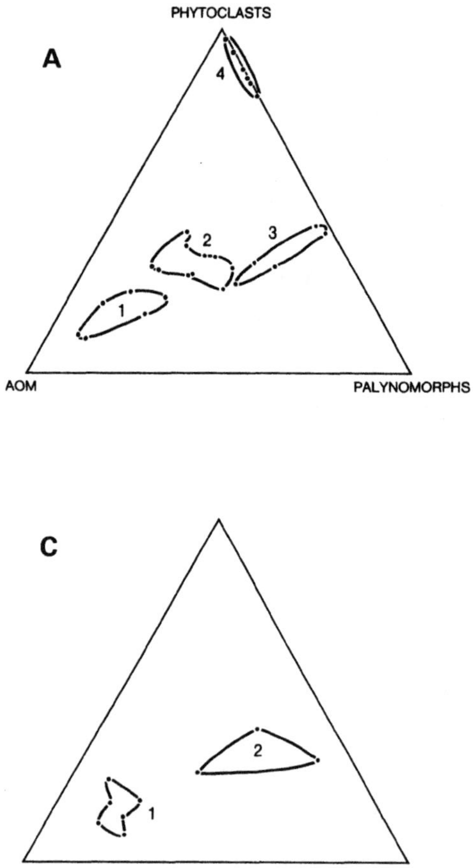

Figure 5.3 (continued)

**E**. Kimmeridge Clay Formation, Aylesbury (W.J. Bailey, unpublished). 1. Lower Kimmeridge Clay (*mutabilis-eudoxus* zones), 2. Upper Kimmeridge Clay (*wheatleyensis* zone).

**F**. Berriasian - Albian pelagic limestones and marls (oxic and anoxic facies) DSDP Site 535, eastern Gulf of Mexico (R.V. Tyson, unpublished). Points are means for different lithological units.

**G**. Speeton Clay, Speeton, Yorkshire (V.O.B. Akinbode, unpublished). 1. D Beds (Ryazanian- early Hauterivian), 2. C Beds (Hauterivian), LB Beds (Barremian).

**H**. Palaeocene (Thanetian), Thanet, Kent (T.S. Sulaeman, unpublished). 1. Thanet Formation, 2. Woolwich-Oldhaven Formations.

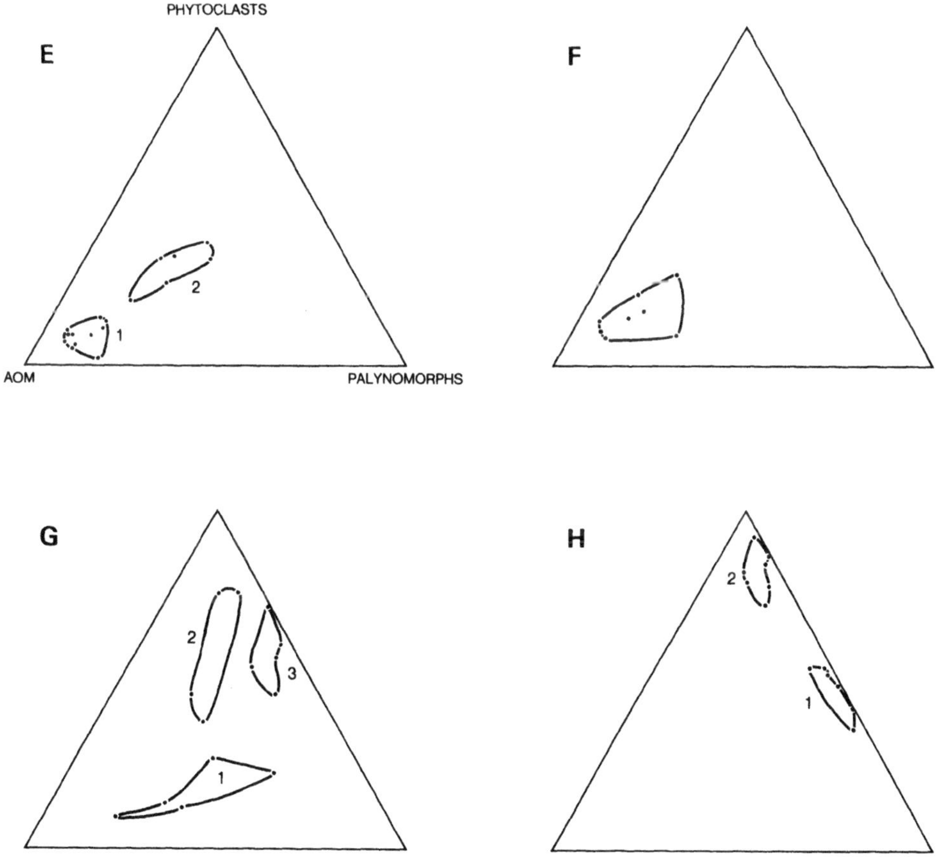

Table 5.4: Parameters for assessing salinity, redox, productivity, and watermass stability. Interpretations should be based on as many of the parameters as possible.

**PARAMETERS FOR ASSESSING SALINITY, REDOX, PRODUCTIVITY AND WATERMASS STABILITY**

| | |
|---|---|
| ***SALINITY*** | |
| **Indirect:** | |
| % phytoclasts (of kerogen) | High values indicate proximity to active fluvio-deltaic source(s) and thus possibility of reduced salinity. |
| Opaque: translucent phytoclasts | Low values indicate good supply of fresh, unoxidised phytoclasts. |
| % cuticle (of kerogen/phytoclasts) | High values suggest strong proximity to active fluvio-deltaci source(s). |
| % marine plankton (of palynomorphs) | Low values may reflect ecological inhibition, or dilution due to high sporomorph supply from runoff. |
| % microspores (of miospores) | High values suggest humid conditions (at least locally), and proximity to active fluvio-deltaic source(s). |
| Sporomorph No./gram | High values indicate strong proximity to active fluvio-deltaic source(s). |
| % reworked palynomorphs | High values generally correlated with active erosion and relatively high runoff. |
| **Direct:** | |
| % chlorococcales (of palynomorphs) | ±100% indicates freshwater. More than a few percent in marine waters suggests some runoff influence. *Botryococcus:Pediastrum* ratio increases with salinity in lake waters. |
| % acritarchs (post-Palaeozoic) | High values suggest inhibition of dinocyst production by abnormal salinity. |
| % prasinophytes (of plankton) | High values suggest inhibition of dinocyst production, but salinity significance unclear. May sometimes reflect stable salinity stratification (depressed surface water salinity). |
| peridinioid:gonyaulacoid cysts | High values ***sometimes*** suggest nearshore and possibly abnormal salinity conditions (post-Jurassic only). |
| dinocyst diversity | Low values (with high dominance), suggest nearshore, stressful environments, and probably often abnormal salinity. |
| foraminiferal test linings | Presence indicates marine conditions in bottomwater. |
| ***PALAEO-OXYGENATION AND REDOX*** | |
| % (fluorescent) AOM (of kerogen) | High values indicate reducing conditions. Distal dysoxic-anoxic shale facies typically have ≥60% (see Fig. 2). Note that % phytoclasts and palynomorphs are low due to dilution by AOM. |
| AOM matrix fluorescence intensity | Increases with preservation, and thus more reducing conditions, but generally less than that of palynomorphs. AOM derived from bacterial mats may be strongly and uniformly fluorescent. |
| % prasinophytes (of plankton) | High relative values are typically associated with anoxic shale facies. Other indices generally indicate only the distal (basinal) nature of dysoxic-anoxic facies (see Tables. 2). |
| ***PRIMARY PRODUCTIVITY*** | |
| Peridinioid:gonyaulacoid ratio | High values occur in upwelling areas (but also influenced by salinity). |
| Dinocyst diversity | Low values with high dominance often associated with upwelling, and/or eutrophic conditions. |
| Dinocyst No./gram | High values often correlated with high productivity (but also influenced by sorting and dilution). |
| Foraminiferal test linings | High numbers/gram (>500?) suggest high productivity; may also be high in dysoxic facies due to reduced grazing pressure. |
| *Pediastrum:Botryococcus* ratio | High values in lakes suggest more eutrophic conditions (but also influenced by many other variables. |
| ***WATER COLUMN STABILITY*** | |
| % prasinophytes (of plankton) | Increases in stable settings (oceans and stratified anoxic basins) as *in situ* dinocyst production declines. |
| % (fluorescent AOM of kerogen) | Increases if stability results in dysoxic or anoxic bottom conditions. |
| | Stable environments generally show distal characteristics (see Tables. 2.) |

Figure 5.4: Ternary diagrams for describing total palynomorph assemblages. Upper: Microplankton-spore-pollen plot (after Federova 1977, and Duringer & Doubinger 1985). Lower: Sporomorph-acritarch-dinocyst plot (after Burger 1980).

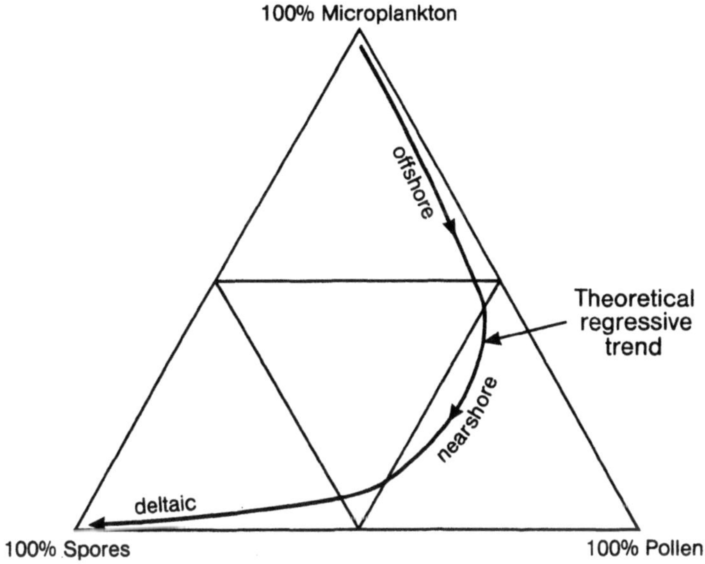


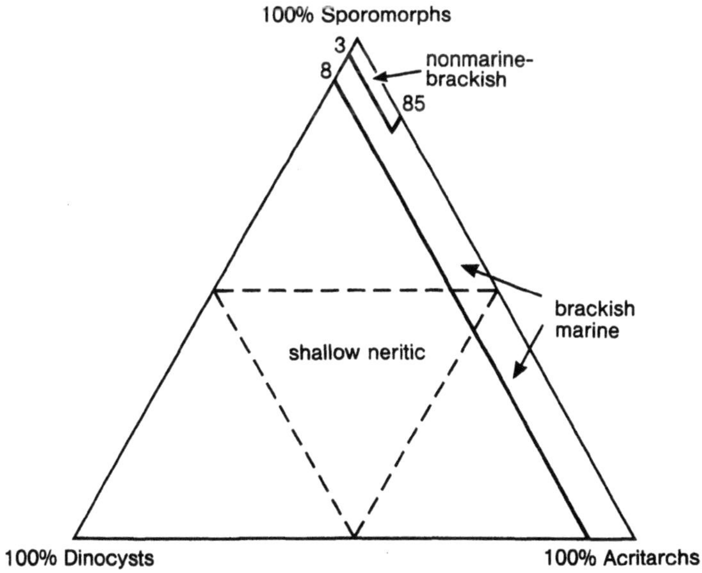

Figure 5.5: Ternary diagrams for describing microplankton assemblages. Arrows indicate the palaeoenvironmental significance of relative displacements of data points.

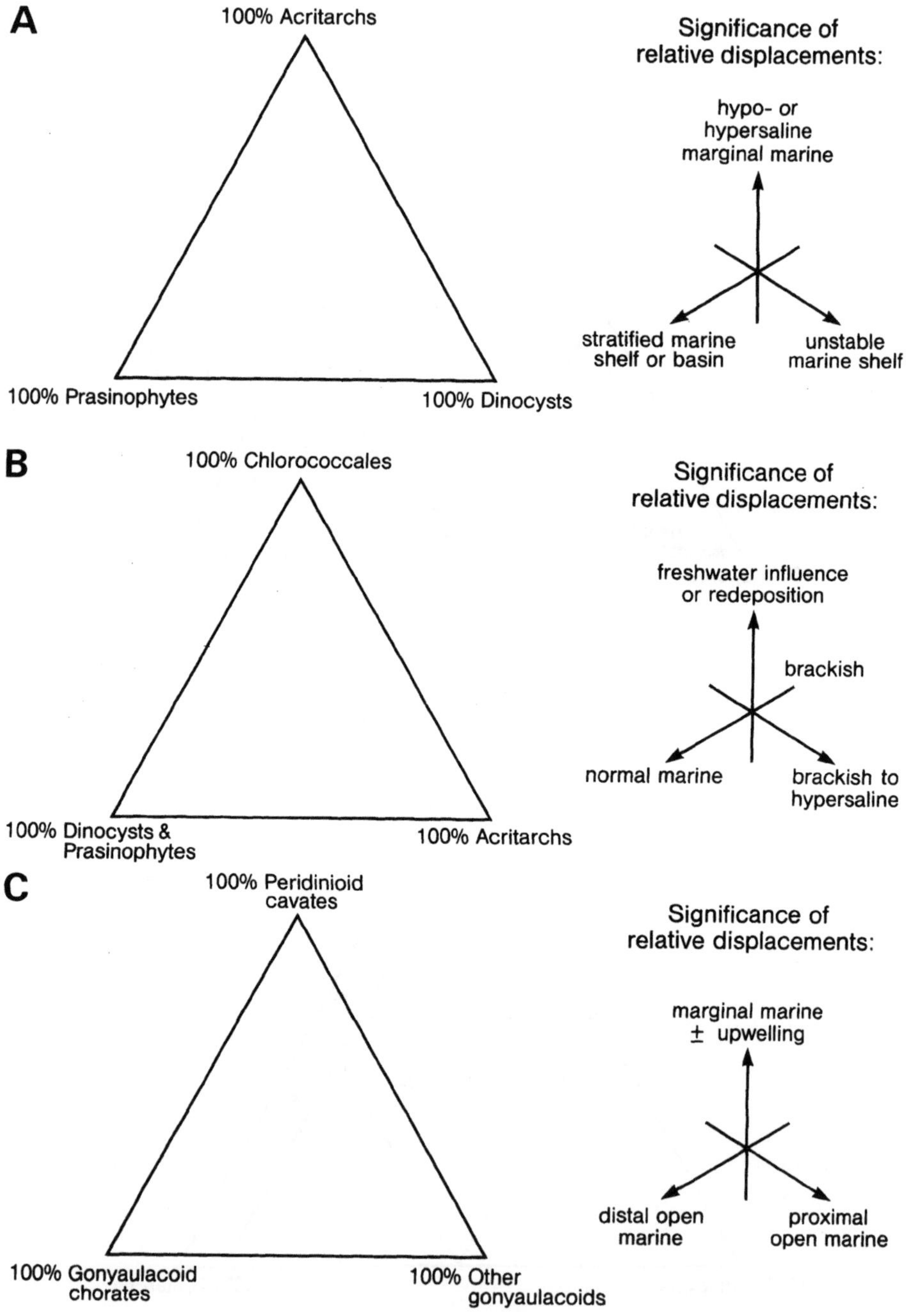

### 5.8.3 Phytoplankton palynomorph plots

A potentially useful plot is that for dinocysts, acritarchs, and prasinophyte algae (as percentages of the total marine microplankton), which may be useful in discriminating marginal marine from normal shelf and stratified shelf or basinal environments.

Another significant plankton plot for use in deltaic to nearshore facies is that for chlorococcale algae, acritarchs, and [dinocysts+prasinophyte algae] as percentages of the total microplankton, which can be potentially used to discriminate freshwater, brackish-marginal, and marine environments.

A plot of gonyaulacoid [proximate + proximochorate + cavate] dinocysts, gonyaulacoid chorate dinocysts, and peridinioid cavate cysts (as percentages of the total dinocysts) can also be potentially useful in palaeoenvironmental diagnosis. It is most useful for mid-Cretaceous to Palaeogene facies.

## 5.9 Conclusions

Palynological analysis of the total kerogen and palynomorph assemblages present in sedimentary rocks can yield a great deal of useful information on depositional environments and hydrocarbon source rock potential. The direct observation of the organic material and the detailed subdivision of its component parts that is possible, gives the technique some considerable advantages over bulk rock methods. However, the integrated use of transmitted white light and incident blue light fluorescence is essential.

The combination of total kerogen and palynomorph parameters described and reviewed in this chapter (see Tables 5.2 and 5.3) has the potential to result in a considerable degree of palaeoenvironmental resolution provided it is based upon proper counting procedures. Of course, detailed analysis is only worthwhile on outcrop and core samples (and thus of limited relevance to much routine industrial work).

The systematic collection of quantitative palynofacies data, along with normal taxonomic data, should eventually permit a much greater understanding of the palaeoecology of individual organic-walled plankton taxa, and thus enhanced palaeoenvironmental resolution. As yet there is very little of this kind of information available. This is largely because most of those engaged in biostratigraphic research have neglected to make palynofacies investigations, or have done so only in a cursory and qualitative manner.

Much more widespread use of "absolute" abundance data is needed; whenever possible, palynomorph variations should be assessed as numbers per gram as well as percentages. It is notable that many of the early "classic" palynofacies studies (such as Muller 1959, Cross et al. 1966) made great use of this type of data. One of the reasons we still rely so heavily on these early studies to guide our interpretations is the lack of more recent investigations using this approach. Only those with a background in Quaternary palynology seem to make adequate use of it. Total kerogen parameters (percentages of phytoclasts and AOM, but not palynomorphs) should also be calibrated by integration with TOC values.

Ternary diagrams are a simple but extremely useful means of presenting palynofacies (percentage) data. Even relatively simple data, when plotted on the appropriate ternary diagram, can demonstrate a significant potential for discriminating temporal and spatial differences in depositional environments. Because all palynofacies parameters are strongly influenced by sedimentological factors, palynofacies studies must be conducted within a sedimentological framework. They should employ sampling strategies that are deliberately designed to statistically assess the influence of any observable sedimentological variables, e.g. sediment colour, grain size, ichnofacies, sedimentary structures, as well as to provide the appropriate stratigraphic coverage.

All of the issues and parameters discussed in this chapter are covered in greater detail in Tyson (in press).

## Acknowledgements

The first draft of this chapter was written in 1989 while the author was a "New Blood" lecturer in palynology and palynofacies at University College London. Thanks go to Janet Baker and Colin Stuart (UCL) for drafting the figures, to Barbara Brown and Christine Jeans (NRG, Newcastle) for preparing the tables, and also to Graham Jenkins for inviting this contribution. The author would also like to thank and acknowledge Victor Akinbode, Bill Bailey, Pippa Jeffcock, Phil Smith, Teteh Sulaeman, and Anne Wylde, past palynology students of the UCL MSc Micropalaeontology Course (1985-1988), some of whose MSc project data is included within Figure 5.3.

### References

Anderson, D.M., Levely, J.J., Reardon, E.M. and Price, C.A. 1985. Sinking characteristics of dinoflagellate cysts. Limnology and Oceanography, 30, 1000-1009.

Batten, D.J. 1973. Use of palynologic assemblage-types in Wealden correlation. Palaeontology, 16, 1-40.

Batten, D.J. 1974. Wealden palaeoecology from the distribution of plant fossils. Proceedings of the Geologic Association, 85, 433-458

Batten, D.J. 1982. Palynofacies, palaeoenvironments, and petroleum. Journal of Micropalaeontology, 1, 107-114.

Batten, D.J. and Lister, J.K. 1988. Evidence of freshwater dinoflagellates and other algae in the English Wealden (Early Cretaceous). Cretaceous Research, 9, 171-179.

Bernier, P. and Courtinat, B. 1979. Le microplankton (Leiosphaeridia) et la matère organique des calcaires d'arrier-recif du Kimmeridgien Superieur dans le Jura Meridional. Systematique, conditions de genese et d'environment. Documents du Laboratoire Géologique, Faculté des Sciences Lyon, 75, 95-117.

Bostick, N.H. 1971. Thermal alteration of clastic organic particles as an indicator of contact and burial metamorphism in sedimentary rocks. Geoscience and Man, 3, 83-92.

Boulter, M.C. and Riddick, A. 1986. Classification and analysis of palynodebris from the Palaeocene sediments of the Forties Field. Sedimentology, 33, 871-886.

Brooks, J. 1981. Organic maturation of sedimentary organic matter and petroleum exploration: a review. In: Brooks, J. (ed) Organic Maturation and Fossil Fuel Exploration. Academic Press, London, 1-37.

Brush, G.S. and Brush, L.M. Jr. 1972. Transport of pollen in a sediment-laden channel: a laboratory study. American Journal of Science, 272, 359-381.

Bryant, I.D., Kantorowicz, J.D. and Love, C.F. 1988. The origin and recognition of laterally continuous carbonate-cemented horizons in the Upper Lias Sands of southern England. Marine and Petroleum Geology, 5, 108-133.

Buchanan, J.B. 1958. The bottom fauna communities across the continental shelf off Accra, Ghana (Gold Coast). Proceedings of the Zoological Society of London, 130, 1-56.

Bujak, J.P. 1984. Cenozoic dinoflagellate cysts and acritarchs from the Bering Sea and northern Pacific, DSDP Leg 19. Micropalaeontology, 30, 180-212.

Bujak, J.P., Barss, M.S. and Williams, G.L. 1977. Offshore east Canada's organic type and color and hydrocarbon potential. The Oil and Gas Journal, 75, 198-201.

Bujak, J.P. and Williams, G.L. 1979. Dinoflagellate diversity through time. Marine Micropalaeontology, 4, 1-12.

Burger, D. 1980. Palynological studies in the Lower Cretaceous of the Surat Basin, Australia. Bureau of Mineral Resources Geology and Geophysics Bulletin, 189, 106pp.

Burgess, J.D. 1974. Microscopic examination of kerogen (dispersed organic matter) in petroleum exploration. Geological Society of America Special Paper, 153, 19-30.

Burns, D.A. 1982. A transmission electron microscope comparison of modern *Botryococcus braunii* with some microfossils previously referred to that species. Revista Espanola de Micropalaeontologia, 14, 165-185.

Bustin, R.M. 1988. Sedimentology and characteristics of dispersed organic matter in Tertiary Niger Delta: origin of source rocks in a deltaic environment. American Association of Petroleum Geologists Bulletin, 72, 277-298.

Cai, D.L., Tan, F.C. and Edmond, J.M. 1988. Sources and transport of particulate organic carbon in the Amazon River. Estuarine, Coastal and Shelf Science, 26, 1-14.

Cane, R.F. 1976. The origin and formation of oil shale. In: Teh Fu Yen and Chilingarian, G.V. (eds) Oil Shale. Developments in Petroleum Science, 5, Elsevier, Amsterdam, 27-60.

Caratini, C., Bellet, J., and Tissot, C. 1983. Les palynofacès: représentation graphique, intérêt de leur étude pour les reconstitutions paléogéographiques. In: Géochimie Organique des Sédiments Marins d'ORGON à MISEDOR. Éditions du Centre National de la Recherche Scientifique, Paris, 327-351.

Chaloner, W.G. and Muir, M. 1968. Spores and floras. In: Murchison, D. and Westoll, T.S. (eds) Coal and Coal-Bearing Strata. Oliver and Boyd, Edinburgh, 127-146.

Claret, J., Jardine, S. and Robert, P. 1981. La diversité des roches meres pétrolières: aspects geologiques et implications économiques à partir de quatre exemples. Bulletin du Centres Recherche Exploration-Production Elf-Aquitaine, 5, 383-417.

Combaz, A. 1964. Les palynofaciès. Revue de Micropaleontologie, 7, 205-218.

Combaz, A. 1980. Les kerogenes vus au microscope. In: Durand, B. (ed) Kerogen: Insoluble Organic Matter From Sedimentary Rocks. Editions Technip, Paris, 55-111.

Correia, M. 1971. Diagenesis of sporopollenin and other comparable organic substances: application to hydrocarbon research. In: Brooks, J., Grant, P., Muir, M.D., Shaw, G. and Van Gijzel, P. (eds) Sporopollenin. Academic Press, London, 569-620.

Courtinat, B. 1989. Les organoclastes des formations lithologiques du Malm dans le Jura Meridional. Systematique, biostratigraphie et elements d'interpretation paleoecologique. Documents de Laboratoire Géologique Lyon, 105, 361pp.

Cross, A.T. 1964. Plant microfossils and geology: an introduction. In: Cross, A.T. (ed) Palynology in Oil Exploration. Society of Economic Paleontologists and Mineralogists Special Publication, 11, 3-13.

Cross, A.T. 1975. Source and distribution of palynomorphs in the Gulf of California. Geoscience and Man, 11, 156.

Cross, A.T., Thompson, G.G. and Zaitzeff, J.B. 1966. Source and distribution of palynomorphs in bottom sediments, southern part of Gulf of California. Marine Geology, 4, 467-524.

Dale, B. 1976. Cyst formation, sedimentation, and preservation: factors affecting dinoflagellate assemblages in recent sediments from Trondheimsfjord, Norway. Review of Palaeobotany and Palynology, 22, 39-60.

Dale, B. 1983. Dinoflagellate resting cysts: "benthic plankton". In: Fryxell, G.A. (ed) Survival Strategies of the Algae. Cambridge University Press, 69-136.

Dale, B. 1986. Life cycle strategies of oceanic dinoflagellates. In: Pierrot-Bults, A.C., Van Der Spoel, S., Zahuranec, B.J., and Johnson, R.K. (eds) Pelagic Biogeography. Unesco Technical Papers in Marine Science, 49, 65-72.

Dance, K.W. 1981. Seasonal aspects of transport of organic and inorganic matter in streams. In: Lock, M.A. and Williams, D.D. (eds) Perspectives in Running Water Ecology, Plenum Press, New York, 69-95.

Darrell II, J.H. and Hart, G.F. 1970. Environmental determinations using absolute miospore frequency, Mississippi River delta. Geological Society of America Bulletin, 81, 2513-2518.

Davey, R.J. 1970. Non-calcareous microplankton from the Cenomanian of England, northern France and North America. Part II. Bulletin of the British Museum (Natural History), Geology, 18, 333-397.

Davey, R.J. 1971. Palynology and palaeo-environmental studies, with special reference to the continental shelf sediments of South Africa. In: Farinacci, A. (ed) Proceedings of the Second Planktonic Conference, Rome 1970. Edizioni Tecnoscienzia-Roma, 331-347.

Davey, R.J. and Rogers, J. 1975. Palynomorph distribution in Recent offshore sediments along two traverses off South West Africa. Marine Geology, 18, 213-225.

Davies, E.H. and Norris, G. 1980. Latitudinal variations in encystment modes and species diversity in Jurassic dinoflagellates. In: Strangway, D.W. (ed) The Continental Crust and Its Mineral Deposits, Geological Association of Canada Special Paper, 20, 361-373.

Davis, R.B. 1967. Pollen studies of near-surface sediments in Maine lakes. In: Cushing, E.J. and Wright, H.E. Jr. (eds) Quaternary Palaeoecology, Yale University Press, New Haven, 143-173.

Degens, E.T. and Mopper, K. 1976. Factors controlling the distribution and early diagenesis of organic material in marine sediments. In: Riley, J.P. and Chester, R. (eds) Chemical Oceanography, 6, 2nd Edition, Academic Press, London, 59-113.

Denison, C. and Fowler, R.M. 1980. Palynological identification of facies in a deltaic environment. In: Proceedings of the Meeting on the Sedimentation of North Sea Reservoir Rocks, Geilo 1980. Norsk Petroleumsforening, Paper XII, 1-22.

Douglas, R.G. 1981. Paleoecology of continental margin basins: a modern case history from the Borderland of Southern California. In: Douglas, R.G., Colburn, I.P. and Gorsline, D.S. (eds) Depositional Systems of Active Continental Margin Basins. Short Course Notes, Pacific Section Society of Economic Paleontologists and Mineralogists, Los Angeles, 121-156.

Downie, C., Hussain, M. and Williams, G.L. 1971. Dinoflagellate cyst and acritarch associations in the Paleogene of southeast England. Geoscience and Man, 3, 29-35.

Dufka, P. 1990. Palynomorphs in the Llandovery black shale sequence of the Prague Basin (Barrandian area, Bohemia). Časopis pro mineralogii a geologii, 35, 15-31.

Durand, B. 1980. Sedimentary organic matter and kerogen. Definition and quantitative importance of kerogen. In: Durand, B. (ed) Kerogen: Insoluble Organic Matter From Sedimentary Rocks. Editions Technip, Paris, 13-34.

Duringer, P. and Doubinger, J. 1985. La palynologie: un outil de characterisation des facies marins et continentaux a la limite Muschelkalk Superieur-Lettenkohle. Science Geologie Bulletin, Strasbourg, 38, 19-34.

Eshet, Y., Druckman, Y., Cousminer, H.L., Habib, D. and Drugg, W.S. 1988. Reworked palynomorphs and their use in the determination of sedimentary cycles. Geology, 16, 662-665.

Evitt, W.R. 1963. Occurrence of freshwater alga Pediastrum in Cretaceous marine sediments. American Journal of Science, 261, 890-893.

Evitt, W.R. 1985. Sporopollenin Dinoflagellate Cysts: Their Morphology and Interpretation. American Association of Stratigraphic Palynologists Foundation, 333pp.

Fauconnier, D. and Slansky, M. 1980. Relations entre le development des Dinoflagelles et la sedimentation phosphate de Bassin de Gafsa (Tunisie). Documents de Bureau du Recherche Géologie et Minieres, 24, 185-204.

Federova, V.A. 1977. The significance of the combined use of microphytoplankton, spores, and pollen for differentiation of multi-facies sediments. In: Samolilovich, S.R. and Timoshina, N.A. (eds) Questions of Phytostratigraphy. Trudy Neftyanoi nauchno-issledovatelskii geologo-razvedochnyi Institus (VNIGRI) Leningrad, 398, 70-88. (in Russian).

Fisher, M.J. 1980. Kerogen distribution and depositional environments in the Middle Jurassic of Yorkshire U.K. In: Bharadwaj, D.C., Singh, H.P. and Tiwari, R.S. (eds) Proceedings of the 4th International Palynological Conference, Lucknow 1976-1977, 2, 574-580.

Fisher, M.J. and Hancock, N.J. 1985. The Scalby Formation (Middle Jurassic, Ravenscar Group) of Yorkshire: reassessment of age and depositional environment. Proceedings of the Yorkshire Geological Society, 45, 293-298.

Glikson, M. and Taylor, G.H. 1986. Cyanobacterial mats; major contributors to the organic matter in Toolebuc Formation oil shales. In: Gravestock, D.I., Moore, P.S. and Pitt, G.M. (eds) Contributions on the Geology and Hydrocarbon Potential of the Eromanga Basin. Geological Society of Australia Special Publication, 12, 273-286.

Goodman, D.K. 1987. Dinoflagellate cysts in ancient and modern sediments. In: Taylor, F.J.R. (ed) The Biology of Dinoflagellates. Botanical Monographs, 21, Blackwell Scientific, Oxford, 649-722.

Gregory, W.A. Jr., Chinn, E.W., Sassen, R. and Hart, G.F. 1991. Fluorescent microscopy of particulate organic matter: Sparta Formation and Wilcox Group, south central Louisiana. Organic Geochemistry, 17, 1-9.

Guy-Ohlson, D. 1988. Developmental stages in the life cycle of Mesozoic Tasmanites. Botanica Marina, 31, 447-456.

Habib, D. 1979. Sedimentology of palynomorphs and palynodebris in Cretaceous carbonaceous facies south of Vigo Seamount. In: Sibuet, J-C., Ryan, W.B.F. et al., Initial Reports of the Deep Sea Drilling Project, U.S. Government Printing Office, Washington D.C. 47(2), 451-467.
Habib, D. 1982. Sedimentary supply origin of Cretaceous black shales. In: Schlanger, S.O. and Cita, M.B. (eds) Nature and Origin of Cretaceous Carbon-Rich Facies, Academic Press, London, 113-127.
Habib, D. 1983. Sedimentation rate dependent distribution of organic matter in the North Atlantic Jurassic-Cretaceous. In: Sheridan, R.E., Gradstein, F.M. et al., Initial Reports of the Deep Sea Drilling Project, U.S. Government Printing Office, Washington D.C., 76, 781-794.
Hancock, N.J. and Fisher, M.J. 1981. Middle Jurassic North Sea deltas with particular reference to Yorkshire. In: Illing, L.V. and Hobson, G.D. (eds) Petroleum Geology of the Continental Shelf of North-West Europe. Heyden and Son Ltd, London, 186-195.
Harland, R. 1973. Dinoflagellate cysts and acritarchs from the Bearpaw Formation (Upper Campanian) of southern Alberta, Canada. Palaeontology, 16, 665-706.
Hart, G.F. 1986. Origin and classification of organic matter in clastic systems. Palynology, 10, 1-23.
Hedges, J.I. and Parker, P.L. 1976. Land-derived organic matter in surface sediments from the Gulf of Mexico. Geochimica et Cosmochimica Acta, 40, 1019-1029.
Hennessee, E.L., Blakeslee, P.J. and Hill, J.M. 1986. The distributions of organic carbon and sulfur in surficial sediments of the Maryland portion of Chesapeake Bay. Journal of Sedimentary Petrology, 56, 674-683.
Herring, J.R. 1985. Charcoal fluxes into sediments of the North Pacific Ocean: the Cenozoic record of burning. In: Sundquist, E.T. and Boecker, W.S. (eds) The Carbon Cycle and Atmospheric $CO_2$: Natural Variations Archean to Present. Geophysical Monographs, 32, American Geophysical Union, Washington D.C., 419-442.
Heusser, L.E. 1983. Pollen distribution in the bottom sediments of the western North Atlantic Ocean. Marine Micropalaeontology, 8, 77-88.
Heusser, L.E. 1988. Pollen distribution in marine sediments on the continental margin off northern California. Marine Geology, 80, 131-147.
Heusser, L.E. and Balsam, W.L. 1985. Pollen sedimentation in the northwest Atlantic: effects of the Western Boundary Undercurrent. Marine Geology, 69, 149-153.
Hoffmeister, W.S. 1954. Microfossil Prospecting for Petroleum. U.S. Patent Office, Washington D.C., Patent No.2, 686,108, 4 pp.
Hoffmeister, W.S. 1960. Palynology has important role in oil exploration. World Oil, 150, 101-104.
Honigstein, A., Lipson-Benitah, S., Conway, B., Flexer, A. and Rosenfeld, A. 1989. Mid-Turonian anoxic event in Israel - a multidisciplinary approach. Palaeogeography, Palaeoclimatology, Palaeoecology, 69, 103-112.
Hooghiemstra, H., Agwu, C.O.C. and Beug, H-J. 1986. Pollen and spore distribution in Recent marine sediments: a record of NW-African seasonal wind patterns and vegetation belts. "Meteor" Forschungsergebnisse, C40, 87-135.
Hopkins, J.S. 1950. Differential flotation and deposition of coniferous and deciduous tree pollen. Ecology, 31, 633-641.
Hughes, N.F. and Moody-Stuart, J.C. 1967. Palynological facies and correlation in the English Wealden. Review of Palaeobotany and Palynology, 1, 259-268.
Hunt, C.O. 1987. Dinoflagellate cyst and acritarch assemblages in shallow-marine and marginal-marine carbonates: the Portland Sand, Portland Stone, and Purbeck Formations (Upper Jurassic/Lower Cretaceous) of southern England and northern France. In: Hart, M.B. (ed) Micropalaeontology of Carbonate Environments. British Micropalaeontological Society Series, Ellis Horwood, Chichester, 208-225.
Hutton, A.C. 1987. Petrographic classification of oil shales. International Journal of Coal Geology, 8, 203-231.
Hutton, A.C. 1988. The lacustrine Condor oil shale sequence. In: Fleet, A.J. Kelts, K. and Talbot, M.R. (eds) Lacustrine Petroleum Source Rocks. Geological Society of London Special Publication, 40, 329-340.
de Jekhowsky, B. 1963. Repartition quantitative des grand groupes de "microorganontes" (spores, hystrichospheres, etc) dans les sediments marins du plateau continental. Compte Rendu Sommaire Seances Société Biogéographie, Paris, 349, 29-47.
Jones, R.W. 1987. Organic facies. In: Brooks, J. and Welte, D. (eds) Advances in Petroleum Geochemistry, 2, Academic Press, London, 1-90.

Lewan, M.D. 1986. Stable carbon isotopes of amorphous kerogens from Phanerozoic sedimentary rocks. Geochimica et Cosmochimica Acta, 50, 1583-1591.

Liengjarern, M., Coast, L. and Downie, C. 1980. Dinoflagellate cysts from the Upper Eocene-Lower Oligocene of the Isle of Wight. Palaeontology, 23, 475-499.

Lister, J.K. and Batten, D.J. 1988. Stratigraphic and palaeo-environmental distribution of Early Cretaceous dinoflagellate cysts in the Hurlands Farm Borehole, West Sussex, England. Palaeontographica, B210, 9-89.

Loh, H., Prauss, M. and Riegel, W. 1986. Primary production, maceral formation and carbonate species in the Posidonia Shale of NW Germany. Mitteilungen der Geologischen Paläontolologischen Institut dem Universität Hamburg, 60, 397-421.

Lorente, M.A. 1990. Textural characteristics of organic matter in several subenvironments of the Orinoco upper delta. Geologie en Mijnbouw, 69, 263-278.

Lund, J.J. and Pedersen, K.J. 1985. Palynology of the marine Jurassic Formations in the Vardekløft Ravine, Jameson Land, East Greenland. Bulletin of the Geological Society of Denmark, 33, 371-399.

Martinez-Hernandez, E., Almeida-Lenero, L., Reyes-Salas, M. and Betancourt-Aguilar, Y. 1980. Estudio palinologico para la determinacion de ambientes en la Cuenca Fuentes-Rio Escondido (Cretacio Superior), Region de Piedras Negreas, Coahulia. Revista Instituto de Geologia, Universidad Nacional Autonoma de Mexico, 4, 167-185.

Masran, T.C. 1984. Sedimentary organic matter of Jurassic and Cretaceous samples from North Atlantic Deep-Sea drilling sites. Bulletin of Canadian Petroleum Geology, 32, 52-73.

Masran, T.C. and Pocock, S.A.J. 1981. The classification of plant-derived particulate organic matter in sedimentary rocks. In: Brooks, J. (ed) Organic Maturation Studies and Fossil Fuel Exploration, Academic Press, London, 145-176.

McLachlan, I.R. and Pieterse, E. 1978. Preliminary palynological results: Site 361, Leg 40, Deep Sea Drilling Project. In: Bolli, H.M., Ryan, W.B.F. et al., Initial Reports of the Deep Sea Drilling Project, U.S. Government Printing Office, Washington D.C., 40, 857-881.

Melia, M.B. 1984. The distribution and relationship between palynomorphs in aerosols and deep-sea sediments off the coast of northwest Africa. Marine Geology, 58, 345-371.

Morzadec-Kerfourn, M.T. 1977. Les Kystes de dinoflagelles dans les sediments Recents des Cotes Bretonnes. Revue de Micropaleontologie, 20, 157-166.

Mudie, P.J. 1982. Pollen distribution in Recent marine sediments, eastern Canada. Canadian Journal of Earth Sciences, 19, 729-747.

Mukhopadhyay, P.K., Hagemann, H.W. and Gormly, J.R. 1985. Characterization of kerogens as seen under the aspect of maturation and hydrocarbon generation. Erdol und Kohle - Erdgas, 38, 7-18.

Muller, J. 1959. Palynology of Recent Orinoco delta and shelf sediments: reports of the Orinoco Shelf expedition; volume 5. Micropalaeontology, 5, 1-32.

Mutterlose, J. and Harding, I. 1987. Phytoplankton from the anoxic sediments of the Barremian (Lower Cretaceous) of North-West Germany. Abhandlungen Geologischen Bundesanstalt, Wien, 39, 177-215.

Nagy, J., Dypvik, H. and Bjaerke, T. 1984. Sedimentological and palaeontological analyses of Jurassic North Sea deposits from deltaic environments. Journal of Petroleum Geology, 7, 169-188.

Nøhr-Hansen, H. 1989. Visual and chemical kerogen analyses of the Lower Kimmeridge Clay, Westbury, England. In: Batten, D.J. and Keen, M.C. (eds) Northwest European Micropalaeontology and Palynology, British Micropalaeontological Society Series, Ellis Horwood, Chichester, 118-131.

Parry, C.C., Whitley, P.K.J. and Simpson, R.D.H. 1981. Integration of palynological and sedimentological methods in facies analysis of the Brent Formation. In: Illing, L.V. and Hobson, G.D. (eds) Petroleum Geology of the Continental Shelf of North-West Europe, Heyden and Son, London, 205-215.

Patterson III, W.A., Edwards, K.J. and Maguire, D.J. 1987. Microscopic charcoal as a fossil indicator of fire. Quaternary Science Reviews, 6, 3-23.

Piasecki, S. 1986. Palynological analysis of the organic debris in the Lower Cretaceous Jydegard Formation, Bornholm, Denmark. Grana, 25, 119-129.

Pocklington, R. and Leonard, J.D. 1979. Terrigenous organic matter in sediments of the St. Lawrence Estuary and the Saguenay Fjord. Journal of the Fisheries Research Board of Canada, 36, 1250-1255.

Pocock, S.A.J. 1982. Identification and recording of particulate sedimentary organic matter. In: Staplin, D.J., Dow, W.G., Milner, C.W.D., O'Connor, D.I., Pocock, S.A.J., van Gijzel, P., Welte, D.H. and Yükler, M.A., How to Assess Maturation and Paleotemparatures. Society of Economic Paleontologists and Mineralogists, Short Course No. 7, 13-131.
Porter, K.G. and Robbins, E.I. 1981. Zooplankton faecal pellets link fossil fuel and phosphate deposits. Science, 212, 931-933.
Powell, T.G. Creaney, S. and Snowdon, L.R. 1982. Limitations of use of organic petrographic techniques for identification of petroleum source rocks. American Association of Petroleum Geologists Bulletin, 66, 430-435.
Reynolds, C.S. 1984. The Ecology of Freshwater Phytoplankton. Cambridge University Press, 384 pp.
Reyre, Y. 1973. Palynologie du Mésozoique Saharien. Traitment des donnés par l'Informatique et apllications à la Stratigraphie et à la Sédimentologie. Mémoires du Muséum National d'Histoire Naturelle, Paris, Serie C, 27, 284 pp.
Rich, F.J. 1989. A review of the taphonomy of plant remains in lacustrine sediments. Review of Palaeobotany and Palynology, 58, 33-46.
Riley, G.A. 1970. Particulate organic matter in sea water. Advances in Marine Biology, 8, 1-118.
Riley, L.A. 1974. A Palynological Investigation of Upper Jurassic-Basal Cretaceous Sediments from England, France, and Iberia. Unpublished PhD Thesis, The Open University, Milton Keynes, 289 pp.
Robert, P. 1988. Organic Metamorphism and Geothermal History: Microscopic Study of Organic Matter and Thermal Evolution of Sedimentary Basins. D. Reidel Publishing Company, Dordrecht. 311 pp.
Rogers, M.A. 1980. Application of organic facies concepts to hydrocarbon source rock evaluation. In: Proceedings of the 10th World Petroleum Congress, Bucharest 1979, Heyden and Son, London, 2, 23-30.
Rupke, N.A. and Stanley, D.J. 1974. Distinctive properties of turbiditic and hemipelagic mud layers in the Algero-Balearic Basin, western Mediterranean Sea. Smithsonian Contributions to the Earth Sciences, 13, 40 pp.
Sander, P.M. and Gee, C.T. 1990. Fossil charcoal: techniques and applications. Review of Palaeobotany and Palynology, 63, 269-279.
Sarjeant, W.A.S., Lacalli, T. and Gaines, G. 1987. The cysts and skeletal elements of dinoflagellates: speculations on the ecological causes for their morphology and development. Micropalaeontology, 33, 1-36.
Schrank, E. 1984. Organic-walled microfossils and sedimentary facies in the Abu Tartur phosphates (Late Cretaceous Egypt). Berliner Geowissenschaftlichen Abhandlungen, ser. a., 50, 177-187.
Scott, R.W. and Kidson, E.J. 1977. Lower Cretaceous depositional systems, west Texas. In: Bebout, D.G. and Loucks, R.G. (eds) Cretaceous Carbonates of Texas and Mexico: Applications to Subsurface Exploration. Reports of Investigations of the Bureau of Economic Geology, University of Texas at Austin, 169-181.
Scull, B.J. Felix, C.J., McCaleb, S.B. and Shaw, W.G. 1966. The inter-discipline approach to paleoenvironmental interpretations. Transactions of the Gulf Coast Association of Geological Societies, 16, 81-117.
Senftle, J.T., Brown, J.H. and Larter, S.R. 1987. Refinement of organic petrographic methods for kerogen characterisation. International Journal of Coal Geology, 7, 105-117.
Showers, W.J. and Angle, D.G. 1986. Stable isotopic characterization of organic carbon accumulation on the Amazon continental shelf. In: Nittrouer, C.A. and DeMaster, D.J. (eds) Sedimentary Processes on the Amazon Continental Shelf. Continental Shelf Research, 6, 227-244.
Smith, D.M., Griffin, J.J. and Goldberg, E.D. 1973. Elemental carbon in marine sediments: a baseline for burning. Nature, 241, 268-270.
Speelman, J.D. and Hills, L.V. 1980. Megaspore paleoecology: Pakowki, Foremost and Oldman Formations (Upper Cretaceous), southeastern Alberta. Bulletin of Canadian Petroleum Geologists, 28, 522-541.
Stach, E., Mackowsky, M.T., Teichmüller, M., Taylor, G.H., Chandra, D., Teichmüller, R., Murchison, D.G. and Zierke, F. 1982. Stach's Textbook of Coal Petrology. 3rd edition. Gebruder Borntraeger, Berlin, 535 pp.

Stanley, D.J. 1986. Turbidity current transport of organic-rich sediments: alpine and Mediterranean examples. Marine Geology, 70, 85-101.
Stanley, E.A. 1965. Abundance of pollen and spores in marine sediments off the eastern coast of the United States. Southeastern Geologist, 7, 25-33.
Stanley, E.A. 1969. Marine palynology. Oceanography and Marine Biology Annual Review, 7, 277-292.
Staplin, F.L. 1969. Sedimentary organic matter, organic metamorphism, and oil and gas occurrence. Bulletin of Canadian Petroleum Geology, 17, 47-66.
Stein, R., Rullkotter, J. and Welte, D.H. 1986. Accumulation of organic-carbon-rich sediments in the Later Jurassic and Cretaceous Atlantic Ocean - a synthesis. Chemical Geology, 56, 1-32.
Summerhayes, C.P. 1983. Sedimentation of organic matter in upwelling regimes. In: Thiede, J. and Suess, E. (eds) Coastal Upwelling: Its Sediments Record. Part B: Sedimentary Records of Ancient Coastal Upwelling. NATO Conference Series, Series IV: Marine Sciences, 10b, Plenum Press, New York, 29-72.
Summerhayes, C.P. 1987. Organic-rich Cretaceous sediments from the North Atlantic. In: Brooks, J. and Fleet, A.J. (eds) Marine Petroleum Source Rocks. Geological Society Special Publication, 26, 301-316.
Summerhayes, C.P. and Masran, T.H. 1983. Organic facies of Cretaceous and Jurassic sediments from DSDP Site 534 in the Blake Bahama Basin, western North Atlantic. In: Sheridan, R.E., Gradstein, F. et al., Initial Reports of the Deep Sea Drilling Project, U.S. Government Printing Office, Washington D.C., 76, 469-480.
Talbot, M.R. and Livingstone, D.A. 1989. Hydrogen Index and carbon isotopes of lacustrine organic matter as lake level indicators. Palaeogeography, Palaeoclimatology, Palaeoecology, 70, 121-137.
Tappan, H. 1980. The Paleobiology of Plant Protists. W.H. Freeman, San Francisco, 1028 pp.
Ten Haven, H.L., Baas, M., De Leeuw, J.W., Schenck, P.A. and Brinkhuis, H. 1987. Late Quaternary Mediterranean sapropels II. Organic geochemistry and palynology of $S_1$, sapropels and associated sediments. Chemical Geology, 64, 149-167.
Tissot, B. and Pelet, R. 1981. Sources and fate of organic matter in ancient sediments. Oceanologica Acta, Special Issue, Actes 26th Congres International de Géologie, Colloque Géologie des Oceans, Paris 1980, 97-103.
Traverse, A. 1988. Paleopalynology. Unwin Hyman, Boston, 600 pp.
Traverse, A. and Ginsburg, R.N. 1966. Palynology of the surface sediments of Great Bahama Bank, as related to water movement and sedimentation. Marine Geology, 4, 417-459.
Tschudy, R.H. 1969. Relationship of palynomorphs to sedimentation. In: Tschudy, R.H. and Scott, R.A. (eds) Aspects of Palynology. Wiley-Interscience, New York, 79-96.
Tschudy, R.H. and Scott, R.A. (eds) 1969. Aspects of Palynology. Wiley-Interscience, New York, 510 pp.
Tyler, M.A., Coats, D.W. and Anderson, D.M. 1982. Encystment in a dynamic environment: deposition of dinoflagellate cysts by a frontal convergence. Marine Ecology - Progress Series, 7, 163-178.
Tyson, R.V. 1984. Palynofacies investigation of Callovian (Middle Jurassic) Sediments from DSDP Site 534, Blake-Bahama Basin, western Central Atlantic. Marine and Petroleum Geology, 1, 3-13.
Tyson, R.V. 1987. The genesis and palynofacies characteristics of marine petroleum source rocks. In: Brooks, J. and Fleet, A.J. (eds) Marine Petroleum Source Rocks, Geological Society Special Publication, 26, 47-67.
Tyson, R.V. 1989. Late Jurassic palynofacies trends, Piper and Kimmeridge Clay Formations, UK onshore and offshore. In: Batten, D.J. and Keen, M.C. (eds) Northwest European Micropalaeontology and Palynology, British Micropalaeontological Society Series, Ellis Horwood, Chichester, 135-172.
Tyson, R.V. 1990. Automated transmitted light kerogen typing by image analysis: 1. General aspects and program description. In: Fermont, W.J.J. and Weegink, J.W. (eds) Proceedings of the International Symposium on Organic Petrology, Zeist, January 1990. Mededelingen Rijks Geologische Dienst, 45, 139-150.
Tyson, R.V. (in press). Sedimentary Organic Matter: Organic Facies and Palynofacies Analysis. Chapman and Hall, London.

Van Der Zwaan, C.J. 1990. Palynostratigraphy and palynofacies reconstruction of the Upper Jurassic to lowermost Cretaceous of the Draugen Field, offshore mid Norway. Review of Palaeobotany and Palynology, 62, 157-186.

Vozzhennikov, T.F. 1965. Vvedenie v izuchenie iskopaemykh Peridineevykh vodoroslei. Akademiya Nauk SSSR, Sibirskae Otedlenie Instituta Geologii i Geofiziki, Izdatel'stvo "Nauka", Moscow, 156 pp. (Introduction to the Study of Fossil Peridinian Algae. Syers, K. (transl.) and Sarjeant, W.A.S. (ed), National Lending Library for Science and Technology, Boston Spa, U.K. (now British Library Document Supply Centre, Boston Spa), 231 pp.)

Wake, L.V. and Hillen, L.W. 1980. Study of a "bloom" of the oil-rich alga *Botryococcus braunii* in the Darwin River reservoir. Biotechnology and Bioengineering, 22, 1637-1656.

Wall, D. 1965. Microplankton, pollen and spores from the Lower Jurassic in Britain. Micropalaeontology, 11, 151-190.

Wall, D., Dale, B., Lohmann, G.P. and Smith, W.K. 1977. The environment and climatic distribution of dinoflagellate cysts in modern marine sediments from regions in the North and South Atlantic Oceans and adjacent seas. Marine Micropalaeontology, 2, 121-200.

Waples, D.W. 1985. Geochemistry in Petroleum Exploration. International Human Resources Development Corporation, Boston. 232 pp.

Weidmann, M. 1967. Petite contribution a la connaisance du flysch. Bulletin des Laboratoires de Geologie, Mineralogie, Geophysique et du Musée géologique de l'Université de Lausanne, 166, 6 pp.

Whitaker, M.F. 1984. The usage of palynology in definition of Troll Field geology. 6th Offshore Northern Seas Conference and Exhibition, Stavanger 1984, Paper G6, 44 pp.

Williams, D.B. and Sarjeant, W.A.S. 1967. Organic-walled microfossils as depth and shoreline indicators. Marine Geology, 5, 389-412.

Williams, L.A. 1984. Subtidal stromatolites in Monterey Formation and other organic-rich rocks as suggested source contributors to petroleum formation. American Association of Petroleum Geologists Bulletin, 68, 1879-1893.

Wollast, R. 1983. Interactions in estuaries and coastal waters. In: Bolin, B. and Cook, R.B. (eds) The Major Biogeochemical Cycles and Their Interactions. Wiley, New York, 385-407.

Woods, R.D. 1955. Spores and pollen - a new stratigraphic tool for the oil industry. Micropaleontology, 1, 368-375.

Richard V. Tyson
Newcastle Research Group (NRG)
Fossil Fuels and Environmental Geochemistry (Postgraduate Institute)
Drummond Building
University of Newcastle
Newcastle upon Tyne
NE1 7RU
United Kingdom

# 6 SEQUENCE STRATIGRAPHY OF BARROW GROUP (BERRIASIAN-VALANGINIAN) SILICICLASTICS, NORTH-WEST SHELF, AUSTRALIA, WITH EMPHASIS ON THE SEDIMENTOLOGICAL AND PALAEONTOLOGICAL CHARACTERIZATION OF SYSTEMS TRACTS

R.W. Jones, P.A. Ventris, A.A.H. Wonders, S. Lowe, H.M. Rutherford, M.D. Simmons, T.D. Varney, J. Athersuch, S.J. Sturrock, R. Boyd and W. Brenner.

**Abstract**

Five Barrow Group (Berriasian to Valanginian) siliciclastic sequences are described from the North-West Shelf, Australia, and calibrated against global third-order (?eustatically-mediated) cycles. Particular emphasis is placed on the sedimentological (core, wireline log) and palaeontological (micropalaeontological, palynological) characterization of constituent systems tracts.

## 6.1 Introduction

### 6.1.1 Present Study

This chapter discusses the sequence stratigraphy of Barrow Group (Berriasian to Valanginian) siliciclastic sequences on the North-West Shelf of Australia. It places particular emphasis on the sedimentological and palaeontological characterization of systems tracts. This is important in that it helps in the constraint of the sequence stratigraphic model. Integration of geological, geophysical, palaeontological and sedimentological data thus enables the prediction of the spatio-temporal distribution of lithofacies (including potential petroleum reservoir lithofacies).

Published sequence stratigraphic, sedimentological (core and wireline log) and palaeontological (micropalaeontological and palynological) data were available from two continuously cored Ocean Drilling Program (ODP) Sites (762 and 763) on the Exmouth Plateau (Boyd et al., 1992; Brenner, 1992; Haq et al., 1992; Jones & Wonders, 1992; Figures 6.1-6.3; Tables 6.1-6.2).

For the purposes of comparison, unpublished stratigraphic, sedimentological (wireline log) and limited palaeontological (palynological) data were available from coeval sequences from a number of petroleum exploration wells in the Barrow Basin (Northern Carnarvon Basin of some authors) (including Barrow, Dampier and Exmouth Sub-Basins) to the east (Figure 6.2) and in the Bonaparte and Papuan Basins to the extreme east. Unfortunately, for reasons of commercial sensitivity, the names and precise locations of these wells cannot be disclosed.

No relevant nannopalaeontological data are available. Barrow Group sections from Leg 122 ODP Sites proved barren of nannofossils (Bralower & Siesser, 1992).



*D. G. Jenkins (ed.), Applied Micropaleontology,* 193–223.
© 1993 *Kluwer Academic Publishers.*

Figure 6.1: Location Map.

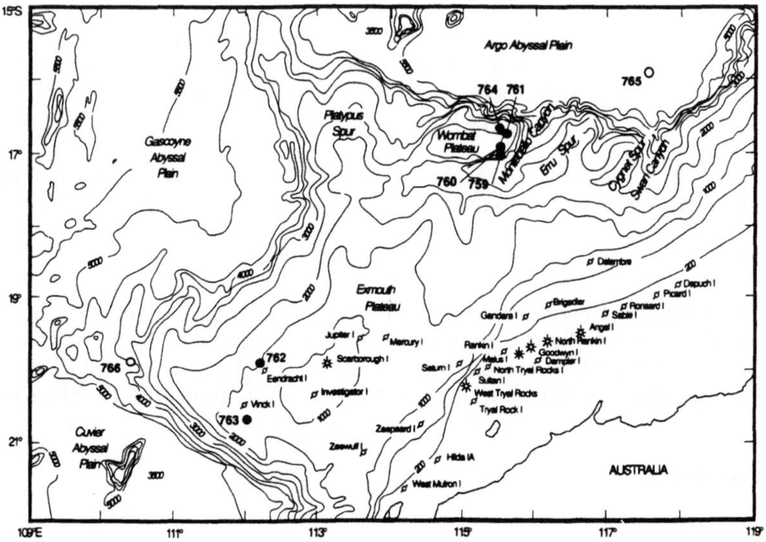


Figure 6.2: Sketch Map of Geological Setting of Barrow Delta Complex. Sediment transport directions shown by arrows. Final progradational front (separating sand-prone proximal delta to south from shale-prone distal delta to north) shown by toothed line. Data from Tait (1985), Boote & Kirk (1989), Boyd et al. (1992), Exon, Borella & Ito (1992), von Rad et al. (1992) and unpublished sources.

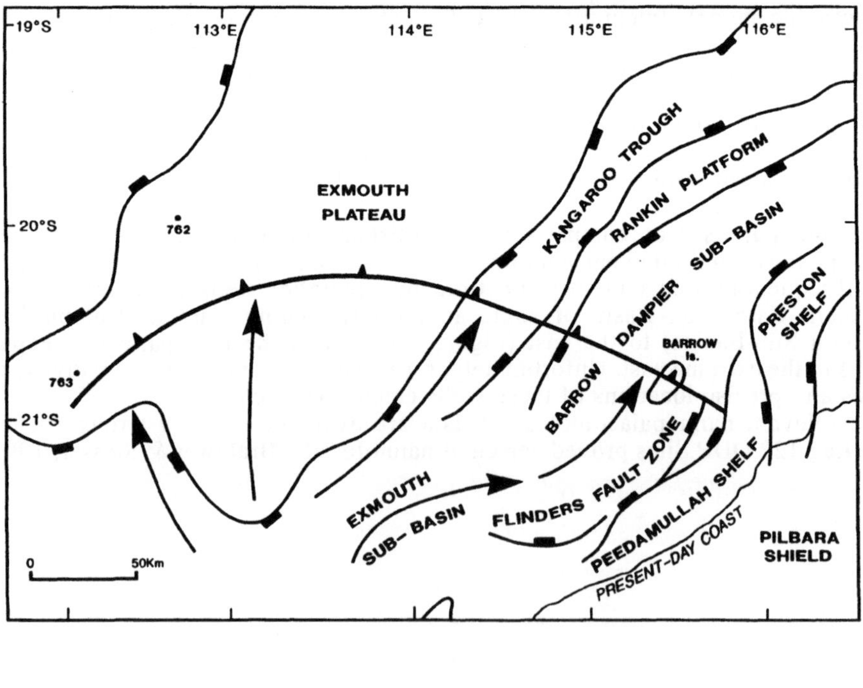

Table 6.1: Semi-quantitative distribution of microfossils in the Barrow Group, Site 763 (after Jones & Wonders, 1992). p - present (1-4 specimens per 10cm³ sample); c - common (>5 specimens per 10cm³ sample).

| Core No. / Microfossils | R | | | | | | | | | | | | | | | | | | | | | | | | | | | | | | | | | | | | | | | | | | | X | | | | | | | | |
|---|---|---|---|---|---|---|---|---|---|---|---|---|---|---|---|---|---|---|---|---|---|---|---|---|---|---|---|---|---|---|---|---|---|---|---|---|---|---|---|---|---|---|---|---|---|---|---|---|---|---|---|---|
| | | 45 R | | | | | 40 R | | | | | 35 R | | | | | 30 R | | | | | 25 R | | | | | 20 R | | | | | 15 R | | | | | 10 R | | | | | 5 R | 4 R | 54 X | | 52 X | | 50 X | | 48 X | | 46 X |
| Saccamminidae Indet. | | | | | | | | | | | | | | | p | | | | | | | | | | | | | | | | | | | | | | | | | | | | | | | | | | | | | |
| *Bathysiphon* sp. | p | | | | p | | c | | p | | p | | | p | | p | | | | | | | | | | | | | | | | | | | | | | | | | | | | | | | | | | | | |
| *Hyperammina* sp. | | | | | | | | | | p | | | | | | | | | | | | | | | | | | | | | | | | | | | | | | | | | | | | | | | | | | |
| *Ammodiscus* sp. | | | | | | | | | | | p | p | | | | | | | | | | | | | | | | | | | | | | | | | | | | | | | | | | | | | | | | |
| *G. gaultina* | | | | | | | | | | | | | | | | | | | | | | | | | | p | | | | | | | | | | p | | | | | | | | | | | | | | | | |
| *Glomospira* sp. | | | | | | | | | | | | | | | | | | | | | | | | | | p | | | | | | | | | | | | | | | | | | | | | | | | | | |
| *A. reophacoides* | c | c | p | p | | | | | p | p | | | p | p | | | p | | | | | | p | c | | | | | | | | | | | | | | | | | | | | | | | | | | | | |
| *Ammomarginulina* sp. | | | | | | | | | | | | | | | | | | | | | | | | | | | | | | | | | | | | | | | | | p | | | | | | | | | | | |
| *H. inconstans* | | | | | | | | | | | | p | | | | | p | | | | | | | | | | | | | | | | | | | | | | | | | | | | | | | | | | | |
| *H. cushmani* | | | | | | | | | | | | | | | | | | | | | | | | | p | p | | | | | | | | | | | | | p | | | | | | | | | | | | | |
| *H.* cf. *infracalloviensis* | p | p | p | | p | p | p | c | c | p | p | | p | p | | p | p | p | | | | | c | | | p | p | p | | p | p | p | | | | p | | | p | | | p | p | | | | | | | | | |
| *Haplophragmoides* sp. | | | | | | | | | | | | | | | | | | | | | | | | | | | | | | | | | | | | | | | p | | | | | | | | | p | p | | | p |
| *Hormosina* sp. | | | | | | | | | | | | | p | | | | | | | | | | | | | | | | | | | | | | | | | | | | | | | | | | | | | | | |
| *Praecystammina* ? sp. | | | | | | | | | | | | | p | | | | | | | | | | p | | | | | | | | | | | | | | | | | | | | | | | | | | | | | |
| *T. bettenstaedti* | | | | p | | | | | | | | p | p | p | | | | | | | | | | | | | | | | | | | | | | | | | | | | | | | | | | | | | | |
| *Textularia* sp. 1 | | | | p | | | | | | | | | | p | | | | | | | | | | | | | | | | | | | | | | | | | | | | | | | | | | | | | | |
| *Textularia* ? sp. | | | | | | | | | | | | | | | | | | | | | | | | | | | | | | | p | | | | | | | | | | | | | | | | | | | | | |
| *Spiroplectammina* ? sp. | | | | | | | | | | | | | | | p | | | | | | | | | | | | | | | | | | | | | | | | | | | | | | | | | | | | | |
| *Trochammina* sp. | | | | | | | | | | | | | | | | | | | | | | | | | | | | | | | | | | | | | | | | p | p | | | | | | | | | | | |
| *V. neocomiensis* | p | p | | | | | | | | | | | p | | p | | | p | | | | | | | | | | | | | | | | | | | | | | p | p | p | | | | | | | | | | |
| Lituolacea Indet. | p | p | p | p | p | p | | p | p | p | p | | p | p | p | p | | p | | | | p | p | | | | p | p | | p | p | p | p | | | p | | | p | p | p | p | | | | | | | | | | p |
| *Astacolus* sp. | | | | | | | | | p | | | | | | | | | | | | | | | | | | | | | | | | | | | | | | | | | | | | | | | | | | | |
| *Dentalina* sp. 1 | | | | | | | | | | | | | | | | | | | | | | | | | | | p | | | | | | | | | | | | | | | | | | | | | | | | | |
| *Dentalina* sp. 2 | | | | | | | | | | | | | | | | | | | | | | | | | | | | | | p | | | | | | | | | | | | | | | | | | | | | | |
| *F.* cf. *hastata* | | | | | | | | | | | | | | | | | | | | | | | | | | p | | | | | | | | | | | | | | | | | | | | | | | | | | |
| *Frondicularia* sp. | | | | | | | | | | | | | | | p | | | | | | | | | | | | | | | | | | | | | | | | | | | | | | | | | | | | | |
| *L. neocomiana* | | | | | | | | | | | | | | p | | | | | | | | | | | | | | | | | | | | | | | | | | | | | | | | | | | | | | |
| *Lagena* sp. 1 | | | | | | | | | | | | | | | | | | | | | | | | | | p | | | | | | | | | | | | | | | | | | | | | | | | | | |
| *Lagena* sp. 2 | p | | | | | | | | | | | | | | | | | | | | | | | | | | | p | | | | | | | | | | | | | | | | | | | | | | | | |
| *Lagena* sp. 3 | | | | | | | | | | | | | | | | | | | | | | | | | | | | | | | p | p | | | | | | | | | | | | | | | | | | | | |
| *L.* ex gr. *muensteri* | | c | | c | | | p | | p | | p | p | p | p | p | c | p | p | | | | p | p | p | | c | p | c | | p | p | p | | | | p | | | p | | p | p | p | | | | | p | | | p | |
| *M.* cf. *robusta* | | | | | | | | | | | | | | | | | | | | | | | | | | | p | | | | | | | | | | | | | | | | | | | | | | | | | |
| cf. *N. sceptrum* | | | | | | | | | | | | | | | | | p | | | | | | | | | | | | | | | | | | | | | | | | | | | | | | | | | | | |
| *P.* cf. *humilis* | | | | | | | | | | | | | | | | | p | | | | | | | | | | | | | | | | | | | | | | p | | | | | | | | | | | | | |
| *S. valanginiana* | | | | | | | | | | | | | | | | | | | | | | | | | | | | | | | | | | | | p | | | | | | | | | | | | | | | | |
| *Saracenaria* sp. | | | | | | | | | | | | | | | | | | | | | | | | | | | | | | | p | | | | | | | | p | | p | | | | | | | | | | | |
| *Vaginulina* sp. | | | | | | | | | | | | | | | | | | | | | | | | | p | | | | | | | | | | | | | | | | | | | | | | | | | | | |
| Nodosariidae indet. | | p | | p | p | | | | p | p | | | | p | | c | | p | | | | p | p | | | p | p | | | p | p | p | | | | | | | | p | | p | | | | | | | | | | |
| *Lingulina* sp. | | | | | | | | | | | | | | | | p | | | | | | | | | | | | | | | | | | | | | | | | | | | | | | | | | | | | |
| Polymorphinidae indet. | | | | p | | p | p | p | p | p | | | | | | p | | | | | | p | | | | p | | p | | | p | | | | | | | | | p | | | | | | | | | | | | |
| *C. valendisensis* | | | | | | | | | | | | | | | | | | | | | | | | | | | | | | | p | | p | | | | | | p | | | | | | | | | | | | | |
| *S. minima* | | | | | | | | | p | | | | | | | | | | | | | | | | | | | | | | | | | | | | | | | | | | | | | | | | | | | |
| Rotaliacea indet. | | | | | | | | | | | | | p | | | | | | | | | | | | | | | | | | | | | | | | | | | | | | | | | | | | | | | |
| *Eucyrtis* ? sp. | | | | | | | | | | | | | | | | | | | | | | | | | | | | | | | | | | | | | | | | | | p | | | | | | | | | | |
| Ostracoda indet. | p | | | p | | p | p | | | | | | p | | | p | | p | | | | | | | | p | | p | | | | p | | | | | | | | p | p | p | | | | | | p | | | | |
| Macrofossil debris | | | | | | | | | | | | | | | | | | | | | | | | | | | | | | | | | | | | | | | | | | | | | | | p | | | | p | |

Table 6.2: Semi-quantitative distribution of microfossils in the Barrow Group, Site 762 (after Jones & Wonders, 1992). p - present (1-4 specimens per 10cm³ sample).

| 91X | 90X | 89X | 88X | 87X | 86X | 85X | 84X | 83X | 82X | 81X | Core No. / Microfossils |
|---|---|---|---|---|---|---|---|---|---|---|---|
| | | p | p | | | | | | p | | *Bathysiphon* sp. |
| | | p | | | | | | | | | Ammodiscidae Indet. |
| | | | | | | | | | p | | *Haplophragmoides* sp. |
| p | | | | | | | | | | | *Textularia* sp. 1 |
| p | | | | | | | | | | | *P. kummi* |
| p | | | | | | | | | | | Lituolacea Indet. |
| p | | p | | p | | | | | | | *L.* ex gr. *muensteri* |
| | | | | p | | | | | | | *S. valanginiana* |
| p | | | | | | | | | | | Nodosariidae Indet |
| | | | | p | | | | | | | *C. valendisensis* |
| p | | | | | | | | | | | *E. caracolla* |
| | p | | p | | | | | | p | | Rotaliacea Indet. |
| | | p | p | | | | | | | | *Cenosphaera* sp. |
| | p | | p | | | | | | | | Macrofossil debris |
| p | p | | | | | | | | | | Ichthyoliths |

Figure 6.3: Interpreted seismic section through Sites 762 and 763. Nos. 1-8 are the seismostratigraphic units of Boyd et al. (1992). Their No. 3 (Berriasian-Valanginian) represents the Barrow Group. Five sequences can be recognized within this unit (see text).

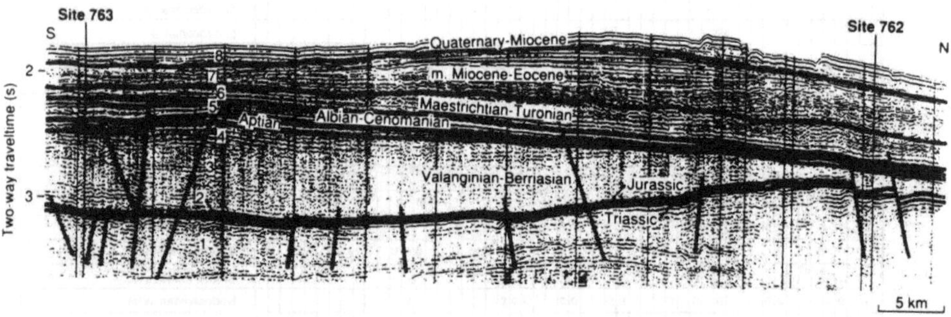

### 6.1.2 Sequence stratigraphy and systems tracts: definitions and discussion

Posamentier (in press) provides a useful overview of sequence stratigraphic concepts and their uses and abuses.

*Sequences*

A sequence is a fundamental building block in stratigraphy. It is defined as "a stratigraphic unit composed of a relatively conformable succession of genetically related strata bounded at its top and base by unconformities or their correlative conformities" (Mitchum et al., 1977). Sequence boundaries (bounding unconformities) are readily recognizable on seismic stratigraphic criteria (erosional truncation, downward shift in facies etc.).

The spatio-temporal distribution of a sequence is dictated by accommodation space, which is defined as "the space ... available for potential sediment accumulation" (Jervey, 1988). Accommodation space is in turn dictated by the complex interplay between three dynamic variables, namely eustasy, tectonism and sediment compaction. Sediment supply fills the available accommodation space. If the rate of sediment supply is greater than that at which accommodation space is created at a given point, water depth will decrease and facies belts will migrate toward the basin centre. This process is termed progradation. Conversely, if the rate of sediment supply is less than that at which accommodation space is created, water depth will increase and facies belts will migrate toward the basin margin. This process is termed retrogradation. If the rates of sediment supply and creation of accommodation space are in equilibrium, there will be no effect on water depth and facies belts will remain static (i.e. sediment will build vertically). This process is termed aggradation.

Haq et al. (1987a) have produced a conceptual two-dimensional model of the development of siliciclastic sequences in relation to eustatic changes in sea-level (the "Exxon Model"). The model can (with the effects of tectonism and sediment supply notwithstanding) be used to predict the spatial distribution of lithofacies (Figure 6.4). Haq et al. (1987a) (and Vail et al., 1977) have generated curves showing variations in eustatic sea-level through geological time, which enhance the predictive utility of the model. Haq (1991) has integrated palaeoclimatic trends with the Cenozoic sea-level curve.

Models of carbonate, carbonate-clastic and carbonate-evaporite sequences are given by Sarg (1988), Jacquin et al. (1991) and Tucker (1991) respectively.

*Genetic stratigraphic sequences*

A genetic stratigraphic sequence is defined as "a package of sediments recording a significant episode of basin-margin outbuilding and basin-filling" (Galloway, 1989). Genetic stratigraphic sequence boundaries (maximum flooding surfaces, see below) are readily recognizable not only on seismic stratigraphic criteria (downlap), but also on sedimentological criteria (lithological breaks in cores or on wireline logs). In terms of sea-level cycles, genetic stratigraphic sequences are 180 degrees out of phase with "Exxon" sequences (see above).

*Systems tracts*

Sequences can be internally subdivided into so-called systems tracts. A systems tract is defined as "a linkage of contemporaneous depositional systems" (e.g., fluvial-deltaic-coastal-shelfal-bathyal) (Brown & Fisher, 1977). The chief types are the low-stand systems tract (LST), the Shelf-Margin Systems Tract or Shelf-Margin Wedge (SMW), the transgressive systems tract (TST), and the high-stand systems tract (HST) (Figure 6.4).

The LST is defined as "the lowermost systems tract in a depositional sequence ... if it lies directly on a type 1 sequence boundary", the SMW as "the lowermost systems tract associated with a type 2 sequence boundary", the TST as "the middle systems tract of both type 1 and type 2 sequences" and the HST as "the upper systems tract in either a type 1 or a type 2 sequence" (van Wagoner et al., 1988). A type 1 sequence boundary is generated by a relative fall in sea-level to a position outboard of the preceding offlap break, and a type 2 sequence boundary by a relative fall in sea-level to a position inboard of the preceding offlap break (Vail et al., 1984).

The LST can be subdivided into a lower Low-Stand Fan (LSF) and an upper Low-Stand Wedge (LSW) (also known as a Prograding Low-Stand Wedge or Complex (LPC)). The LST (or SMW) and TST are separated by the Transgressive Surface (TS), which is defined as "the first significant marine flooding surface ... within a sequence (van Wagoner et al., 1988). The TST and HST are separated by the Maximum Flooding Surface (MFS), which is defined as "the surface corresponding to the time of maximum flooding" (van Wagoner et al., op. cit.).

Condensed sections commonly occur at the TS and HFS. Condensed sections are defined as "thin marine stratigraphic units consisting of pelagic to hemipelagic sediments characterized by very low sedimentation rates", which are "most areally extensive at the time of maximum regional transgression ... " (Loutit et al., 1988). They commonly contain "abundant and diverse planktonic and benthonic microfossil assemblages", which enable correlation with the open oceans and calibration against the global standard biostratigraphic zonation schemes established there (Loutit et al., op. cit.). As such, they are of great importance in providing the chronostratigraphic framework for sequence analysis, and in delineating systems tracts (see also below).

### 6.1.3 Previous work on the Early Cretaceous stratigraphy of the North-West Shelf

Previous publications on the stratigraphy and/or petroleum geology of the North-West Shelf (including those on the Barrow Group and associated rock units) include those of Symonds & Cameron (1977), Wiseman (1979), Exon, von Rad & von Stackelberg (1982), Powell (1982), Tait (1985), Erskine & Vail (1988), Hocking et al. (1988), Woodside Offshore Petroleum (1988), Boote & Kirk (1989), BMR Palaeogeographic Group (1990), Struckmeyer et al. (1990), Posamentier & Erskine (1991), Boyd & Bent (1992), Boyd et al. (1992), Exon, Borella & Ito (1992), Exon, Haq & von Rad (1992), Haq et al. (1992), von Rad et al. (1992) and Boyd et al. (in press).

Many of the aforementioned authors discussed the sequential development of the Barrow Delta Complex in the context of the evolution of the passive margin of the Exmouth Plateau. The consensus is that "Gondwanan" rifting phases took place in the Carboniferous to Permian, and Late Triassic, prior to a major "break-up" phase within the Late Jurassic to Early Cretaceous (see also Gorur & Sengor, 1992).

Figure 6.4: Siliciclastic Sequence Model (after Haq et al., 1987a). Applies to Type I Sequence (see text).

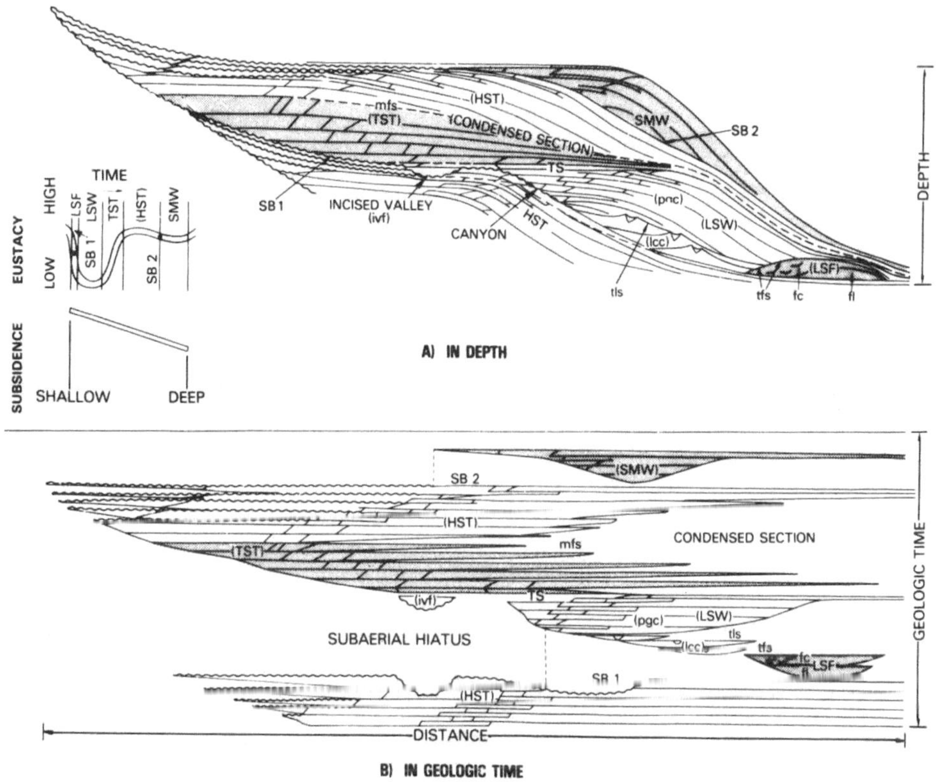


**LEGEND**

SURFACES

(SB) SEQUENCE BOUNDARIES
- (SB 1) = TYPE 1
- (SB 2) = TYPE 2

(DLS) DOWNLAP SURFACES
- (mfs) = maximum flooding surface
- (tfs) = top fan surface
- (tls) = top leveed channel surface

(TS) TRANSGRESSIVE SURFACE
- (First flooding surface above maximum regression)

Z FACIES BOUNDARIES
(Proximal at left; Distal at right)

SYSTEMS TRACTS

HST = HIGHSTAND SYSTEMS TRACT
TST = TRANSGRESSIVE SYSTEMS TRACT
LSW = LOWSTAND WEDGE SYSTEMS TRACT
- ivf = incised valley fill
- pgc = prograding complex
- lcc = leveed channel complex

LSF = LOWSTAND FAN SYSTEMS TRACT
- fc = fan channels
- fl = fan lobes

SMW = SHELF MARGIN WEDGE SYSTEMS TRACT

Barrow Group deltaic and associated sediments were laid down during the Earliest Cretaceous (Berriasian to Valanginian). The delta complex was fed by (chiefly sandy) sediment derived from uplifted areas to the south (possibly in the Gascoyne Sub-Basin), and was channelled to the north. It prograded into the rapidly subsiding Barrow Sub-Basin and onto the southern part of the Exmouth Plateau, widening into a thick (locally up to 2km thick), broadly fan-shaped package of sediment effectively confined only by the rift-margin (Flinders Fault Zone) to the east, and to some extent by the Rankin Platform and the Exmouth Plateau to the north (Figure 6.2). At its maximum extent, it formed an east-west trending arcuate front crossing the Exmouth Plateau only some 7.5km to the south-east of the location of Site 763 and some 50km south-east of Site 762. In fact, Sites 762 and 763 were deliberately located on the palaeoprodelta for safety reasons, so as to avoid any risk of penetrating potentially gas-bearing sandy sediments on the palaeodelta (note that the *Sedco Explorer* is not equipped with a blow-out preventer).

Up to five progradational sequences have been recognized within the Barrow Delta Complex (see also "Sequence Stratigraphy of the Barrow Group" Section below). Some contain LSFs derived from erosion of earlier sequences exposed at sequence boundaries, though only LSWs (LPCs) were penetrated at the ODP Sites. The LSFs may be mounded in both strike and dip orientation, and their tops may be marked by laterally continuous high-amplitude reflectors. These features indicate that they are sand-prone, and constitute attractive petroleum exploration targets (cf. Weimer & Link, 1991).

### 6.1.4 Previous work on the sedimentological characteristics of systems tracts

There is a plethora of previous publications on the sedimentological (wireline log) characteristics of systems tracts. Some of the most recent include those of van Wagoner et al. (1990), Armentrout (1991a,b), Armentrout et al. (1991), Mitchum et al. (1991), Pacht et al. (1991, and in press), Sangree et al. (1991), Vail & Wornardt (1991) and Gregory & Hart (1992) (and references therein).

In their work on the sedimentological and palaeontological characteristics of the systems tracts of Palaeogene sequences in the Gulf Coast, Gregory & Hart (1992) noted progradational wireline log stacking patterns during LSTs, and generally retrogradational stacking patterns during both TSTs and HSTs. However, in areas associated with deltaic sedimentation, they also noted progradational stacking patterns during both TSTs and HSTs.

Characteristic wireline log expressions of systems tracts within the "Exxon Model" (see above) are shown on Figure 6.5.

### 6.1.5 Previous work on the palaeontological characteristics of systems tracts

Previous publications on the palaeontological characteristics of systems tracts in siliciclastic sequences include those of Stewart (1987), van Gorsel (1988), Poumot (1989), McNeil et al. (1990), Rosen & Hill (1990), Shaffer (1990), Armentrout (1991b), Armentrout et al. (1991), Hill & Rosen (1991), Vail & Wornardt (1991), Gregory & Hart (1992), and Valicenti et al. (in prep.) (abstract, Valicenti et al., 1991) (see also Loutit et al., 1988)). The only previous publication known to the authors on the palaeontological characteristics of systems tracts in carbonate sequences is that of Cubaynes et al. (1991) (but see also Rioult et al. (1991)).

Figure 6.5: Characteristic wireline log expressions of systems tracts within the "Exxon Model" (Haq et al., 1987a). Hypothetical well in approximate location of ODP Sites 762 and 763. Legend as per Figure 6.4. MFS may be represented by a gamma-log peak at proximal sites and a gamma-log trough at distal sites (see text).

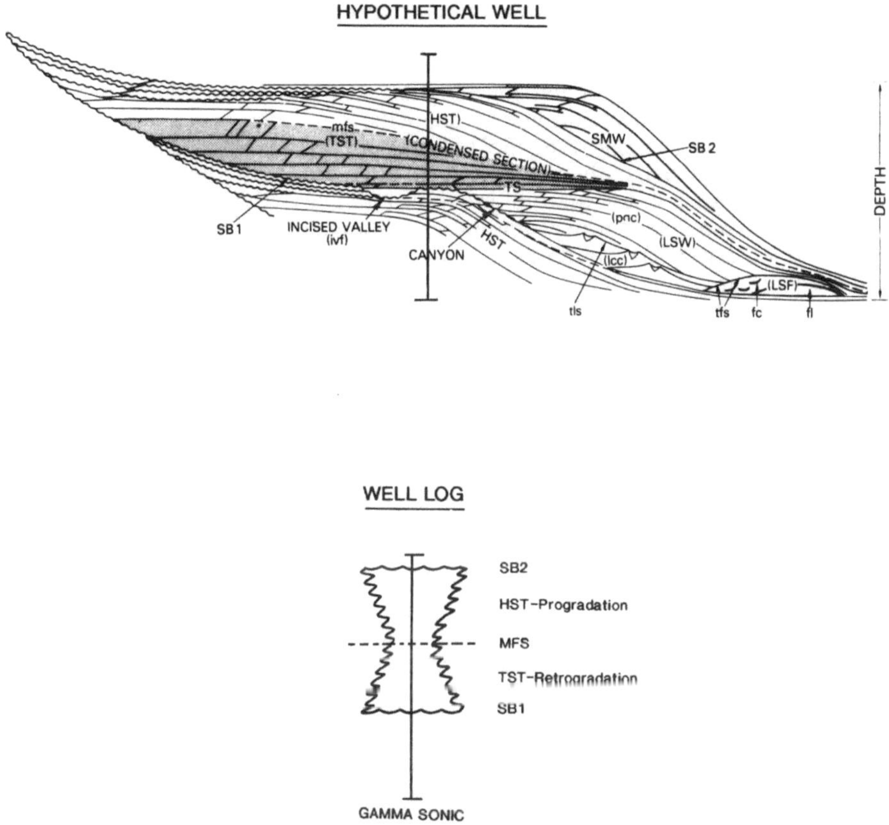


In one of the earliest studies of its kind, Stewart (1987) described the integrated bio- and sequence- stratigraphy of the Central North Sea. He noted that the microfossil assemblage associated with the Late Palaeocene Forties Low-Stand Fan (his Sequence 7) is impoverished, and consists predominantly of agglutinating (benthonic) foraminifera (and rare planktonic diatoms). Agglutinating taxa present include *Ammodiscus, Glomospira, Haplophragmoides,* hormosinids, *Hyperammina, Lagenammina, Recurvoides, Reticulophragmium, Rhabdammina, Saccammina* and *Usbekistania* (RWJ's unpublished observations; see also Charnock & Jones (1990)). Significantly, all of these genera are among the "morphogroups" interpreted by Jones & Charnock (1985) as epifaunal (surface- rather than sediment- dwelling). As such, they might be expected to be among the pioneer colonizers of submarine fans during interturbiditic intervals.

In a wide-ranging paper dealing with the Neogene (essentially benthonic foraminiferal) biostratigraphy of Indonesia, van Gorsel (1988) noted that:
"Although there are no microfossils that are unique to certain systems tracts, there are some indirect criteria that help identify position within a sequence:
-common reworked shallow marine fauna in deep marine beds are suggestive of updip erosion and turbidite transport, and thus point to lowstand fan deposits;
-abundant large arenaceous forams ("flysch faunas") are thought to reflect relatively rapid mud deposition in a deep marine facies, which most likely occur in the lowstand fan or distal lowstand wedge deposits;
-in the transgressive systems tract, most clastics are trapped in aggradational alluvial and coastal plain deposits. Shorelines receive comparatively little sediment, so these tend to be dominated by marine processes (waves, tides). Water turbidity is relatively low, and "clear water" marine microfaunas like *Cellanthus, Operculina, Amphistegina*, larger foraminifera and sea-grass species like planoconvex *Cibicides* and *Discorbis* are most frequent here. Also glauconite-rich and calcareous beds, including reefal buildups, are most frequent in transgressive deposits;
-the highstand systems tract is characterized by relatively rapidly prograding shorelines. Influence of rivers and deltas is relatively high, and "turbid water" marine foraminifera tend to dominate here (*Ammonia, Pseudorotalia* in the shallow, *Uvigerina* and *Bolivina* in the deeper parts)."

In another wide-ranging paper dealing this time with the Neogene palynostratigraphy of the tropical realm, Poumot (1989) documented variations in palynomorph populations through idealized depositional sequences. Of particular note are his observations of strong fluvial activity (erosion) during sea-level falls at sequence boundaries, widespread development of spores during LSTs, of Palmae (palms) during TSTs, of mangroves at MFSs, and of Rubiaceae and Euphorbiaceae (open forest and swamp pollen groups) during early HSTs, and Gramineae or *Casuarina* (littoral pollen groups) during late HSTs.

In their paper on the Cenozoic stratigraphy of the Beaufort-Mackenzie Basin, Arctic Canada, McNeil et al. (1990) documented variations in benthonic foraminiferal populations in sequences in wells along a transect from the palaeo-Mackenzie Delta (shelf) to the palaeo-prodelta (slope). In the well closest to the initial shelf-edge, they noted a transition within the Oligocene to Middle Miocene (?second-order) Kugmallit-Mackenzie Bay Sequence from a shelf-slope microfauna in the essentially retrogradational foreset facies of the ?LPC and TST to a shelf microfauna in the progradational topset facies of the HST. Their ?LPC and TST microfaunas are characterized by the calcareous benthonic genera *Gyroidina, Nuttallides*, *Rotaliatina* and *Turrilina* and the agglutinating genera *Bathysiphon* and *Trochammina*. Their HST microfaunas are characterized by the calcareous benthonic genera *Asterigerina, Cibicidoides, Cyclogyra, Elphidium, Elphidiella, Ehrenbergina, Globobulimina, Globocassidulina, Melonis, Miliolinella, Pullenia, Quinqueloculina, Sphaeroidina* and *Valvulineria* and the agglutinating genera *Ammodiscus, Ammolagena, Bathysiphon, Cystammina, Gravellina, Haplophragmoides, Recurvoides, Reticulophragmium, Rhabdammina, Spirosigmoilinella* and *Trochammina*.

In their papers on the Neogene and Pleistocene sequence stratigraphy of the Gulf of Mexico, Rosen & Hill (1990), Shaffer (1990), Armentrout (1991b), Armentrout et al. (1991), Hill & Rosen (1991) and Vail & Wornardt (1991) noted calcareous nannoplankton and/or foraminiferal abundance and diversity maxima in condensed sections and used them to constrain systems tract and/or sequence boundaries.
In their paper on the systems tracts of Palaeogene sequences in the Gulf Coast, Gregory & Hart (1992) noted "terrigenous domination" of palynofloras and "above

average numbers of recycled palynomorphs" during LSTs, and generally "marine dominated" palynofloras during both TSTs and HSTs. However, in areas associated with deltaic sedimentation, they also noted "terrigenous dominated" and locally "low diversity" palynofloras during TSTs, and "marine dominated" and locally "high marine diversity" palynofloras (retrogradational) to "terrigenous dominated" palynofloras (progradational) during HSTs.

Most recently, in their work on the sedimentological and palaeontological characteristics of the systems tracts of an Early Cretaceous (Albian) sequence in the Bredasdorp Basin, South Africa, Valicenti et al. (in prep.) (abstract, Valicenti et al., 1991) noted palynomorph abundance maxima and diversity minima in the condensed sections atop LSFs (two to three species) and within TSTs (often only one species), and abundance and diversity maxima in the condensed sections at MFSs (up to forty species). Valicenti (pers. comm., 1991) attributes the abundance maximum within the TST to "blooming" of opportunistic species in newly-created niches (with nutrient availability enhanced by increased fresh-water run-off), and the diversity maximum at the MFS to increased environmental stabilization and intraspecific competition.

## 6.2 Sequence stratigraphy of the Barrow Group

### 6.2.1 Introduction

Five third-order sequences (BD1-BD5) from the Barrow Group of the Exmouth Plateau are defined and discussed below. The discussion concerns age, gross depositional environment, systems tracts, sedimentological characteristics of systems tracts, and palaeontological characteristics of systems tracts. Comments on coeval sequences in the Barrow Basin to the east and in the Bonaparte and Papuan Basins to the extreme east are added as appropriate.

The sequences correspond to the five unnamed sequences described by Haq et al. (1992) from the Barrow Group of the Exmouth Plateau. Tectonic complications notwithstanding, Haq et al. (op. cit.) used the palynostratigraphic datings of their sequences (Brenner, 1992) to infer calibrations against the global (?eustatically-mediated) third-order cycles of Haq et al. (1987a), the biostratigraphic calibration of which is discussed by Haq et al. (1987b, 1988). They calibrated their equivalents of Sequences BD1-BD3 against Third-Order Cycles 1.4-1.6 respectively (from Second-Order Supercycle LBZ-1), and their equivalents of BD4-BD5 against 2.1 and 2.2 respectively (from Second-Order Supercycle LBZ-2) (see Figure 6.6; see also Exon, Borella & Ito, 1992).

There is some supportive biostratigraphic evidence for the ages assigned on the basis of this calibration above and beyond that presented by Haq et al. (1992). Sequences BD1, BD2 and BD3 (calibrated against the Berriasian Cycles 1.4, 1.5 and 1.6 respectively) all contain the benthonic foraminifer *Haplophragmoides* cf. *infracalloviensis*, which ranges no younger than the basal part of the Albidum Zone of the Boreal Ryazanian (Boisseri Zone of the Tethyan Berriasian) (Jones & Wonders, 1992, and references therein; unpublished observations). Sequence BD4 (calibrated against the latest Berriasian to earliest Valanginian Cycle 2.1) contains the dinoflagellate palynomorph *Egmontodinium torynum*, which ranges no younger than the Paratollia Zone of the Boreal Valanginian (Roubaudi Zone of the Tethyan Valanginian) (unpublished observations).

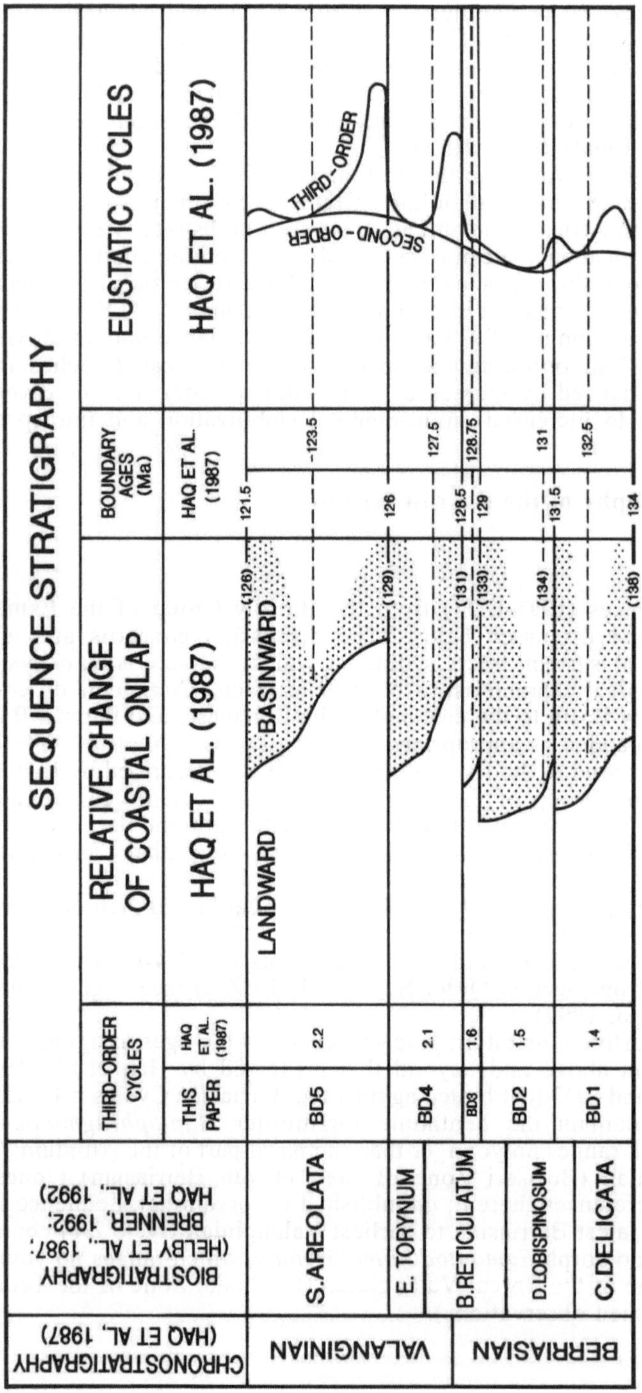


Figure 6.6: Sequence stratigraphic summary.

There is also some supportive radiometric evidence. K/Ar dating of authigenic glauconite grains from Sequence BD4 (calibrated against the latest Berriasian to earliest Valanginian Cycle 2.1) yielded an age of 132 +/- 2 Ma (Berriasian) (Haq et al., 1992).

Our own unpublished observations indicate that third-order sequences BDl, BD2 and BD3 can be regarded as representing a second-order LSF (essentially aggradational), and BD4 and BD5 a second-order LSW (essentially progradational). A chronostratigraphic diagram of the Barrow Group siliciclastic sequences at Sites 762 and 763 is given on Figure 6.7 and a wireline log correlation between the two sites is given on Figure 6.8.

### 6.2.2 Use of palaeontology in palaeoenvironmental interpretation

The use of micropalaeontology in palaeoenvironmental interpretation is well established (see, for instance, Jones & Wonders (1992), and references therein). The use of palynology in palaeoenvironmental interpretation, as opposed to stratigraphic studies, is less well established. This is presumably because terrestrially-derived palynomorphs (plant cuticle, pollen, spores, wood fragments) and marine pelagic palynomorphs (acritarchs, dinocysts) provide no direct information on marine benthonic environments. However, the percentage of terrestrially-derived palynomorphs provides a useful measure of "continentality"; see, for instance, Williams & Serjeant (1967). Moreover, the ratio of "deep" marine dinocysts (heterotrophic peridinaceans) to "shallow" marine dinocysts (autotrophic (photosynthetic) gonyaulacaceans) provides some measure of "oceanicity" (or upwelling) (see, for instance, Lewis et al. (1990), and references therein).

Figure 6.7: Chronostratigraphic Diagram, Barrow Group Siliciclastic Sequences, Sites 762 and 763 (modified after Haq et al., 1992). LPC - Low-Stand Prograding Complex; TST - Transgressive Systems Tract; MFS - Maximum Flooding Surface; HST - High-Stand Systems Tract; B - *Bathysiphon*; C - *Cenosphaera*; H - *Haplophragmoides* sp.; Hi - *Haplophragmoides* cf. *infracalloviensis*; Lm - *Lenticulina* ex gr. *muensteri*.

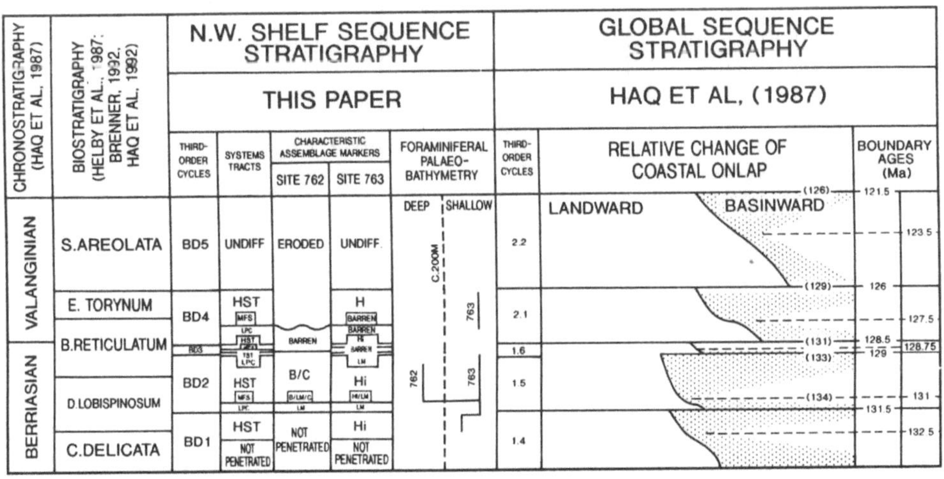

Figure 6.8: Wireline log correlation, barrow group siliciclastic sequences, sites 762 and 763. SB - Sequence Boundary; LPC - Low-Stand Prograding Complex; TST - Transgressive Systems Tract; MFS - Maximum Flooding Surface; HST - High-Stand Systems Tract; Char. Assem. Markers - Characteristic Assemblage Markers; B - *Bathysiphon*; C - *Cenosphaera*; H - *Haplophragmoides* sp.; Hi - *Haplophragmoides* cf. *infracalloviensis*; Lm - *Lenticulina* ex gr. *muensteri*. BD1 is penetrated at Site 763, but not shown (no wireline logs). The BD3 TST at Site 763 (proximal palaeo-prodelta) condenses to merge with the MFS at Site 762 (distal palaeo-prodelta). The borehole sites are approximately 100km apart.

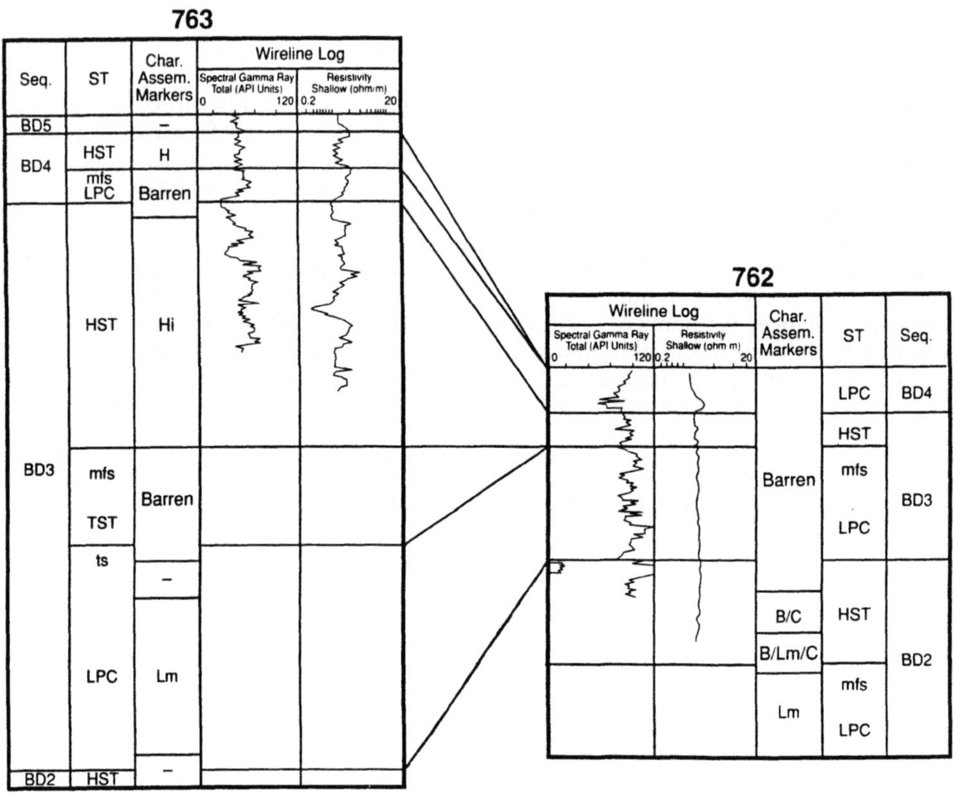

### 6.2.3 Sequence BD1

*Definition*

Base undefined; top defined by "reflector 34" of Haq et al. (1992), calibrated against 131.5Ma sequence boundary unconformity of Haq et al. (1987a).

*Age*

Berriasian, *C. delicata* to *D. lobispinosum* Zones of Helby et al. (1987) (Brenner, 1992).

*Gross depositional environment*

Haq et al. (1992) interpreted the depositional environment of this sequence as prograding prodeltaic at Site 763 (proximal palaeo-prodelta). It is not penetrated at Site 762.

*Systems tracts*

Only a high-stand systems tract (HST) is penetrated at Site 763 (Cores 46R-37R) (Haq et al., 1992).

*Sedimentological characteristics of systems tracts*

On core evidence (no wireline log data being available), the HST at Site 763 (proximal palaeo-prodelta) consists of locally shaly siltstone (with siderite) (Haq et al., 1992).

*Micropalaeontological characteristics of systems tracts*

The HST at Site 763 (proximal palaeo-prodelta) is characterized by low to high microfaunal abundance and diversity: diversity ranges from 3-9 (mean 6). In terms of taxonomic composition, agglutinating foraminifera predominate, and the agglutinating species *Haplophragmoides* cf. *infracalloviensis* is the most characteristic assemblage marker. The associated calcareous benthonic foraminiferal species *Lenticulina* ex gr. *muensteri* and *Spirillina minima* probably indicate palaeodepths of the order of 50-150m and 100-200m respectively (Jones & Wonders, 1992, and references therein).

*Palynological characteristics of systems tracts*

The HST at Site 763 (proximal palaeo-prodelta) is characterized by low to high palynofloral diversity: diversity indices range from 0.19-0.54 (mean 0.36). The percentage of terrestrially derived palynomorphs (plant cuticle, pollen, spores, wood fragments) is generally 60-90% (though it is >90% in Cores 46R-45R at the base of the HST and <60% in Cores 38R-37R at the top of the HST) (WB's unpublished observations).

*Comments*

In terms of sedimentological characteristics, BD1 in one of the wells in the Barrow Sub-Basin and in several wells in the Dampier Sub-Basin (on the palaeo-prodelta) is similar to BD1 at Site 763 on the Exmouth Plateau (also on the palaeo-prodelta). In both, it is represented by thin shales and siltstones. In the Barrow Sub-Basin well, the HST exhibits a weakly progradational gamma-log motif (coarsening-upward from shale to siltstone). BD1 in the Bonaparte and Papuan Basins to the extreme east is sand-prone.

6.2.4 Sequence BD2

*Definition*

Base defined by "reflector 34" of Haq et al. (1992), calibrated against 131.5Ma sequence boundary unconformity of Haq et al. (1987a); top by "reflector 36" of Haq et al. (1992), calibrated against 129Ma sequence boundary unconformity of Haq et al. (1987a). Maximum flooding surface (MFS) marked by "reflector 35" of Haq et al. (1992), calibrated against 131Ma downlap surface of Haq et al. (1987a).

*Age*

Berriasian, *D. lobispinosum* to *B. reticulatum* Zones of Helby et al. (1987) (Brenner, 1992).

*Gross depositional environment*

Haq et al. (1992) interpreted the depositional environment of this sequence as prograding prodeltaic at Site 763 (proximal palaeo-prodelta) and prodeltaic at Site 762 (distal palaeo-prodelta).

*System tracts*

Both a low-stand prograding complex (LPC) and a high-stand systems tract (HST) are penetrated at Sites 763 (Cores 35R-31R and Cores 29R-21R respectively) and 762 (Cores 91X-90X and Core 88X respectively) (Haq et al., 1992). They are separated by MFSs (Core 30R at Site 763; Core 89X at Site 762) (Haq et al., op. cit.).

*Sedimentological characteristics of systems tracts*

On core evidence (no wireline log data being available), the LPC and HST at Site 763 (proximal palaeo-prodelta) consist of locally shaly siltstone and subordinate shale (with pyrite and siderite), and the MFS of siltstone (with siderite) (Haq et al., 1992). The MFS is characterized by comparatively high shipboard $CaCO_3$ content (10%). On core and limited wireline log evidence, the LPC and HST at Site 762 (distal palaeo-prodelta) consist of siltstone, and the MFS of shale (Haq et al., 1992).

*Micropalaeontological characteristics of systems tracts*

The LPC at Site 763 (proximal palaeo-prodelta) is characterized by moderate to high microfaunal abundance and diversity: diversity ranges from 4-9 (mean 7). Agglutinating and calcareous benthonic foraminifera occur in approximately equal numbers, though as at Site 762, the calcareous benthonic species *Lenticulina* ex gr. *muensteri* is the most characteristic assemblage marker. The assemblage probably indicates palaeodepths in the range 50-150m. The MFS is characterized by moderate abundance and diversity (6 species). Agglutinating and calcareous benthonic foraminifera again occur in approximately equal numbers, and both the agglutinating species *Haplophragmoides* cf. *infracalloviensis* and the calcareous benthonic species *Lenticulina* ex gr. *muensteri* are characteristic assemblage markers. The assemblage again indicates a palaeodepth in the range 50-150m. The HST is characterized by zero to high abundance and diversity: diversity ranges from 0-10 (mean 3). Agglutinating and calcareous benthonic foraminifera again occur in approximately equal numbers, though the agglutinating species *Haplophragmoides* cf. *infracalloviensis* is the most characteristic assemblage marker. The assemblage yet again indicates a palaeodepth in the range 50-150m (Jones & Wonders, 1992, and references therein).

The LPC at Site 762 (distal palaeo-prodelta) is characterized by low to moderate microfaunal abundance and diversity: diversity ranges from 1-6 (mean 4). Calcareous benthonic foraminifera predominate, and the calcareous benthonic species *Lenticulina* ex gr. *muensteri* is the most characteristic assemblage marker. The assemblage probably indicates a palaeodepth in the range 50-150m. The MFS is characterized by moderate abundance and diversity (4 species). Agglutinating and calcareous benthonic foraminifera occur in approximately equal numbers. The agglutinating foraminiferal species *Bathysiphon* sp., the calcareous benthonic foraminiferal species *Lenticulina* ex gr. *muensteri*, and the actinommid radiolarian species *Cenosphaera* sp. are all characteristic assemblage markers. The assemblage probably indicates a palaeodepth in the range 200-500m. The HST is characterized by low abundance and diversity (3 species). Agglutinating foraminifera predominate. The agglutinating foraminiferal species *Bathysiphon* sp. and the actinommid radiolarian species *Cenosphaera* sp. are both characteristic assemblage markers. The assemblage again probably indicates a palaeodepth in the range 200-500m (Jones & Wonders, 1992, and references therein).

*Palynological characteristics of systems tracts*

The LPC at Site 763 (proximal palaeo-prodelta) is characterized by low to high palynofloral diversity: diversity indices range from 0.03-0.47 (mean 0.26). The percentage of terrestrially-derived palynomorphs (plant cuticle, pollen, spores, wood fragments) is generally >90% (though it is <60% in Core 35R). The MFS is characterized by moderate diversity: diversity indices range from 0.24-0.35 (mean 0.30). The percentage of terrestrially-derived palynomorphs is >90%. The HST is characterised by low to high diversity: diversity indices range from 0.06-0.46 (mean 0.24). The percentage of terrestrially-derived palynomorphs is generally 60-90%, though it is >90% in Cores 23R-22R at the top of the HST (WB's unpublished observations). BD2 at site 762 (distal palaeo-prodelta) contains 60-90% terrestrially-derived palynomorphs (WB's unpublished observations).

*Comments*

In terms of sedimentological characteristics, BD2 in wells in the Barrow Sub-Basin (on the palaeo-delta) is rather different from BD2 in wells in the Dampier Sub-Basin and at Sites 762 and 763 on the Exmouth Plateau (on the palaeo-prodelta). In the Barrow Sub-Basin wells, it is represented by thick aggradational sandstones, in the Dampier Sub-Basin and Exmouth Plateau wells by thin shales and siltstones.

BD2 in the Bonaparte and Papuan Basins to the extreme east is sand-prone, particularly in the LST.

6.2.5 Sequence BD3

*Definition*

Base defined by "reflector 36" of Haq et al. (1992), calibrated against 129Ma sequence boundary unconformity of Haq et al. (1987a); top by "reflector 39" of Haq et al. (1992), calibrated against 128.5Ma sequence boundary unconformity of Haq et al. (1987a). Transgressive surface (TS) at Site 763 marked by "reflector 37" of Haq et al. (1992). Maximum flooding surfaces (MFSs) at Sites 762 and 763 marked by "reflector 38" of Haq et al. (1992), calibrated against 128.75Ma downlap surface of Haq et al. (1987a).

*Age*

Berriasian, *B. reticulatum* Zone of Helby et al. (1987) (Brenner, 1992).

*Gross depositional environment*

Haq et al. (1992) interpreted the depositional environment of this sequence as prograding prodeltaic (Cores 46R-4R) to ?turbiditic (Cores 54X-52X) at Site 763 (proximal palaeo-prodelta) and prodeltaic at Site 762 (distal palaeo-prodelta).

*Systems tracts*

A low-stand prograding complex (LPC), a transgressive systems tract (TST) and a high-stand systems tract (HST) at Site 763 (Cores 19R-15R, Cores 13R-12R and Cores 11R-53X respectively), and a low-stand prograding complex (LPC) and a high-stand systems tract (HST) are penetrated at Site 762 (Cores 86X-85X and Core 83X respectively) (Haq et al., 1992). The TST and HST at Site 763 and the LPC and HST at Site 762 are separated by MFSs (Core 12R at Site 763; Core 84X at Site 762) (Haq et al., op. cit.). The LPC and TST at Site 763 are separated by a TS (Core 14R).

*Sedimentological characteristics of systems tracts*

On core and limited wireline log evidence, the LPC at Site 763 (proximal palaeo-prodelta) consists of locally shaly siltstone (with pyrite and siderite), the TST of locally shaly siltstone and subordinate shale (with siderite), the MFS of siltstone, and the HST of siltstone and subordinate shale and carbonate (with pyrite and siderite) (Haq et al., 1992). The TST and MFS are both characterized by comparatively high shipboard calcium carbonate contents (10-80% and 40% respectively). The HST exhibits a weakly progradational gamma-log motif (coarsening-upward).

On core and wireline log evidence, the LPC at Site 762 (distal palaeo-prodelta) consists of siltstone, the MFS of shale, and the HST of shale and subordinate siltstone (Haq et al., 1992). The LPC exhibits an essentially aggradational gamma-log motif. The MFS exhibits a comparatively low gamma-log response. The HST exhibits a weakly progradational gamma-log motif.

*Micropalaeontological characteristics of systems tracts*

The LPC at Site 763 (proximal palaeo-prodelta) is characterized by zero to high microfaunal abundance and diversity: diversity ranges from 0-9 (mean 5). Agglutinating and calcareous benthonic foraminifera occur in approximately equal numbers, though the calcareous benthonic species *Lenticulina* ex gr. *muensteri* is the most characteristic assemblage marker. The assemblage probably indicates palaeodepths in the range 50-150m. The TST and the MFS are both barren. The HST is characterized by zero to high abundance and diversity: diversity ranges from 0-8 (mean 3). Agglutinating and calcareous benthonic foraminifera occur in approximately equal numbers, though the agglutinating species *Haplophragmoides* cf. *infracalloviensis* is the most characteristic assemblage marker. The assemblage probably indicates palaeodepths in the range 50-150m. The occurrence of the radiolarian species *Eucyrtis*? sp. in Core 5R is arguably indicative of palaeodepths of >200m (Jones & Wonders, 1992, and references therein). This core can be interpreted as representing part of a submarine fan complex. A case can therefore be made for placing it within the BD4 LST rather than the BD3 HST.

The LPC, MFS and HST at Site 762 (distal palaeo-prodelta) are all barren of microfauna (Jones & Wonders, 1992).

*Palynological characteristics of systems tracts*

The LPC at Site 763 (proximal palaeo-prodelta) is characterized by zero to moderate palynofloral diversity: diversity indices range from 0.00-0.35 (mean 0.17). The percentage of terrestrially-derived palynomorphs (plant cuticle, pollen, spores, wood fragments) is generally >90%. The TST is characterized by zero to moderate diversity: diversity indices range from 0.00-0.25 (mean 0.19). The percentage of terrestrially-derived palynomorphs is >90%. The MFS is characterized by low to moderate diversity: diversity indices range from 0.15-0.22 (mean 0.19). The percentage of terrestrially-derived palynomorphs is >90%. The HST is characterized by zero to high diversity: diversity indices range from 0.00-0.40 (mean 0.20). The percentage of terrestrally derived palynomorphs is generally 60-90 %, though it is > 90% in core 11R at the base of the HST (WB's unbuplished observations).

The LPC at Site 762 (distal palaeo-prodelta) contains 60-90% terrestrially-derived palynomorphs at its base and <60% at its top. The MFS and HST contain <60% terrestrially-derived palynomorphs (WB's unpublished observations).

*Comments*

In terms of sedimentological characteristics, BD3 in wells in the Barrow Sub-Basin (on the palaeo-delta) is rather different from BD3 at Sites 762 and 763 on the Exmouth Plateau (on the palaeoprodelta). In the Barrow Sub-Basin wells, it is represented by locally thick, essentially aggradational, sandstones and siltstones, and in the Exmouth Plateau wells by thin, locally (HST) weakly progradational shales and siltstones.

BD3 cannot be confidently identified in the Dampier Sub-Basin. This could be because it is very thin, owing to slow deposition over a short time-span (0.5Ma), and thus unresolvable.

### 6.2.6 Sequence BD4

*Definition*

Base defined by "reflector 39" of Haq et al. (1992), calibrated against 128.5Ma sequence boundary unconformity of Haq et al. (1987a); top by "reflector 3" of Haq et al. (1992), calibrated against 126Ma sequence boundary unconformity of Haq et al. (1987a).

*Age*

Berriasian to Valanginian, *B. reticulatum* to *E. torynum* Zones of Helby et al. (1987) (Brenner, 1992).

*Gross depositional environment*

Haq et al. (1992) interpreted the depositional environment of this sequence as turbiditic? (Cores 52X-49X) to prodeltaic (Cores 49X-48X) at Site 763 (proximal palaeo-prodelta) and prodeltaic at Site 762 (distal palaeo-prodelta).

*Systems tracts*

A low-stand prograding complex (LPC) and a high-stand systems tract (HST) are penetrated at Site 763 (Core 51X and Cores 50-49X respectively) (Haq et al., 1992). They are separated by a maximum flooding surface (MFS) calibrated against the 127.5 Ma downlap surface of Haq et al. (1987a) (Core 51X) (Haq et al., 1992). Only a low-stand prograding complex (LPC) is penetrated at Site 762 (Core 81X) (Haq et al., 1992).

*Sedimentological characteristics of systems tracts*

On core and wireline log evidence, the LPC at Site 763 (proximal palaeo-prodelta) consists of siltstone and subordinate carbonate (with pyrite and glauconite), the MFS and HST of siltstone (the MFS with pyrite, the HST with pyrite and glauconite) (Haq et al., 1992). The MFS is characterized by a comparatively high shipboard calcium carbonate content (50%), and a correspondingly low gamma-log response.

On core and wireline log evidence, the LPC at Site 762 (distal palaeo-prodelta) consists of shale and subordinate siltstone and carbonate (with pyrite) (Haq et al., 1992). It exhibits an aggradational to weakly retrogradational gamma-log motif (fining-upward).

*Micropalaeontological characteristics of systems tracts*

The LPC and MFS at Site 763 (proximal palaeo-prodelta) (Core 51X) are barren of microfauna. The HST is characterized by low microfaunal abundance and diversity: diversity ranges from 1-2. Agglutinating and calcareous benthonic foraminifera occur in approximately equal numbers, though the agglutinating species *Haplophragmoides* sp. is the most characteristic assemblage marker. The assemblage probably indicates palaeodepths in the range 50-150m (Jones & Wonders, 1992, and references therein).

The LPC at Site 762 (distal palaeo-prodelta) (Core 81X (part)) is barren of microfauna (Jones & Wonders, 1992).

*Palynological characteristics of systems tracts*

BD4 at Site 763 (proximal palaeo-prodelta) contains 60-90% terrestrially-derived palynomorphs (WB's unpublished observations).

The LPC at Site 762 (distal palaeo-prodelta) contains <60% terrestrially-derived palynomorphs (WB's unpublished observations).

*Comments*

The unconformity at the base of Sequence BD4 appears to be regionally correlatable around the North-West Shelf of Australia. LST sandstones might at least locally be developed in reservoir facies.

In terms of sedimentological characteristics, BD4 in wells in the Barrow Sub-Basin (on the palaeo-delta) is rather different from BD4 in wells in the Dampier Sub-Basin and at Sites 762 and 763 on the Exmouth Plateau (on the palaeo-prodelta). In the Barrow Sub-Basin wells, it is represented by thick, initially (LST) aggradational, later (TST) aggradational to retrogradational and finally (HST) progradational, sandstones and siltstones, and in the Dampier Sub-Basin and Exmouth Plateau wells predominantly by thin, locally (LPC) retrogradational shales and siltstones and local sandstones.

6.2.7 Sequence BD5

*Definition*

Base defined by "reflector 3" of Haq et al. (1992), calibrated against 126Ma sequence boundary unconformity of Haq et al. (1987a); top by "break-up unconformity" of Haq et al. (1992), tentatively calibrated against 121.5Ma sequence boundary unconformity of Haq et al. (1987a).

*Age*

Valanginian, *S. areolata* Zone of Helby et al. (1987) (Brenner, 1992).

*Gross depositional environment*

Haq et al. (1992) interpreted the depositional environment of this sequence as prodeltaic at Site 763 (proximal palaeo-prodelta). It is not penetrated at Site 762.

*Systems Tracts*

Only undifferentiated BD5 is penetrated at Site 763 (Core 48X) (Haq et al., 1992).

*Sedimentological characteristics of systems tracts*

On core and wireline log evidence, BD5 at Site 763 (proximal palaeo-prodelta) consists of siltstone (with pyrite) (Haq et al., 1992).

*Micropalaeontological characteristics of systems tracts*

BD5 at Site 763 (proximal palaeo-prodelta) is barren of microfauna (Jones & Wonders, 1992).

*Palynological characteristics of systems tracts*

No detailed palynofloral data are available.

*Comments*

The unconformity at the base of Sequence BD5 appears to be regionally correlatable around the North-West Shelf of Australia. The "break-up" unconformity at the top of Sequence BD5 is also regionally correlatable. Mardie Greensand and Huderong Shale sediments deposited during the succeeding thermal subsidence phase are of a predominantly fine-grained clastic lithology, and form regional seals to reservoirs developed in the Barrow Delta Complex.

In terms of sedimentological characteristics, BD5 in wells in the Barrow Sub-Basin (on the palaeo-delta) is rather different from BD5 in wells in the Dampier Sub-Basin and at Site 763 on the Exmouth Plateau (on the palaeo-prodelta). In the Barrow Sub-Basin, it is represented by thick, aggradational (LST-TST) and progradational (HST), sandstones and siltstones and in the Dampier Sub-Basin and Exmouth Plateau wells by thin siltstones.

In terms of palaeontological characteristics, the BD5 MFS in wells in the Papuan Basin is similar to the BD2 MFS on the Exmouth Plateau. Both contain actinommid radiolarians (*Cenosphaera* spp.).

## 6.3 Sedimentological characteristics of systems tracts

### 6.3.1. Low-stand systems tracts (LSTS)

LSTs in studied exploration wells on the palaeo-delta (Barrow Basin) are represented by predominantly coarse- and medium- grained clastics. The characteristic gamma-log motif is aggradational.

LSTs (LPCs) in the ODP boreholes on the palaeo-prodelta (Exmouth Plateau) are represented by predominantly fine-grained clastic lithologies (shales, shaly siltstones, and siltstones with pyrite, siderite and locally glauconite) and subordinate carbonates. The characteristic gamma-log motif is aggradational, the less characteristic retrogradational (BD4).

### 6.3.2 Transgressive systems tracts (TSTs)

TSTs in exploration wells on the palaeo-delta are represented by predominantly coarse- and medium- grained clastics. The characteristic gamma-log motifs are aggradational and retrogradational.

The TST in the ODP borehole on the proximal palaeo-prodelta (Site 763) is represented by shaly siltstone and subordinate shale (with siderite). It is characterized by a comparatively high $CaCO_3$ content (10-80%).

### 6.3.3 Maximum flooding surfaces (MFSs)

MFSs in exploration wells on the palaeo-delta are represented by siltstones and argillaceous sandstones. They are characterized by high gamma-log responses.

MFSs in the ODP boreholes on the palaeo-prodelta are represented by predominantly fine-grained clastic lithologies (shales and siltstones with pyrite and siderite). As in many other mixed-component (clastic-carbonate) sedimentary systems, they are characterized by comparatively high $CaCO_3$ contents (10-50%) (due to clastic starvation), and correspondingly low gamma-log responses.

### 6.3.4 High-stand systems tracts (HSTs)

HSTs in exploration wells on the palaeo-delta are represented by predominantly coarse-grained clastics. The characteristic gamma-log motifs are aggradational to progradational in BD1-BD3 (second-order LSF) and progradational in BD4-BD5 (second-order LSW).

HSTs in the exploration well and in the ODP boreholes on the palaeo-prodelta are represented by predominantly fine-grained clastic lithologies (shales, shaly siltstones, siltstones with pyrite, siderite and locally glauconite) and subordinate carbonates. Their characteristic gamma-log motif is progradational (coarsening-upward).

## 6.4 Palaeontological characteristics of systems tracts

Characteristic microfaunal assemblage markers for LSTs (LPCs), TSTs, MFSs and HSTs at Sites 763 (proximal palaeo-prodelta) and 762 (distal palaeo-prodelta) are given on Figure 6.7.

### 6.4.1 Low-Stand systems tracts (LSTs)

*Micropalaeontology*

LST (LPC) assemblages from the palaeo-prodelta sites are characterized by zero to moderate microfaunal abundances, and diversities in the range 0-10 (mean 4). In terms of taxonomic composition, agglutinating and calcareous benthonic foraminifera occur in approximately equal numbers, though calcareous benthonic species locally predominate. The calcareous benthonic species *Lenticulina* ex gr. *muensteri* is the most characteristic LPC assemblage marker at both proximal and distal palaeo-prodelta sites: significantly, this species is regarded as an indicator of regressive environments in the Jurassic. LPC assemblages from both sites probably indicate palaeodepths in the range 50-150m (Jones & Wonders, 1992).

LPC assemblages from both palaeo-prodelta sites are similar to the *Dentalina-Nodosaria* Assemblage described by Nagy (1985a, b) from the Jurassic Dunlin and Brent Formations of the Statfjord of the Northern North Sea, interpreted by him as indicative of distal deltaic environments characterized by comparatively little clastic input. They are also similar to the "clear-water" assemblages described by van Gorsel (1988) from the Neogene of Indonesia, interpreted by him as indicative of TSTs. The apparent similarity between distal LST (LPC) assemblages and TST assemblages is not surprising. Sediment is expected to be effectively sequestered in proximal parts of the shelf during both LSTs (LPCs) and TSTs.

*Palynology*

LST (LPC) assemblages from the proximal palaeo-prodelta site are characterized by low to high palynofloral diversity: diversity indices range from 0.03-0.47. Mean values range from 0.26 in Sequence BD2 to 0.17 in Sequence BD3. The percentage of terrestrially-derived palynomorphs (plant cuticle, pollen, spores, wood fragments) is generally >90%. The percentage of terrestrially-derived palynomorphs in LPC assemblages from the distal palaeo-prodelta site ranges from <60% to 60-90%.

Penecontemporaneously reworked palynomorphs are common in provisionally interpreted LPCs in the Dampier Sub-Basin wells (PAV's unpublished observations).

6.4.2 Transgressive systems tracts (TSTs)

*Micropalaeontology*

The TST from the proximal palaeo-prodelta site is barren of microfauna. No data are available from the distal palaeo-prodelta site.

*Palynology*

The TST from the proximal palaeo-prodelta site is characterized by low to moderate palynofloral diversity: diversity indices range from 0.06-0.25 (mean 0.19). The percentage of terrestrially-derived palynomorphs (plant cuticle, pollen, spores, wood fragments) is >90%. No data are available from the distal palaeo-prodelta site.

6.4.3 Maximum flooding surfaces (MFSs)

*Micropalaeontology*

MFS assemblages from the palaeo-prodelta sites are characterized by zero to moderate microfaunal abundances, and diversities in the range 0-6 (mean 2). Agglutinating and calcareous benthonic foraminifera occur in approximately equal numbers. The agglutinating species *Haplophragmoides* cf. *infracalloviensis* and the calcareous benthonic species *Lenticulina* ex gr. *muensteri* are characteristic MFS assemblage markers at the proximal palaeo-prodelta site, and the agglutinating foraminiferal species *Bathysiphon* sp., the calcareous benthonic foraminiferal species *Lenticulina* ex gr. *muensteri*, and the radiolarian species *Cenosphaera* sp. are characteristic MFS assemblage markers at the distal palaeo-prodelta site. The proximal palaeo-prodelta MFS assemblage probably indicates a palaeodepth in the range 50-150m, and the distal palaeo-prodelta MFS assemblage a palaeodepth in the range 200-500m (Jones & Wonders, 1992, and references therein).

*Palynology*

MFS assemblages from the proximal palaeo-prodelta site are characterized by low to moderate palynofloral diversity: diversity indices range from 0.15-0.35. Mean values range from 0.30 in Sequence BD2 to 0.19 in Sequence BD3. The percentage of terrestrially-derived palynomorphs (plant cuticle, pollen, spores, wood fragments) is generally >90%. The percentage of terrestrially-derived palynomorphs in MFS assemblages from the distal palaeo-prodelta site ranges from <60% to 60-90%.

In any given sequence, palynofloral diversity is generally at a maximum at the MFS. The unusually high percentages of terrestrially-derived palynomorphs in the BD2 and BD3 MFSs may be due to artifacts of sampling (admixturing of systems tracts within cores) (but see also "Significance of Barren or Impoverished Samples" section below).

6.4.4 High-stand systems tracts (HSTs)

*Micropalaeontology*

HST assemblages from the palaeo-prodelta sites are characterized by zero to moderate microfaunal abundances, and diversities in the range 0-10 (mean 4). Values decrease from the lower to the upper sequences at both sites. Agglutinating and calcareous benthonic foraminifera occur in approximately equal numbers, though agglutinating species locally predominate. The agglutinating species *Haplophragmoides* cf. *infracalloviensis* and *H.* sp. are the most characteristic HST assemblage markers at the proximal palaeo-prodelta site, and the agglutinating species *Bathysiphon* sp. and the radiolarian species *Cenosphaera* sp. are the most characteristic HST assemblage markers at the distal palaeo-prodelta site. The proximal palaeo-prodelta HST assemblage probably indicates a palaeodepth in the range 50-150m (or, in Sequence BD1, in the range 100-150m), and the distal palaeo-prodelta HST assemblage a palaeodepth in the range 200-500m (Jones & Wonders, 1992, and references therein).

HST assemblages from both palaeo-prodelta sites are similar to the *Ammodiscus-Dentalina* Assemblage described by Nagy (1985a, b), interpreted by him as indicative of distal deltaic environments characterized by considerable clastic input (probably due to deltaic progradation). They are also similar to the "Thecamoebian" assemblage (which also contains agglutinating foraminifera such as *Trochammina* and *Ammobaculites*) described by Carbonel & Moyes (1987) from the Pleistocene of the Mahakam Delta, Kalimantan, Indonesia, interpreted by them as indicative of "prograding phases" and to the "turbid-water" assemblage described by van Gorsel (1988) from the Neogene of Indonesia, interpreted by him as indicative of HSTs.

*Palynology*

HST assemblages from the proximal palaeo-prodelta site are characterized by low to high palynofloral diversity: diversity indices range from 0.06-0.54. Mean values range from 0.36 in Sequence BD1 through 0.24 in Sequence BD2 to 0.20 in Sequence BD3. The percentage of terrestrially-derived palynomorphs (plant cuticle, pollen, spores, wood fragments) ranges is generally 60-90%. The percentage of terrestrially-derived palynomorphs in HST assemblages from the distal palaeo-prodelta site ranges from <60% to 60-90%.

6.4.5 Significance of barren or impoverished samples

Microfaunal abundance and diversity values are comparatively low in both TST (zero) and MFS (zero to moderate) assemblages from the palaeo-prodelta sites. This could be an artefact of the unrepresentativeness of the samples. Alternatively, it could be attributable to unfavorable environmental conditions (dysoxia) especially in the MFS case. Microfaunal abundance and diversity values tend to be lower in the upper sequences at both sites. This could be associated with increased terrigenous dilution due to increased sedimentation rate in these increasingly progradational sequences.

Palynofloral diversity values also tend to be lower in the upper sequences at the proximal palaeo-prodelta site. This could be associated with a decline in primary phytoplankton productivity due to increased turbidity and decreased light penetration inhibiting photosynthesis. This could in turn account for the high proportions of terrestrially-derived palynomorphs observed in the BD2 and BD3 MFS samples at the proximal palaeo-prodelta site (though these could equally be caused by sampling artefacts, see above).

## 6.5 Conclusions

### 6.5.1 Sequence stratigraphy

Five Barrow Group siliciclastic sequences are described from the North-West Shelf, Australia, corresponding to the five unnamed sequences described by Haq et al. (1992) from the same area. They can be calibrated against five of the global (?eustatically-mediated) third-order cycles described by Haq et al. (1987a). The age-range represented by the sequences is Berriasian to Valanginian. Third-order sequences BD1, BD2 and BD3 can be regarded as representing a second-order LSF (aggradational), and BD4 and BD5 a second-order LSW (progradational).

Sedimentological and palaeontological characterization of constituent systems tracts (see below) helps in the constraint of the sequence stratigraphic model, and ultimately in the prediction of the spatio-temporal distribution of lithofacies, including potential petroleum reservoir lithofacies.

### 6.5.2 Sedimentological characteristics of systems tracts

*Low-stand systems Tracts (LSTs)*

LSTs in the study area characteristically exhibit aggradational gamma-log motifs. Those on the palaeo-delta (Barrow Sub-Basin) are characteristically sand-prone, while those on the palaeo-prodelta (Dampier Sub-Basin and Exmouth Plateau) (LPCs) are shale-prone.

*Transgressive systems tracts (TSTs)*

TSTs in the study area characteristically exhibit aggradational and retrogradational gamma-log motifs. Those on the palaeo-delta are sand-prone, while those on the palaeo-prodelta are shale-prone.

*Maximum flooding surfaces (MFSs)*

MFSs on the palaeo-delta are characterized by gamma-log peaks, while those on the palaeo-prodelta are characterized by gamma-log troughs.

*High-stand systems tracts (HSTs)*

HSTs in the study area characteristically exhibit aggradational to progradational gamma-log motifs in Sequences BD1-BD3 (Second-Order LSF) and progradational gamma-log motifs in Sequences BD4-BD5 (Second-Order LSW). Those on the palaeo-delta are sand-prone, those on the palaeo-prodelta shale-prone.

### 6.5.3 Palaeontological characteristics of systems tracts

*Low-stand systems tracts (LSTs)*

LSTs (LPCs) from the palaeo-prodelta (Exmouth Plateau) are characterized by zero to high microfaunal abundances and diversities. Agglutinating and calcareous benthonic foraminifera occur in approximately equal numbers, though calcareous benthonic species locally predominate. LPC assemblages from both sites probably indicate palaeodepths in the range 50-150m.

LPCs from the proximal palaeo-prodelta site are characterized by zero to high palynofloral diversities. Mean diversity index values decrease from Sequence BD2 to Sequence BD3. Terrestrially-derived palynomorph percentages are generally high at the proximal palaeo-prodelta site, and moderately high at the distal palaeo-prodelta site.

*Transgressive systems tracts (TSTs)*

The TST from the proximal palaeo-prodelta (Site 763) is characterized by zero microfaunal abundance and diversity. The TST from the proximal palaeo-prodelta site is characterized by zero to moderate palynofloral diversities. Terrestrially-derived palynomorph percentages are high.

*Maximum flooding surfaces (MFSs)*

MFSs from the palaeo-prodelta are characterized by zero to moderate microfaunal abundances and diversities. Agglutinating and calcareous benthonic foraminifera occur in approximately equal numbers. The proximal palaeo-prodelta MFS assemblage probably indicates a palaeodepth in the range 50-150m, and the distal palaeo-prodelta MFS assemblage a palaeodepth in the range 200-500m.

MFSs from the proximal palaeo-prodelta site are characterized by low to moderate palynofloral diversities. Mean diversity index values decrease from Sequence BD2 to Sequence BD3. In any given sequence, diversity is generally at a maximum at the MFS. Terrestrially derived palynomorph percentages are generally high at the proximal palaeo-prodelta site, and range from moderately low to moderately high at the distal palaeo-prodelta site.

*High-stand systems tracts (HSTs)*

HSTs from the palaeo-prodelta are characterized by zero to high moderate microfaunal abundances and diversities. Values decrease from the lower to the upper sequences at both proximal and distal sites. Agglutinating and calcareous benthonic foraminifera occur in approximately equal numbers, though agglutinating species locally predominate. The proximal palaeo-prodelta HST assemblage probably indicates a palaeodepth in the range 50-150m (or, in Sequence BD1, in the range 100-150m), and the distal palaeo-prodelta HST assemblage a palaeodepth in the range 200-500m.

HSTs from the proximal palaeo-prodelta site are characterized by zero to high palynofloral diversities. Mean diversity index values decrease from Sequence BD1 to Sequence BD3. Terrestrially-derived palynomorph percentages are generally moderately high at the proximal palaeo-prodelta site, and range from moderately low to moderately high at the distal palaeo-prodelta site.

## Acknowledgements

BP granted permission for RWJ, PAV, AAHW, SL, MMR, MDS, TDV, JA and SJS to publish, and also provided drafting facilities. BP Australia released well data from the Barrow and Bonaparte Basins. Nick Milton of BP Glasgow provided us with an integrated seismic and wireline log correlation of several wells in the Barrow Basin (from palaeo-delta to palaeo-prodelta), and also allowed us to reproduce in modified form a figure (Figure 6.5) from an unpublished BP report. Bil Haq of the Division of Ocean Sciences, National Science Foundation, Washington DC is thanked for allowing us to reproduce a figure (Figure 6.4) from one of his publications (Haq et al., 1987), and for allowing us to use data from another (Haq et al., 1992).


### References


Armentrout, J.M. 1991a. Research Conference Overview. In: Armentrout, J.M. & Perkins, B.F. (eds): Sequence Stratigraphy as an Exploration Tool: Concepts and Practices in the Gulf Coast. Proceedings of the Society of Economic Palaeontologists and Mineralogists Foundation (Gulf Coast Section) Eleventh Annual Research Conference, Houston 1991, 1-10.

Armentrout, J.M. 1991b. Paleontologic Constraints on Depositional Modelling. Weimer, P. & Link, M.H. (eds): Seismic Facies and Sedimentary Processes of Submarine Fans and Turbidite Systems. Springer-Verlag, 137-170.

Armentrout, J.M., Echols, R.J. & Lee, T.D. 1991. Patterns of Foraminiferal Abundance and Diversity: Implications for Sequence Stratigraphic Analysis. In: Armentrout, J.H. & Perkins, B.F. (eds): Sequence Stratigraphy as an Exploration Tool: Concepts and Practices in the Gulf Coast. Proceedings of the Society of Economic Palaeontologists and Mineralogists Foundation (Gulf Coast Section) Eleventh Annual Research Conference, Houston 1991, 53-58.

Boote, R.D. & Kirk, R.B. 1989. Depositional Wedge Cycles on Evolving Plate Margin, Western and Northwestern Australia. American Association of Petroleum Geologists Bulletin, 73(2): 216-243.

Boyd, R. and Bent, A. 1992. Triassic-Lower Cretaceous Wireline Log Data from Sites 759 through 764. In: von Rad, U., Haq, B.U., et al.: Proceedings of the Ocean Drilling Program, Scientific Results, 122, 377-391.

Boyd, R., Williamson, P. & Haq, B.U. 1992. Seismic Stratigraphy and Passive Margin Evolution of the Southern Exmouth Plateau. In: von Rad, U., Haq, B.U., et al.: Proceedings of the Ocean Drilling Programme, Scientific Results, 122, 39-59.

Boyd, R., Williamson, P. & Haq, B.U., in press. Seismic Stratigraphy and Passive Margin Evolution of the Southern Exmouth Plateau. In: Posamentier, H.W., Summerhayes, C.P., Haq, B.U. & Allen, G.P. (eds): Stratigraphic and Facies Associations in a Sequence Stratigraphic Framework. International Association of Sedimentologists Special Publication.

Bralower, T.J. & Siesser, W.G. 1992. Cretaceous Calcareous Nannofossil Biostratigraphy of Sites 761, 762 and 763, Exmouth and Wombat Plateaus, Northwest Australia. In: von Rad, U., Haq, B.U., et al.: Proceedings of the Ocean Drilling Programme, Scientific Results, 122, 529-556.

Brenner, W. 1992. Dinoflagellate Cyst Stratigraphy of the Lower Cretaceous Sequence at Sites 762 and 763, Exmouth Plateau, Northwest Australia. In: von Rad, U., Haq, B.U., et al.: Proceedings of the Ocean Drilling Program, Scientific Results, 122, 511-528.

Brown, L.F. & Fisher, W.L. 1977. Seismic Stratigraphic Interpretation of Depositional Systems: Examples from Brazil Rift and Pull-Apart Basins. American Association of Petroleum Geologists Memoir, 26: 213-248.

BMR Palaeogeographic Group 1990. Australia: Evolution of a Continent. Bureau of Mineral Resources, Geology and Geophysics.

Carbonel, P. & Moyes, J. 1987. Late Quaternary Paleoenvironments of the Mahakam Delta (Kalimantan, Indonesia). Palaeogeography, Palaeoclimatology, Palaeoecology, 61: 265-284.

Charnock, M.A. & Jones, R.W. 1990. Agglutinated Foraminifera from the Palaeogene of the North Sea. In: Hemleben, Ch., Kaminski, M.A., Kuhnt, W. & Scott, D.B. (eds): Paleoecology, Biostratigraphy, Paleoceanography and Taxonomy of Agglutinated Foraminifera. Kluwer Academic Publishers, 139-244.

Cubaynes, R., Rey, J., Ruget, C., Courtinat, B. & Bodergat, A.-M. 1991. Relations between Systems Tracts and Micropaleontological Assemblages on a Toarcian Carbonate Shelf (Quercy, Southwest France). Bulletin de la Societe Geologique de France, Ser. 8, 6(6): 989-993.

Erskine, R.D. & Vail, P.R. 1988. Seismic Stratigraphy of the Exmouth Plateau. In: Bally, A.W. (ed): Atlas of Seismic Stratigraphy, Volume 2. American Association of Petroleum Geologists, Studies in Geology Series, No. 27, 167-173.

Exon, N.F., Borella, P.E. & Ito, M. 1992. Sedimentology of Marine Cretaceous Sequences in the Central Exmouth Plateau (Northwest Australia). In: von Rad, U., Haq, B.U., et al.: Proceedings of the Ocean Drilling Program, Scientific Results, 122, 233-257.

Exon, N.F., Haq, B.U. & von Rad, U. 1992. Exmouth Plateau Revisited: Scientific Drilling and Geological Framework. In von Rad, U., Haq, B.U., et al.: Proceedings of the Ocean Drilling Program, Scientific Results, 122, 3-20.

Exon, N.F., von Rad, U. & von Stackelberg, U. 1982. The Geological Development of the Passive Margins of the Exmouth Plateau off Western Australia. American Association of Petroleum Geologists Bulletin, 62: 40-72.

Galloway, W.E. 1989. Genetic Stratigraphic Sequences in Basin Analysis I: Architecture and Genesis of Flooding-Surface Bounded Depositional Units. American Association of Petroleum Geologists Bulletin, 73(2): 125-142.

Gorur, N. & Sengor, A.M.C. 1992. Paleogeography and Tectonic Evolution of the Eastern Tethysides: Implications for the Northwest Australian Margin Breakup History. In: von Rad, U., Haq, B.U., et al.: Proceedings of the Ocean Drilling Program, Scientific Results, 122, 83-106.

Gregory, W.A. & Hart, G.F. 1992. Towards a Predictive Model for the Palynologic Response to Sea-Level Changes. Palaios, 7: 3-33.

Haq, B.U. 1991. Sequence Stratigraphy, Sea-Level Change, and Significance for the Deep Sea. Special Publications of the International Association of Sedimentologists, 12: 3-39.

Haq, B.U., Boyd, R., Exon, N.F. & von Rad, U. 1992. Evolution of the Central Exmouth Plateau: a Post-Drilling Perspective. In: von Rad, U., Haq, B.U., et al.: Proceedings of the Ocean Drilling Program, Scientific Results, 122, 801-816.

Haq, B.U., Hardenbol, J. & Vail, P.R. 1987a. Chronology of Fluctuating Sea-Levels Since the Triassic. Science, 235: 1156-1166.

Haq, B.U., Hardenbol, J. & Vail, P.R. 1987b. The New Chronostratigraphic Basis of Cenozoic and Mesozoic Cycles. In: Ross, C.A. & Haman, D. (eds): Timing and Depositional History of Eustatic Sequences: Constraints on Seismic Stratigraphy. Cushman Foundation for Foraminiferal Research Special Publication, No. 24, 7-13.

Haq, B.U., Hardenbol, J. & Vail, P.R. 1988. Mesozoic and Cenozoic Chronostratigraphy and Eustatic Cycles. In: Wilgus, C., et al. (eds): Sea-Level Change- an Integrated Approach. Society of Economic Paleontologists and Mineralogists Special Publication, No. 42, 71-108.

Helby, R., Morgan, R. & Partridge, A.D. 1987. A Palynological Zonation of the Australian Mesozoic. In: Jell, P.A. (ed): Studies in Australian Mesozoic Palynology. Memoirs of the Association of Australasian Palaeontologists, 1-94.

Hill, W.A. & Rosen, R.N. 1991. Concepts and Models to Pliocene-Miocene Sequence Stratigraphy, Central Gulf of Mexico. In: Armentrout, J.M. & Perkins, B.F. (eds): Sequence Stratigraphy as an Exploration Tool: Concepts and Practices in the Gulf Coast. Proceedings of the Society of Economic Palaeontologists and Mineralogists Foundation (Gulf Coast Section) Eleventh Annual Research Conference, Houston 1991, 193-197.

Hocking, R.M., Moors, M.T. & van der Graaff, W.G.E. 1988. The Geology of the Carnarvon Basin. Western Australia. Bulletin of the Survey of Western Australia, 133.

Jacquin, T., Arnaud-Vanneau, A., Arnaud, H., Ravenne, C. & Vail, P.R. 1991. Systems Tracts and Depositional Sequences in a Carbonate Setting: a Study of Continuous Outcrops from Platform to Basin at the Scale of Seismic Lines. Marine and Petroleum Geology, 8: 122-139.

Jervey, M.T. 1988. Quantitative Geological Modelling of Siliciclastic Rock Sequences and their Seismic Expression. Society of Economic Paleontologists and Mineralogists Special Publication, 42: 47-69.

Jones, R.W. & Charnock, M.A. 1985. "Morphogroups" of Agglutinating Foraminifera, their Life Positions and Feeding Habits and Potential Applicability in (Paleo)Ecological Studies. Revue de Paleobiologie, 4(2): 311-320.

Jones, R.W. & Wonders, A.A.H. 1992. Benthic Foraminifers and Paleobathymetry of Barrow Group (Berriasian-Valanginian) Deltaic Sequences, Sites 762 and 763, Northwest Shelf, Australia. In: von Rad, U., Haq, B.U., et al.: Proceedings of the Ocean Drilling Program, Scientific Results, 122, 557-568.

Lewis, J., Dodge, J.D. & Powell, A.J. 1990. Quaternary Dinoflagellate Cysts from the Upwelling System Offshore Peru, Hole 686B, ODP Leg 112, In: Proceedings of the Ocean Drilling Program, Scientific Results, 112, 323-328.

Loutit, T.S., Hardenbol, J., Vail, P.R. & Baum, G.R. 1988. Condensed Sections: the Key to Age Determinations and Correlation of Continental Margin Sequences. Society of Economic Paleontologists and Mineralogists Special Publication, 42: 183-213.

McNeil, D.H., Dietrich, J.R. & Dixon, J. 1990. Foraminiferal Biostratigraphy and Seismic Sequences: Examples from the Cenozoic of the Beaufort-Mackenzie Basin, Arctic Canada. In: Hemleben, Ch., Kaminski, M.A., Kuhnt, W. & Scott, D.B. (eds): Paleoecology, Biostratigraphy, Paleoceanography and Taxonomy of Agglutinated Foraminifera. Kluwer Academic Publishers, 859-882.

Mitchum, R.M., Jr., Sangree, J.B., Vail, P.R. & Wornardt, W.W. 1991. Sequence Stratigraphy in Late Cenozoic Expanded Sections, Gulf of Mexico. In: Armentrout, J.M. & Perkins, B.F. (eds): Sequence Stratigraphy as an Exploration Tool: Concepts and Practices in the Gulf Coast. Proceedings of the Society of Economic Palaeontologists and Mineralogists Foundation (Gulf Coast Section) Eleventh Annual Research Conference, Houston 1991, 237-256.

Mitchum, R.M., Jr., Vail, P.R. & Sangree, J.B. 1977. Stratigraphic Interpretation of Seismic Reflection Patterns in Depositional Sequences. American Association of Petroleum Geologists Memoir, 26: 117-143.

Nagy, J. 1985a. Jurassic Foraminiferal Facies of the Statfjord Area, Northern North Sea, I. Journal of Petroleum Geology, 8: 273-295.

Nagy, J. 1985b. Jurassic Foraminiferal Facies of the Statfjord Area, Northern North Sea, II. Journal of Petroleum Geology, 8: 389-404.

Pacht, J.A., Bowen, B., Shaffer, B.L. & Pottorf, W.R. 1991. Sequence Stratigraphy of Plio-Pleistocene Strata in the Offshore Louisiana Gulf Coast: Applications to Hydrocarbon Exploration. In: Armentrout, J.M. & Perkins, B.F. (eds): Sequence Stratigraphy as an Exploration Tool: Concepts and Practices in the Gulf Coast. Proceedings of the Society of Economic Palaeontologists and Mineralogists Foundation (Gulf Coast Section) Eleventh Annual Research Conference, Houston 1991, 269-285.

Pacht, J.A., Bowen, B., Shaffer, B.L. & Pottorf, W.R., in press. Systems Tracts, Seismic Facies, and Attribute Analysis within a Sequence Stratigraphic Framework - Example from the Offshore Louisiana Gulf Coast. In: Marine Clastic Reservoirs. Springer-Verlag.

Posamentier, H.W., in press. An Overview of Sequence Stratigraphic Concepts: Uses and Abuses. International Association of Sedimentologists Special Publication.

Posamentier, H.W. & Erskine, R.D. 1991. Seismic Expression and Recognition Criteria of Ancient Submarine Fans. In: Weimer, P. & Link, M.H. (eds): Seismic Facies and Sedimentary Processes of Submarine Fans and Turbidite Systems. Springer-Verlag, 197-222.

Poumot, C. 1989. Palynological Evidence for Eustatic Events in the Tropical Neogene. Bulletin des Centres Recherches Exploration-Production Elf-Aquitaine, 13(2): 437-453.

Powell, D.E. 1982. The Northwest Australian Continental Margin. Philosophical Transactions of the Royal Society of London, A305: 45-62.

Rioult, M., Dugue, O., Jan du Chene, R., Ponsot, C., Fily, G., Moron, J.-M. & Vail, P.R. 1991. Outcrop Sequence Stratigraphy of the Anglo-Paris Basin, Middle to Upper Jurassic (Normandy, Maine, Dorset). Bulletin des Centres Recherches Exploration-Production Elf-Aquitaine, 15: 101-194.

Rosen, R.N. & Hill, W.A. 1990. Biostratigraphic Application to Pliocene-Miocene Sequence Stratigraphy of the Western and Central Gulf of Mexico and its Integration to Lithostratigraphy. Transactions- Gulf Coast Association of Geological Societies, 60: 737-743.

Sangree, J.B., Vail, P.R. & Mitchum, R.M., Jr. 1991. A Summary of Exploration Applications of Sequence Stratigraphy. In: Armentrout, J.M. & Perkins, B.F. (eds): Sequence Stratigraphy as an Exploration Tool: Concepts and Practices in the Gulf Coast. Proceedings of the Society of Economic Palaeontologists and Mineralogists Foundation (Gulf Coast Section) Eleventh Annual Research Conference, Houston 1991, 321-327.

Sarg, J.F. 1988. Carbonate Sequence Stratigraphy. In: Wilgus, C., et al. (eds): Sea-Level Change-an Integrated Approach. Society of Economic Paleontologists and Mineralogists Special Publication, No. 42, 115-181.

Shaffer, B.L. 1990. The Nature and Significance of Condensed Sections in Gulf Coast Late Neogene Sequence Stratigraphy. Transactions - Gulf Coast Association of Geological Societies, 60, 767-776.

Stewart, I.J. 1987. A Revised Stratigraphic Interpretation of the Early Palaeogene of the Central North Sea. In: Brooks, J. & Glennie, K. (eds): Petroleum Geology of North-West Europe. Graham & Trotman, 557-576 .

Struckmeyer, H.I.M., Yeung, M. & Bradshaw, M.T. 1990. Mesozoic Palaeogeography of the Northern Margin of the Australian Plate and its Implications for Hydrocarbon Exploration. Proceedings of the First Papua New Guinea Petroleum Convention: 137-152.

Symonds, P.A. & Cameron, P.J. 1977. The Structure and Stratigraphy of the Carnarvon Terrace and Wallaby Plateau. Australian Petroleum Exploration Association Journal, 17: 30-41.

Tait, A.M. 1985. A Depositional Model for the Dupuy Member and the Barrow Group in the Barrow Sub-Basin, North-Western Australia. Australian Petroleum Exploration Association Journal, 25: 282-290.

Tucker, M.E. 1991. Sequence Stratigraphy of Carbonate-Evaporite Basins: Models and Applications to the Upper Permian (Zechstein) of Northeast England and Adjoining North Sea. Journal of the Geological Society, 148: 1019-1036.

Vail, P.R., Hardenbol, J. & Todd, R.G. 1984. Jurassic Unconformities, Chronostratigraphy, and Sea-Level Changes from Seismic Stratigraphy and Biostratigraphy. In Schlee, J.S. (ed): Interregional Unconformities and Hydrocarbon Accumulation. American Association of Petroleum Geologists Memoir, 36, 129-144.

Vail, P.R., Mitchum, R.H., Jr., Thompson, M., III, Todd, R.G. Sangree, J.B., Widmier, J., Bubb, N.N. & Natelid, W.G. 1977. Seismic Stratigraphy and Global Sea-Level Changes. American Association of Petroleum Geologists Memoir, 26: 49-212.

Vail, P.R. & Wornardt, W.W. 1991. Well Log-Seismic Sequence Stratigraphy: an Integrated Tool for the 90's. In: Armentrout, J.M. & Perkins, B.F. (eds): Sequence Stratigraphy as an Exploration Tool: Concepts and Practices in the Gulf Coast. Proceedings of the Society of Economic Palaeontologists and Mineralogists Foundation (Gulf Coast Section) Eleventh Annual Research Conference, Houston 1991, 379-397.

Valicenti, H., Benson, J.M., Petrie, H.S.P., Pringle, A.A & MacMillan, I.K. 1991. Correlation of Depositional Systems Tract Boundaries within the Albian 14A Sequence, Bredasdorp Basin, South Africa, using Condensed Sections. 1er Colloque de Stratigraphie et de Paleogeographie des Bassins Sedimentaires Ouest-Africains/11e Colloque Africain de Micropaleontologie, Libreville, Gabon 1991, Abstracts.

Van Gorsel, J.T. 1988. Biostratigraphy in Indonesia: Methods, Pitfalls and New Directions. Proceedings of the Indonesian Petroleum Association Seventeenth Annual Convention, October 1988: 275-300.

Van Wagoner, J.C., Mitchum, R.M., Campion, K.M. & Rahmanian, V.D. 1990. Siliciclastic Sequence Stratigraphy in Well Logs, Cores and Outcrops: Concepts for High-Resolution Correlation of Time and Facies. American Association of Petroleum Geologists, Methods in Exploration Series, No. 7.

Van Wagoner, J.C., Posamentier, H.W., Mitchum, R.M., Vail, P.R., Sarg, J.F., Loutit, T.S. & Hardenbol, J. 1988. An Overview of Sequence Stratigraphy and Key Definitions. Society of Economic Paleontologists and Mineralogists Special Publication, 42: 39-45.

Von Rad, U., Exon, N.F. & Haq, B.U. 1992. Rift-to-Drift History of the Wombat Plateau, Northwest Australia: Triassic to Tertiary Leg 122 Results. In: von Rad, U., Haq, B.U., et al.: Proceedings of the Ocean Drilling Program, Scientific Results, 122, 765-800 .

Von Rad, U., Haq, B.U., et al. 1992. Proceedings of the Ocean Drilling Program, Scientific Results, 122. Ocean Drilling Program (College Station, Texas).

Weimer, P. & Link, M.H. 1991. Global Petroleum Occurrences in Submarine Fans and Turbidite Systems. In: Weimer, P. & Link, M.H. (eds): Seismic Facies and Sedimentary Processes of Submarine Fans and Turbidite Systems. Springer-Verlag, 9-70.

Williams, D.B. & Serjeant, W.A.S. 1967. Organic-Walled Microfossils as Depth and Shoreline Indicators. In: Hallam, A. (ed): Depth Indicators in Marine Sedimentary Environments. Elsevier (Marine Geology, Special Issue), 389-412.

Wiseman, J.F. 1979. Neocomian Eustatic Changes-Biostratigraphic Evidence from the Carnarvon Basin. Australian Petroleum Exploration Association Journal 19(1): 66-73.

Woodside Offshore Petroleum 1988. A Review of the Petroleum Geology and Hydrocarbon Potential of the Barrow-Dampier Sub-Basin and Environs. In: Purcell, P.G. & Purcell, R.D. (eds): The North West Shelf Australia. Petroleum Exploration Society of Australia, 115-128.

R.W. Jones and P.A. Ventris
BP Exploration (Frontier & International)
4-5 Longwalk Road
Stockley Park
Uxbridge
Middlesex UB11 1BP.

# 7 ENGINEERING & ECONOMIC GEOLOGY

Malcolm B. Hart

## 7.1 Introduction

Many of the geological problems associated with engineering works and the extraction of industrial minerals can be solved using standard, or somewhat innovative, micropalaeontological techniques. This is especially true in cases where the micropalaeontological input is needed to provide precise stratigraphical control within a particular geological setting.

The roles of micropalaeontology in this area can be identified as follows:

- where detailed stratigraphical knowledge is required;
- where subsurface/borehole material is being employed and, as a result, sample size may be limited;
- where it is necessary to identify the presence of faults and/or repeated successions;
- where stratigraphical knowledge is required in an area of complex facies relationships; and
- where there is a need to identify the presence or absence of reworked or allochthonous material.

Figure 7.1: Geological map of south-east England and the location of engineering and other sites mentioned in the text.

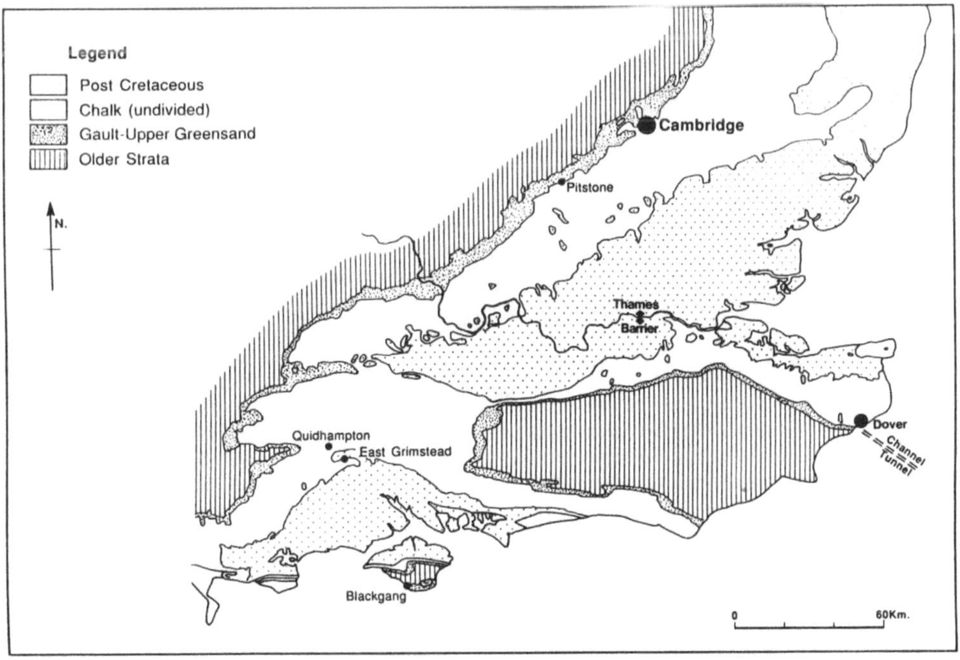




*D. G. Jenkins (ed.), Applied Micropaleontology, 225–262.*
© 1993 *Kluwer Academic Publishers.*

The types of investigation chosen to exemplify the geological situations listed above include:-

1. provision of precise stratigraphical control - the Channel Tunnel site investigation;
2. the identification of a lack of geological problems in the form of faults, folds, etc. - the Thames Barrier site investigation;
3. the identification of exotic elements and/or allochthonous materials - the Site investigation for the M10 Cambridge by-pass;
4. the identification of repetitive successions and/or displacements - the Blackgang landslip on the Isle of Wight; and
5. the identification of stratigraphically controlled facies for the exploitation of industrial minerals - examples from Pitstone, Quidhampton and East Grimstead.

Figure 7.2: Geological succession represented in the various sites indicated in Figure 7.1: Column 1 is the succession represented at the site of the Thames Barrier; Column 2 is the Channel Tunnel; Column 3 is Blackgang, Isle of Wight; Column 4 is East Grimstead/Quidhampton; Column 5 is Pitstone; and Column 6 is Cambridge.

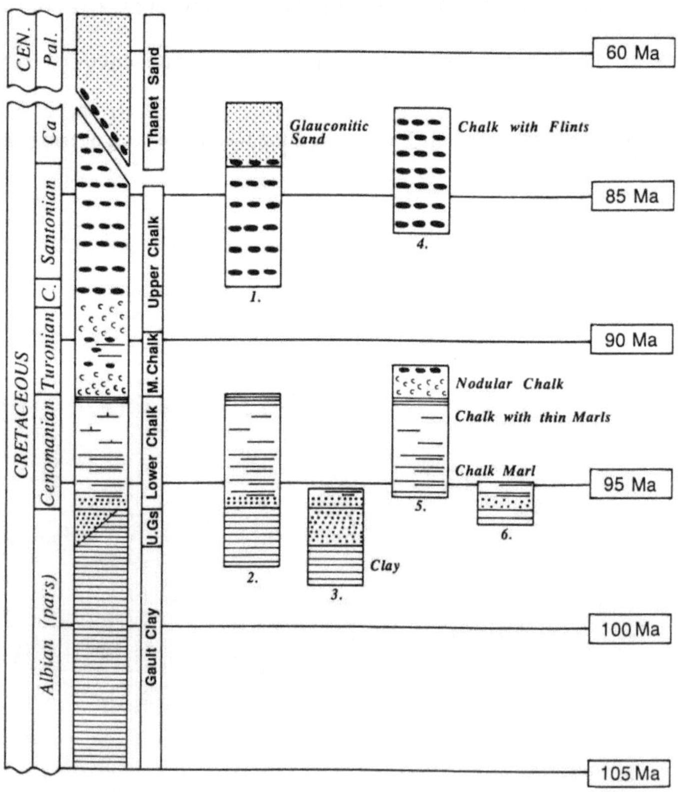

All the examples chosen to illustrate this application of micropalaeontology are from Southern England (Figures 7.1 and 7.2) and involve strata of Cretaceous and Palaeocene age. This closely reflects the author's interests but there are other reasons. Whenever the chalk is involved in any investigation that involves borehole material, microfossils are normally the only way in which a typical engineering description of "structureless white chalk" can be interpreted stratigraphically. Only in exceptionally well cored material and exposed faces could this be done on lithological characteristics alone. Because of its areal extent in Southern England, numerous geological activities involve chalk in one way or another. In all these applications it has been necessary to acquire large amounts of data for rapid correlation with the *minimum* of effort (= time = money). In some cases the data generated are enormous and, because of that, graphical techniques have been employed. This is especially true of those examples requiring the resolution of stratigraphical problems. This type of work is made all the more difficult by the size of the "site". Most geologists are trained in, and are familiar with, inter-regional and inter-basinal correlation. In such cases it is normal for a few taxa, common to both areas, to be relied upon for the creation of an appropriate stratigraphy. However, in engineering and economic geology, the site may only be 1km or less across and the accuracy required is usually far beyond that which the client is prepared to pay for in the form of boreholes and other data gathering methods (e.g. geophysical surveys). In some cases, however, this is not necessarily true and the data obtained for prestige projects (e.g. Channel Tunnel and Thames Barrier) are generally adequate.

In every case it must be understood that the client does not want his succession fitted into a global stratigraphic scheme but he wants precise local correlation. This is frequently very difficult. While extinctions or first appearances of taxa provide such guidelines to the international biostratigrapher it is impossible to guarantee that the (usually) inadequate sampling will pick up all such bio-events in boreholes only metres apart. The accuracy desired is often greater than the sample spacing. Figure 7.3 shows how this problem manifests itself in a series of imaginary boreholes.

In any preliminary research the published literature will almost certainly be consulted and, while this often provides a boost to confidence in one's awareness of the fauna, it can be misleading. Published, academically-oriented research will often depend on a large number of closely spaced samples, with re-sampling undertaken at key boundaries. Taxonomic splitting, often using biometrics, may well be beyond what is possible with either the material or the time available. In many cases it is probably safer to use an informal, specifically generated taxonomy appropriate for the job in hand. This was done by D.J. Carter in the early (1958/59) Channel Tunnel Site Investigation, with nearly all the species being given informal names and/or numbers. This is almost the traditional utilitarian approach used by the oil industry in the 1950's. This was modified over a period of years to in-house taxonomic decisions, many of which may follow the research literature after a period of time. Although quite obvious, it is probably well worth stating again that the client, especially in one-off engineering investigations, probably has not the slightest interest in the fauna being employed by the consultant, or in its significance.

The individual examples are described below and their special problems identified. Each example is very different and all had to be tackled in a variety of ways, some of which worked, while others were recognised as imperfect as work progressed. A flexible approach is the key to a successful outcome. The cost of such work is inevitably a minute proportion of any site investigation programme, which in itself is often a low percentage of the final cost or economic value of the commodity. In the case of the Thames Barrier Site the costs of the micropalaeontological investigation were only about 0.0001% of the final costs of the structure.

Figure 7.3: This diagram represents three boreholes (with various types of samples) at an engineering site together with an imaginary research section in a remote area. Bioevents 'X', 'Y' and 'Z' in the boreholes show how the samples from the engineering site are almost more difficult to correlate than to correlate one borehole with the remote, separate basin.

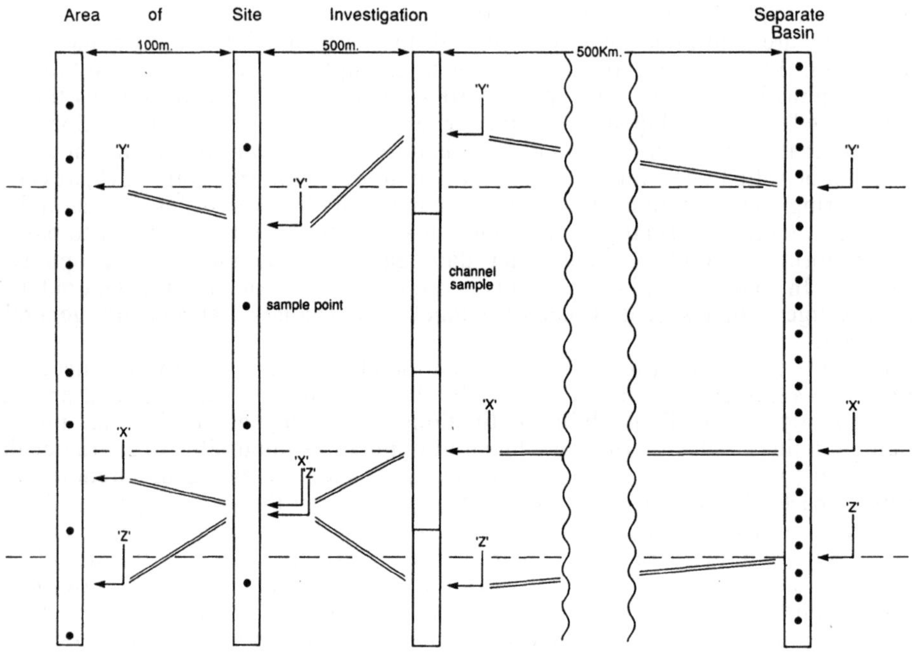


## 7.2 The Channel Tunnel site investigation

In many ways this must be regarded as one of the engineering feats of the 20th Century and it is interesting to consider the role of micropalaeontology in the project. Much of the work for the project was done in 1958/59 and 1964/66 by D.J. Carter (ex-Imperial College, London), with the author being involved in the refinement of the zonation and stratigraphy from 1965 onwards, leading through to on-site work with Transmanche Link in 1988/91. The micropalaeontological work involves the *precise* determination of the stratigraphy and the *accurate* correlation of the successions involved in the route.

The route of the bored tunnels (two running tunnels and one central service tunnel) involves the Lower Chalk - Gault Clay succession, and while most of the route is through the lower part of the Lower Chalk it is the boundary between this and the Gault Clay which is of particular significance. Detailed work on the Gault foraminifera has been undertaken by Berthelin (1880), Chapman (1891-98), Hart (1970, 1973a), Magniez-Jannin (1975), Carter & Hart (1977a), Price (1977) and Harris (1982). There is therefore a large data base on the foraminiferal fauna, with two workable zonations (Carter & Hart, 1977a and Price, 1977) currently available. The Carter & Hart (1977a) scheme has been used in the correlation of the subsurface sections and boreholes. The Albian/Cenomanian boundary (Figure 7.4) shows a number of very distinct foraminiferal changes. Several of the taxa are not formally identified, while the names of others (e.g. *Arenobulimina* spp.) probably should be changed in line with the recent proposals of Barnard & Banner (1980). When the zones, so defined (see Carter & Hart,1977a, for formal designations) are applied to the succession below East Wear Road (Folkestone, Kent) and a series of boreholes (see Figure 7.6 for locations) along the route of the tunnel towards Dover, a number of significant hiatuses are identified. Of major importance are the erosional surfaces seen at the base of Zones 6, 6A and 7. The base of Zone 6 is the lowermost surface of "Bed XII", using the Price (1874) nomenclature. The base of Zone 7 is the base of the Glauconitic Marl and is a widespread hiatus across Southern England (Hart, 1970; Carter & Hart, 1977a). Zone 6A is unknown from surface outcrop, except for a few centimetres in the coastal sections of East Wear Bay, Folkestone. Westwards (e.g. the Glyndebourne Borehole, Sussex) it is not recorded (Figure 7.5), while eastward towards Dover and under the English Channel (Figure 7.6) it takes on a greater significance. Lithologically, Zone 6A appears as a highly micaceous mudstone and it is likely to have very different engineering properties compared to the normal Gault Clay. Under the English Channel it becomes extensively developed and attains its maximum thickness, before thinning and finally disappearing in mid-Channel. From approximately the same point on the route, the basal part of the chalk succession becomes progressively lost at the level of the "base of Zone 7" hiatus. This can be particularly seen if one considers the correlation of the Lower Chalk across the area.

Carter & Hart (1977a) formally established a series of zones based on benthonic and planktonic Foraminiferida (Figure 7.7). These have been updated by Hart et al. (1989) but for ease of subsurface correlation the original zonal numbers and definitions are retained. When published in a final form all the data will be fully updated to the appropriate terminology. Two methods of correlation have been employed for the part of the succession involving the Lower Chalk. As shown in Figure 7.7 there is a workable zonal scheme (Carter & Hart, 1977a) based on benthonic foraminifera (Zones 7, 8, 9, 10, 11(i), 11(ii), 12, 13, 14). Coupled with this is the use of planktonic:benthonic ratios (also described by Carter & Hart, 1977a) for correlation. While these graphical plots cannot be reliably used in wide, regional, correlation it is possible to use the graphs to identify correlation lines (Figures 7.7 and 7.8) for use between closely-spaced boreholes.

The planktonic:benthonic ratios were calculated (for the purpose of this work) on the 250μm-500μm grain-size fraction, and involved counts of over 300 specimens per sample. Samples were normally taken at a 1m or less spacing, with additional samples close to key boundaries. The planktonic component of all the counts was separated into three categories, viz. *Rotalipora* spp., *Praeglobotruncana/Dicarinella* spp., and *Hedbergella/Whiteinella* spp. The graphs produced by this analysis look like the one shown in Figure 7.7.

Figure 7.4: Foraminiferal zonation of the Albian/Cenomanian boundary in the area of the Channel Tunnel. The zonation (Zones 5-9) is that described in Carter & Hart (1977a) and the taxonomy of all the species illustrated is also covered in that publication. The Glauconitic Marl (shown as Zone 7 on this diagram) is not always the same age in Southern England (see Figure 7.5).

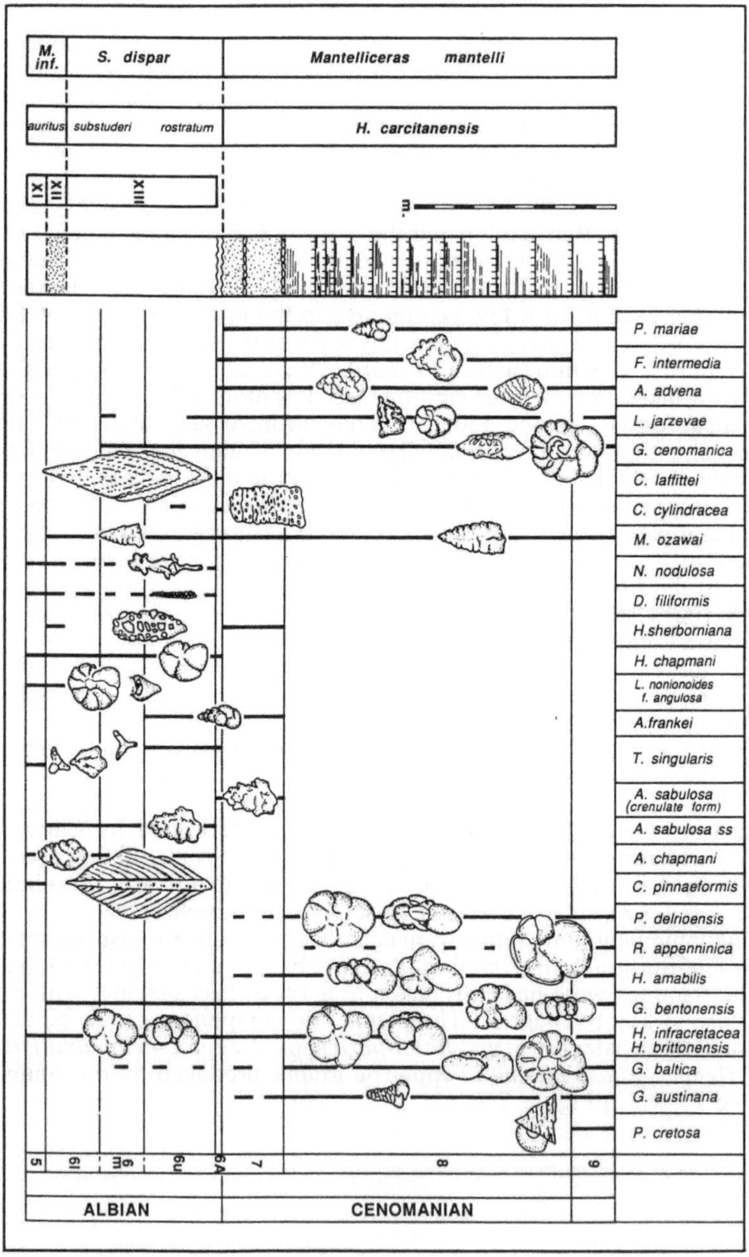

Figure 7.5: Correlation of the Albian/Cenomanian from Folkestone (East Wear Road) and Dover (P.000). The borehole and field sites are shown in Figure 7.6. The Glyndebourne borehole is located a few kilometres east of Lewes in Sussex. At this locality the Glauconitic Marl is of Zone 8 age, compared with Zone 7 in the Folkestone/Dover area.

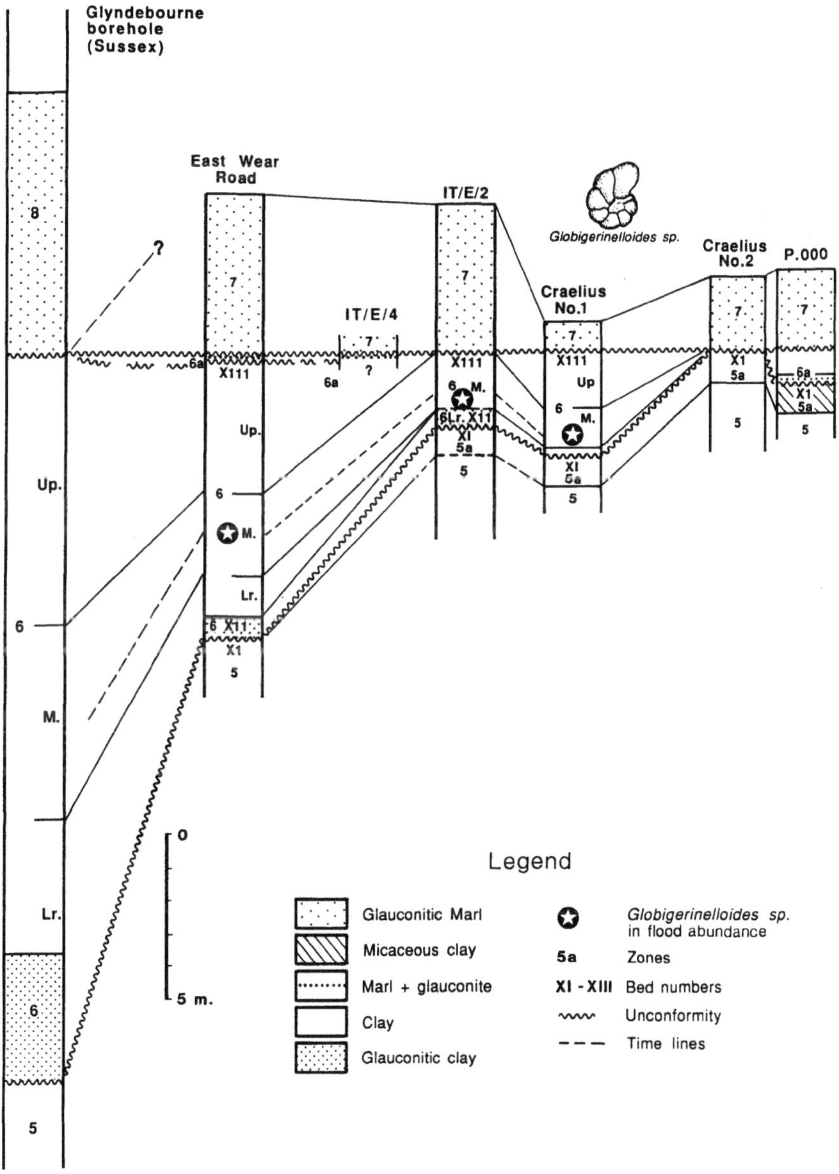

Figure 7.6: Correlation of the Albian/Cenomanian boundary across the English Channel. The top of the Glauconitic Marl has been used as a datum simply as a convenient line across the section. The lower diagram shows how levels of abundance of *Hedbergella* spp. can be used in the correlation of Zone 6a, and Zone 7.

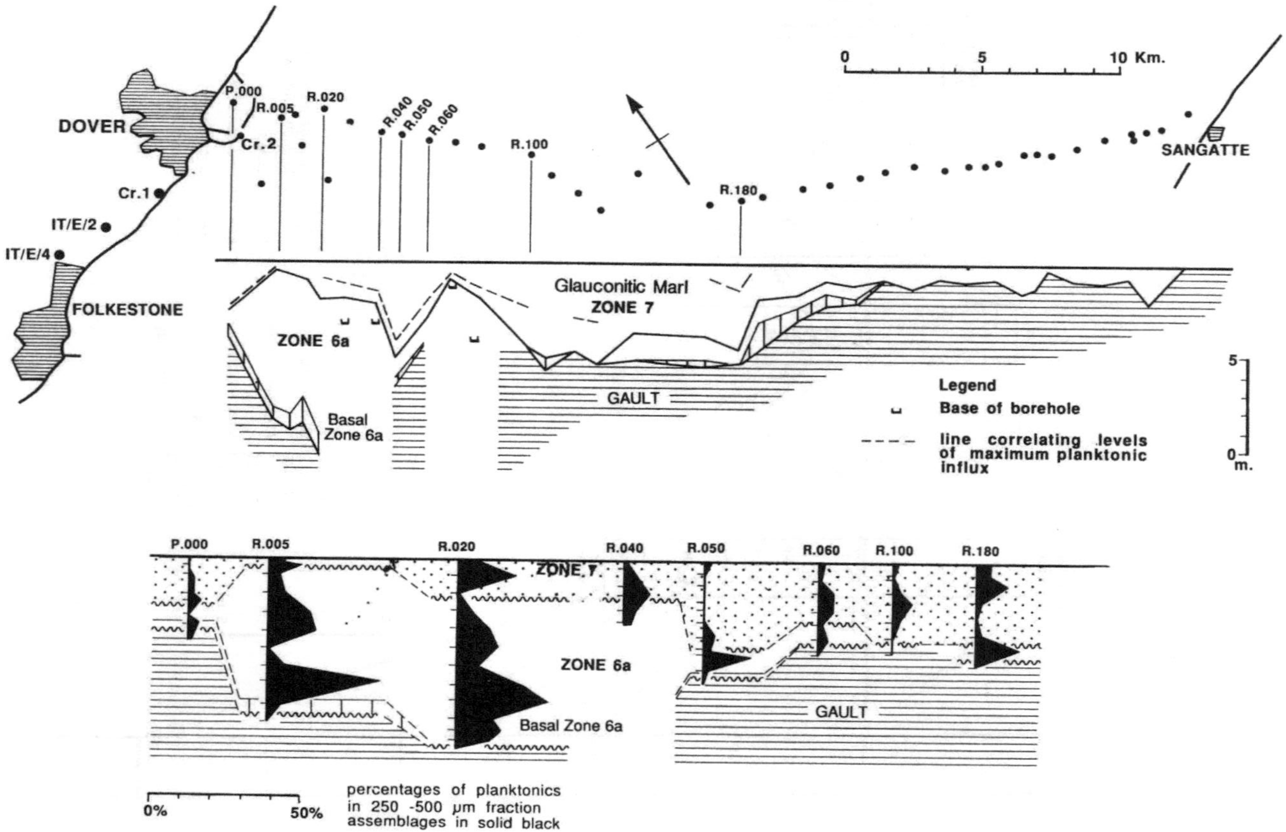

Figure 7.7: Schematic log of borehole P.000, which was located in Dover Harbour. The sample points are shown in the left-hand column, while the foraminiferal zones of Carter & Hart (1977a) are shown immediately to the right of the lithological log. Graph A shows the simple planktonic:benthonic ratio based on the 500-250μm grain size fraction. Graph B shows the same graph but with the planktonic taxa identified. The mid-Cenomanian non-sequence is highlighted as is the so-called 'C' line of correlation. The lines used in the cross-Channel graphical correlation are also indicated. Another example of this type of correlation is shown in Figure 7.20.

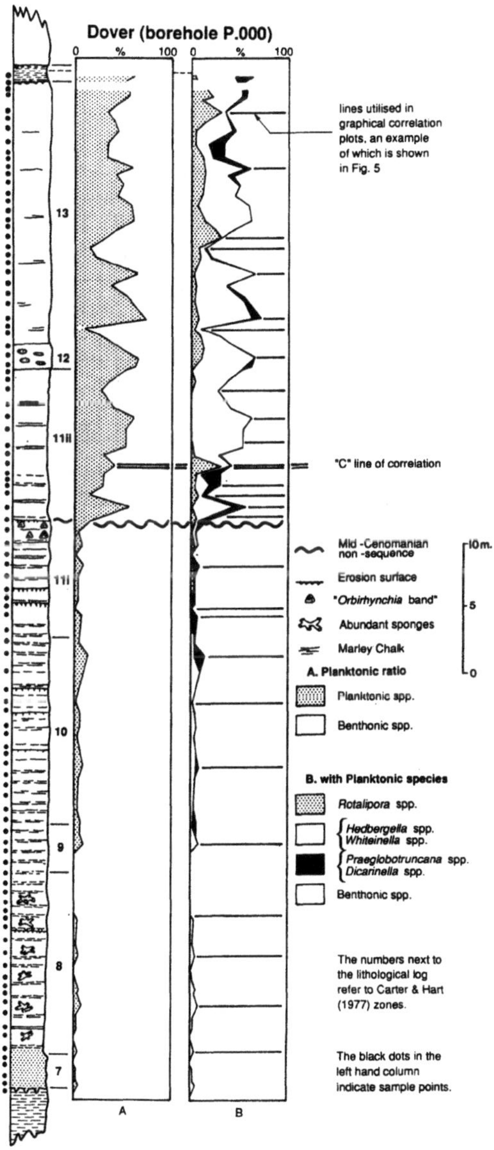

Just above the 11(i)/11(ii) boundary (the mid-Cenomanian non-sequence) there is a marked peak in the occurrence of *Rotalipora cushmani* (Morrow). This is the so-called "c-peak" which can be followed across the English Channel (see Figure 7.8). All the other features of these graphs were identified by letter and symbol, and then used to construct correlation lines which can be followed, in many cases, across the section. While publication of the whole profile is not possible at the present time, Figure 7.8 shows four un-named boreholes and the correlation that was generated between them. One is close to Dover Harbour, two are in mid-Channel, while the last one is near to the French coast. It must be remembered that Figure 7.8 has been drawn using the "c-peak" as the horizontal datum; this is simply for convenience but it also removes the effects of all post-Cenomanian folding etc. There are several features of interest in this correlation.

1. Everything above the "c-peak" of *Rotalipora cushmani* is essentially horizontal, indicating an almost layer-cake depositional environment.
2. The mid-Cenomanian non-sequence becomes a distinct unconformity as one correlates this hiatus across the Channel. This coincides with the base of the *R.cushmani* Taxon Range Zone and the 11(i)/11(ii) boundary of the benthonic zonation. Followed westwards from Dover, towards SW England, this unconformity takes on an even greater significance (Kennedy, 1970; Hart, 1970; Carter & Hart, 1977a).
3. The base of the chalk becomes progressively younger towards the French coast and this can be confirmed using the data from the Boulonnais provided by Amedro et al., (1978) although a number of benthonic taxa have slightly different names. This loss of the very clay-rich lowermost Lower Chalk changes quite significantly the engineering properties of the materials encountered *en route*.

Using a combination of standard benthonic foraminiferal zonation, simple graphical data plots, and a very primitive form of graphical correlation (done in 1965/66) it has been possible to provide an appropriate stratigraphical data base for the project. Subsequent improvements in seismic surveys and downhole logging techniques have enhanced the resolution in key areas but these were not readily available when the greater part of the micropalaeontological work was done.

### 7.3 The Thames Barrier site investigation

The Thames Barrier spans the River Thames between Woolwich and Silvertown and is designed to prevent flooding in Central London. It consists of ten moveable steel gates which close to the incoming flow when abnormally high tides (aided by a North Sea surge) are forecast. Each gate is supported by concrete piers which house the operating machinery. The four central gates each weight 3300 tonnes. The design of the structures, gates and operating machinery assumes sound foundations. The site investigation was conducted between December 1971 and May 1972 by Foundation Engineering Ltd under the direction of Rendel Palmer & Tritton, the consulting engineers to the Greater London Council. The consulting engineers advised Foundation Engineering Ltd that a micropalaeontological and biostratigraphical investigation of the chalk across the site would be the best way of assessing the site and D.J. Carter, assisted by the author, undertook the investigation. The final report was presented to the GLC in May 1972, after which time construction began. The first use of the completed barrier was on 31st October 1982.

Figure 7.8: Correlation of the Albian/Cenomanian successions in the area of the Channel Tunnel showing the methods developed using the data in Figure 7.7. The diagram shows the succession at either end of the Channel Tunnel and two unspecified boreholes from the mid-Channel area. The 'C' line of correlation has been used as this may represent a synchronous, basin-wide event. The base of the chalk succession on the French coast can be seen to be markedly younger than that on the English coast.

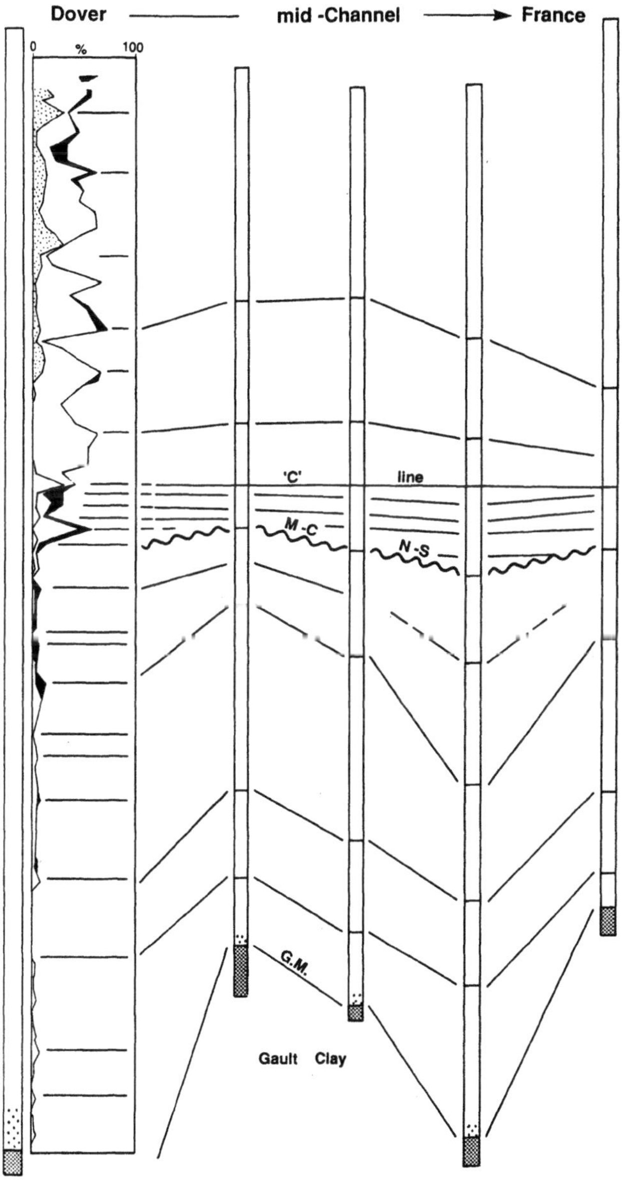

The geology of the area can be seen from the Geological Survey Sheet 271. It is underlain by Upper Chalk which dips north-north-east below the Palaeocene Thanet Sands. Across the site, the unconformable Thanet Sand and the Upper Chalk are unconformably overlain by river alluvium and some late glacial flood plain gravels. The river bed is generally covered in mud, although in some places chalk was easily exposed by dredging. Apart from near the north bank, where some of the small structures were founded in Thanet Sands, the pier foundations in the river cut through the Tertiary and younger sediments and rest in the Upper Chalk.

68 boreholes were drilled to investigate the Upper Chalk. There was some *in situ* engineering testing as well as a number of tests on samples in the laboratory. A full marine geophysical survey was undertaken by Wimpey Laboratories Ltd. There were 49 over-water borings, three of which were done from staging on the south bank and 46 from barges. There were 19 land boreholes. All the chalk boreholes yielded abundant microfaunas and most of the chalk cores showed very little drilling disturbance. The main cause of core disruption was an occasional flint jammed in the drill-bit. Allowance had to be made for data obtained from such material.

The engineering properties of a bed of fresh, undisturbed chalk show relatively little lateral variation when the area involved, like that covered by the Thames Barrier Site, is small (Carter & Mallard, 1974). However, appreciable changes can occur when beds from different levels within the chalk sequence are considered. In the absence of weathering and frost shattering, for example, the uniformity or otherwise of the engineering properties of chalk depend on whether it is, or is not, all of closely similar age. This was not known with certainty prior to the investigation. The two objectives of the micropalaeontological study were:

(i) the production of a correlation sufficiently detailed to permit the construction of geological sections through the site, showing the stratigraphical horizon of each pier founding level and the position of faults which might leak or affect the stability of the pier foundations;

(ii) the location of boundaries between undisturbed *in situ* chalk and disturbed solifluction and/or frost shattered chalk in the borehole sections by the identification of sedimentary grains and other material not normally associated with the chalk.

Although the Lower Chalk was studied in detail in the Channel Tunnel site investigation (Carter, 1966) some information on the distribution of microfossils in the Middle Chalk and the upper part of the Upper Chalk was available (Williams-Mitchell, 1948; Barnard & Banner, 1953; Barr, 1962, 1966; Hart, 1970; Owen, 1970). In the lower part of the Upper Chalk, however, no study of the microfaunal sequence was sufficiently comprehensive to be of use in the Thames Barrier site investigation. Much basic research was therefore required to determine the microfaunal succession, and to apply it to obtain fine, accurate correlations.

The sampling and processing procedures have been described by Carter & Hart (1977b) but in most boreholes, samples were taken from undifferentiated chalk at 1.5m intervals. For cores showing variable lithology the interval was reduced to 1.0m and additional samples were taken immediately below horizons of marked lithological change. Other samples (from U4 or U100 percussion cores) were taken carefully following extrusion, always removing the disturbed material round the edge of the core. The majority of samples had to be crushed carefully, under water, using a heavy pestle and mortar. They were then washed very carefully, dried, stored and catalogued.

Figure 7.9: Outline geological map of the Thames Barrier site. The area of this map is indicated on the inset map of SE England. Pegwell Bay is the nearest surface exposure of the Chalk/Thanet Sands succession and this was used as a reference section during the investigation.

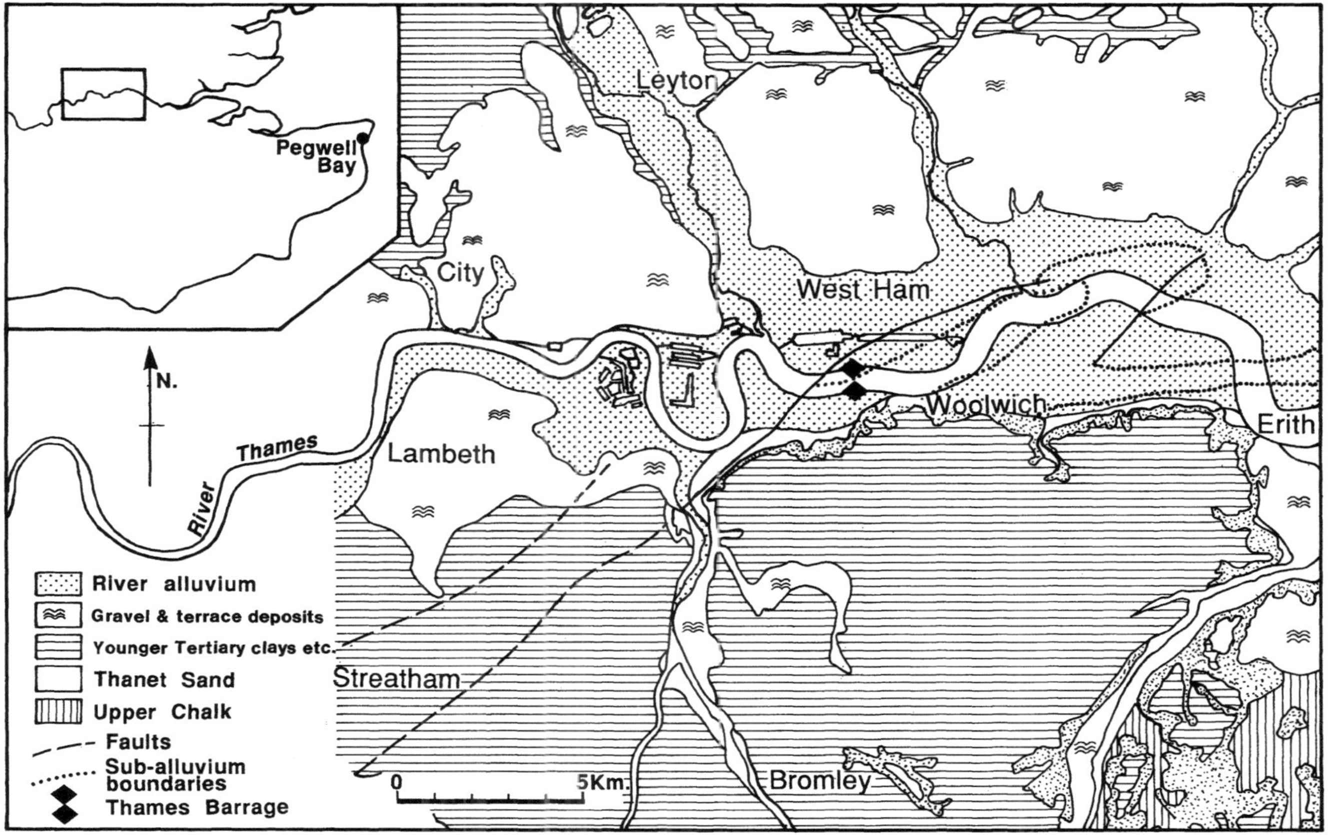

Borehole 25, located on the southern bank (Figure 7.9), was used as a reference section, and studied first. Three grain-size fractions were studied >500μm, 500-250μm and 250-65μm, but counts were only done on the middle fraction as it contained the majority of the diagnostic fauna that was present in that size range. As only a minority of species encountered were described in the literature, all species - except those immediately identifiable - were placed in genus and given an "informal" letter (e.g. *Gaudryina* sp. B) or number. Certain species known to have very long or, ranges for other reasons, little stratigraphic significance, were not counted individually but in groups. In all 71 species of benthonic and 8 species of planktonic foraminifera were identified and over 250,000 individuals counted.

Faunal distribution charts for borehole 25 were prepared showing the stratigraphic range and abundance of the species encountered. Different methods of presentation of the statistics and grouping of species were employed for different charts. The following types of chart were found to be useful:

(i) a simple range chart showing the total vertical range of all species in all grain-size fractions;

(ii) a distribution chart for the 500-250μm grain-size fraction showing the percentage abundance of species and groups of species within one particular benthonic superfamily (Cassidulinacea - *sensu* Loeblich & Tappan, 1964) plotted as a cumulative histogram (Figure 7.11A);

(iii) a distribution chart for the 500-250μm grain-size fraction showing the percentage abundance of species groups of planktonic foraminifera in the total assemblage, expressed as a cumulative percentage graph (Figure 7.11B);

(iv) a distribution chart for the 500-250μm grain-size fraction showing the percentage abundance of the four superfamilies represented in the benthonic fauna, expressed as a cumulative percentage graph (Figure 7.11C - Buliminacea only).

The foraminiferal zonation was based on (i) and (ii). The recognition of intrazonal sub-divisions is based on (ii), (iii) and (iv). Foraminiferal zones were erected for borehole 25 and the overlapping borehole 19 (Figure 7.11). To avoid the use of complex biological nomenclature these were given the names of colours (which were later plotted on the geological sections across the site, Figure 7.12). In 1971/72 the relationship of these zones to the accepted macrofossil zonation in the lower part of the Upper Chalk was not clear, but subsequent work by Bailey (1978), Bailey & Hart (1979) and Hart et al. (1989) has clarified the situation.

**Red Zone** - upper part of the *Micraster coranguinum* Zone (by comparison with the Kent coast sections).
**Blue Zone** - equivalent to the middle and lower parts of the *M. coranguinum* Zone.
**Orange Zone** - equivalent to the very lowest part of the *M. coranguinum* Zone and the larger part of the *Micraster cortestudinarium* Zone.
**Violet Zone** - equivalent to the *Sternotaxis planus* Zone and possibly the basal part of the *M. cortestudinarium* Zone.
**Yellow Zone** - equivalent to the top of the Middle Chalk (cf. *Terebratulina lata* Zone).

Figure 7.10: Microfaunal analysis of Borehole 25 - Thames Barrier site investigation. The informal zonation used in the site investigation is compared with the zonation (A-E) established by Bailey (1978). The macrofossil zonation is based on extrapolation of the zonal boundaries and not the distribution of macrofossils in the borehole cores.

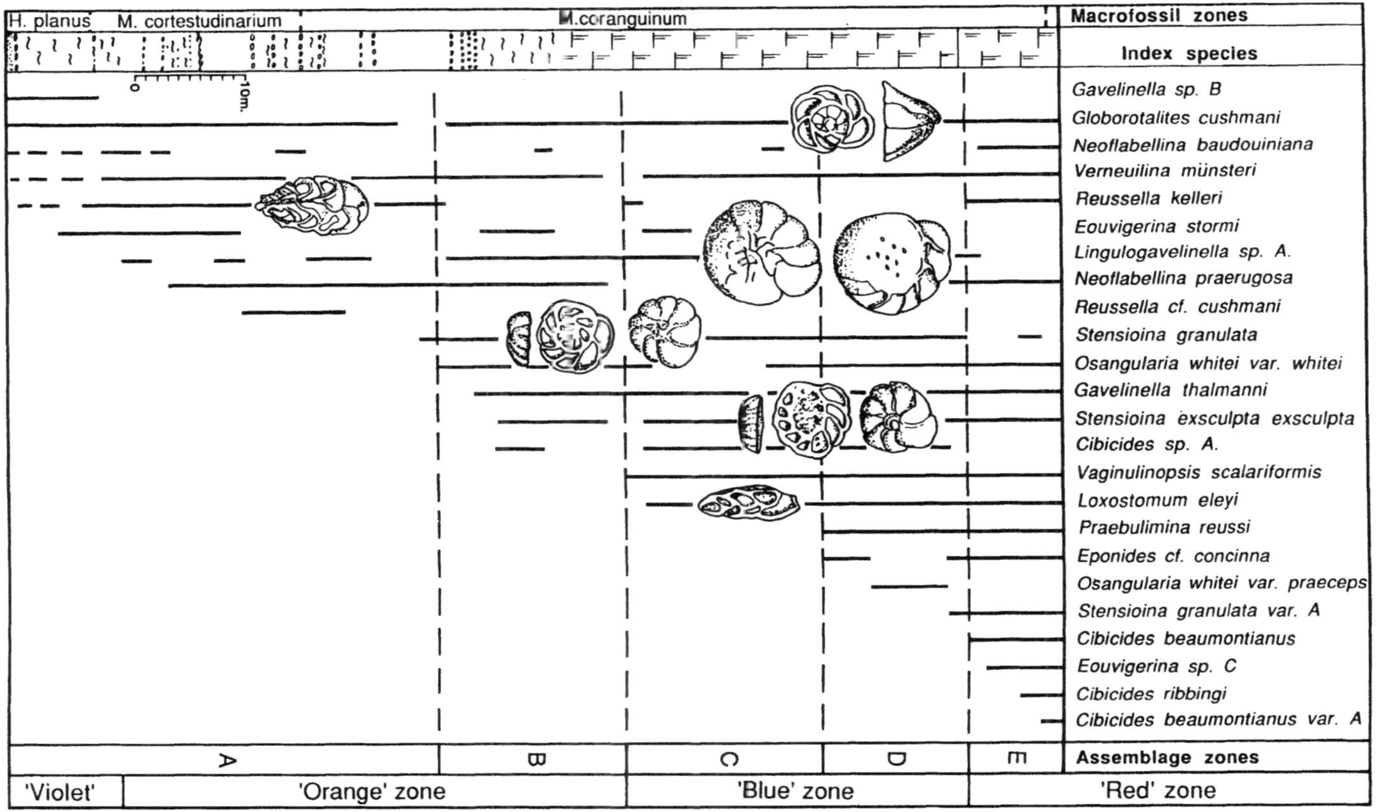

Figure 7.11: Raw micropalaeontological data from boreholes 25, 19 and 52 - Thames Barrier site investigation. The 60-30 fr is the 500-250μm size fraction which was used as a standard in the project. Graphs A, B and C were used to effect a correlation. An example of the problems explained in Figure 7.3 can be seen clearly if one considers the marker species shown against Boreholes 25 and 52. It will be noticed that the first appearances of *L. eleyi* and *S. exsculpta exsculpta* are reversed. It is also clear that correlation lines 'A', 'II' and 'III' provide a good correlation, as does the sharp reduction of the *L.* aff. *L. vombensis* population at the Blue/Red boundary.

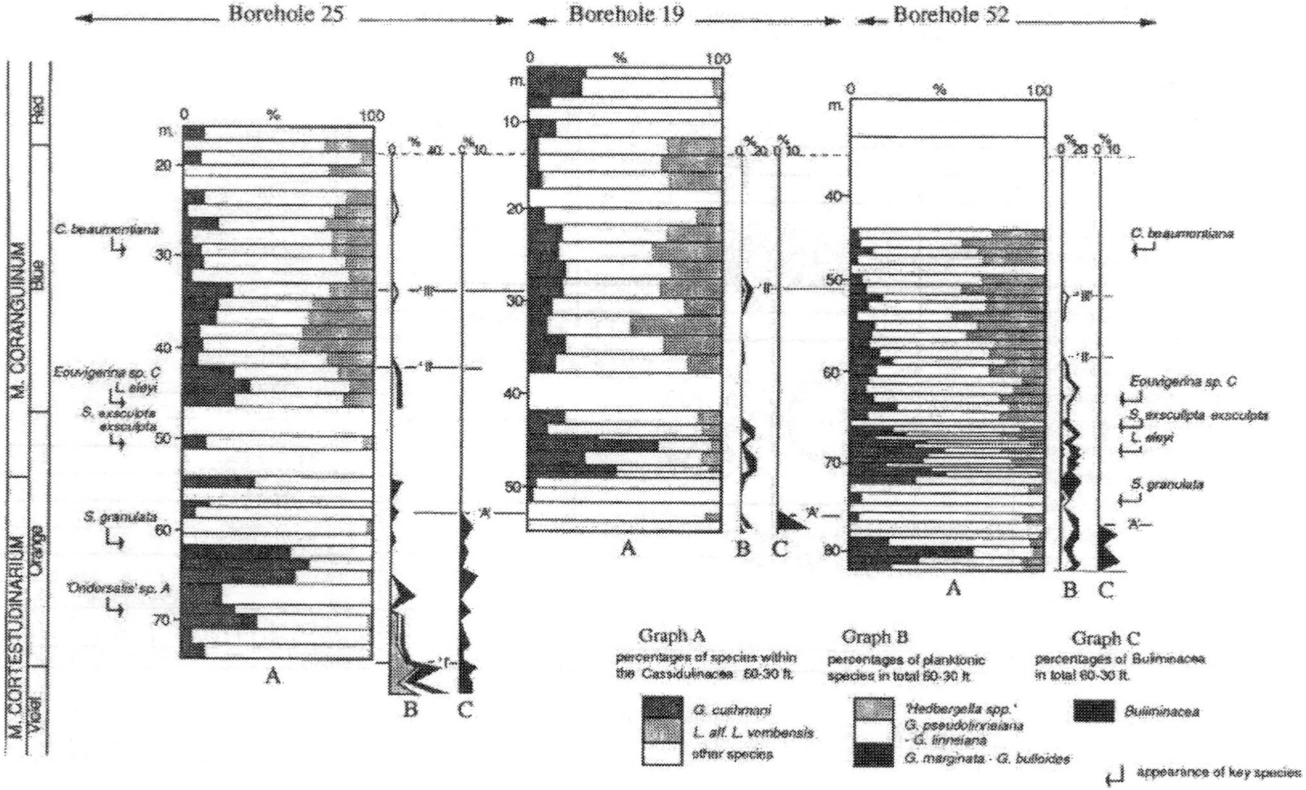

After all the palaeontological data were extracted from the samples, several different methods of correlation were employed.

Microfossil zones are based on the vertical ranges or overlap of vertical ranges of species. Any species near the end of its vertical range tends to be rare and therefore irregularly distributed laterally; for this reason species can appear and disappear at slightly different levels in closely-spaced different boreholes (e.g. *Stensioina exsculpta exsculpta* appears before *Loxostomum eleyi* up-section in borehole 25, but the order of appearance is reversed in Borehole 52 [see Figure 7.11]). For this reason the positions of zonal boundaries can vary slightly from borehole to borehole and correlations based on them are not always completely accurate. The "Red/Blue" zonal boundary, however, is sharp and easily recognisable. The microfaunal change marking it occurs within the stratigraphical thickness of 0.01m and the accuracy with which it is located in the boreholes depends solely on the sampling interval (Figure 7.3). The change downwards from the "Red" to the "Blue" zone involves the sudden appearance in abundance of the very striking *Lingulogavelinella* sp.cf. *L. vombensis* (Brotzen). Subsequent work by Bailey (1978), Robaszynski et al. (1980) and Hart et al. (1989) has shown the value of this taxon in correlation, although it is now accepted as not being the "*vombensis*" of the original author (a Maastrichtian species from Sweden).

Correlations based on zonal boundaries were supplemented by others based on different criteria.

### *(i) Fluctuations in the abundance of benthonic species*

Many long-ranging benthonic species fluctuate in abundance through the drilled section. These fluctuations are particularly well-marked in the Superfamily Cassidulinacea and are best demonstrated using cumulative percentage histograms of species or groups of species within the Superfamily in the 500-250µm grain-size fraction. The cumulative histograms for Boreholes 25, 19 and 52 are compared in Figure 7.11, Graph A. The overall distribution patterns are governed largely by those of two of the constituent species, *Globorotalites micheliniana* (d'Orbigny) and *L.* sp.cf. *L. vombensis.*

Similar histograms were prepared for 18 boreholes. Differences were found to occur both as the distance between the boreholes and between samples in the same boreholes increases. The former reflects increasing difference induced by lateral change of environment between assemblages living on the bottom, and the latter shows increasing distortion of the true distribution pattern caused by increasingly incomplete representation. Correlation lines based on the connection of features of these intrinsically unstable patterns must be treated with considerable reserve and considered as guides to correlation rather than firm ties.

### *(ii) Planktonic:benthonic ratio*

Planktonic foraminifera in general evolved rapidly, but in these assemblages they show little differentiation and the same species are sporadically distributed throughout the section. However, studies of those in Borehole 25 reveal changes in their abundance at various levels. Graphs for Boreholes 25, 19 and 52 are shown in Figure 7.11. Three features are easy to recognize: the sudden fall in planktonic percentages up-section at the "Orange"/"Violet" boundary (Level "I"); the sudden disappearance of planktonic species up-section in the lower part of the "Blue" zone (Level "II"), and their sudden, temporary reappearance in fair abundance near its middle (Level "III"). The unequivocal identification of these and other features depends on the availability of long, tightly sampled sequences from the boreholes to be correlated. In the majority of holes recovery was too poor to fulfil these requirements.

*(iii) Percentage abundance of superfamilies*

When the 250-500μm grain-size fraction of benthonic assemblages from Borehole 25 were separated into their constituent superfamilies, the overall pattern was not sufficiently distinctive to serve as a basis for correlation. However, one superfamily (Buliminacea - Loeblich & Tappan, 1964), which is well represented in the <250μm grain-size fraction, contains a few species which attain large size in some parts of the section and therefore also appear in the 250-500μm grain-size fraction. These large-sized specimens appear sporadically in the lower part of the "Violet" zone, increasing in abundance upwards, and may constitute up to 15% of the 250-500μm grain-size fraction in the lower part of the "Orange" zone. About two-thirds of the way up the "Orange" zone these large specimens abruptly disappear at a level thought to represent a slight erosion break. This horizon (Level A) which might approximate to the *M. coranguinum/M. cortestudinarium* zonal boundary, is sharply defined and a good datum; its position in Boreholes 25, 19 and 52 is indicated in Figure 7.11. Because of its low position in the succession it is not represented in the majority of boreholes.

It is generally accepted on evidence from Kent and Surrey that the Thanet Sand, often with the Bull Head Bed at its base, overlies chalk that has been very gently folded and tilted prior to erosion and therefore is slightly different in age in different places. Profiles of identifiable horizons within the Chalk on site demonstrate that although the erosion plane cuts only one Zone (the *M. coranguinum* Zone) this has been thrown into a series of shallow flexures before bevelling occurred. The river alluvium and flood plain gravels were shown to rest on Thanet Sand only to the north of Borehole 12; to the south of it they rest directly on chalk. Post-Thanetian structures were investigated by plotting a stratum contour map of the sub-Thanet erosion surface (top of the Chalk) based on triangulation of site borehole levels. This was fully discussed in Carter & Hart (1977b) who identified an en-echelon suite of minor NE-SW faults (of Miocene age?) across the area (Figure 7.12).

In order to determine the nature of any pre-Thanetian structures Carter & Hart (1977b) used the "Red"/"Blue" zonal boundary for a second contouring exercise. This identified a series of minor folds (also shown in Figure 7.12) that must post-date deposition of the "Red" zone and be pre-Thanetian. They probably relate to inversion of the Chalk at the close of the Cretaceous (Laramide movements?). Failure to differentiate frost-shattered and soliflucted materials from firm *in situ* bedrock can lead both to stratigraphical miscorrelations in site investigations and difficulties during construction. These deposits are difficult to recognise, especially in poorly sampled successions, since the one often passes imperceptibly into the other. Both can pass downwards or laterally into *in situ* chalk and even in surface sections the placing of boundaries is largely subjective. However, solifluction chalk that has shifted appreciably can often be recognized micropalaeontologically since it contains fine-grained exotic material (e.g. Tertiary sand grains, glauconite, vegetable material, small snail shells etc.) introduced during movement.

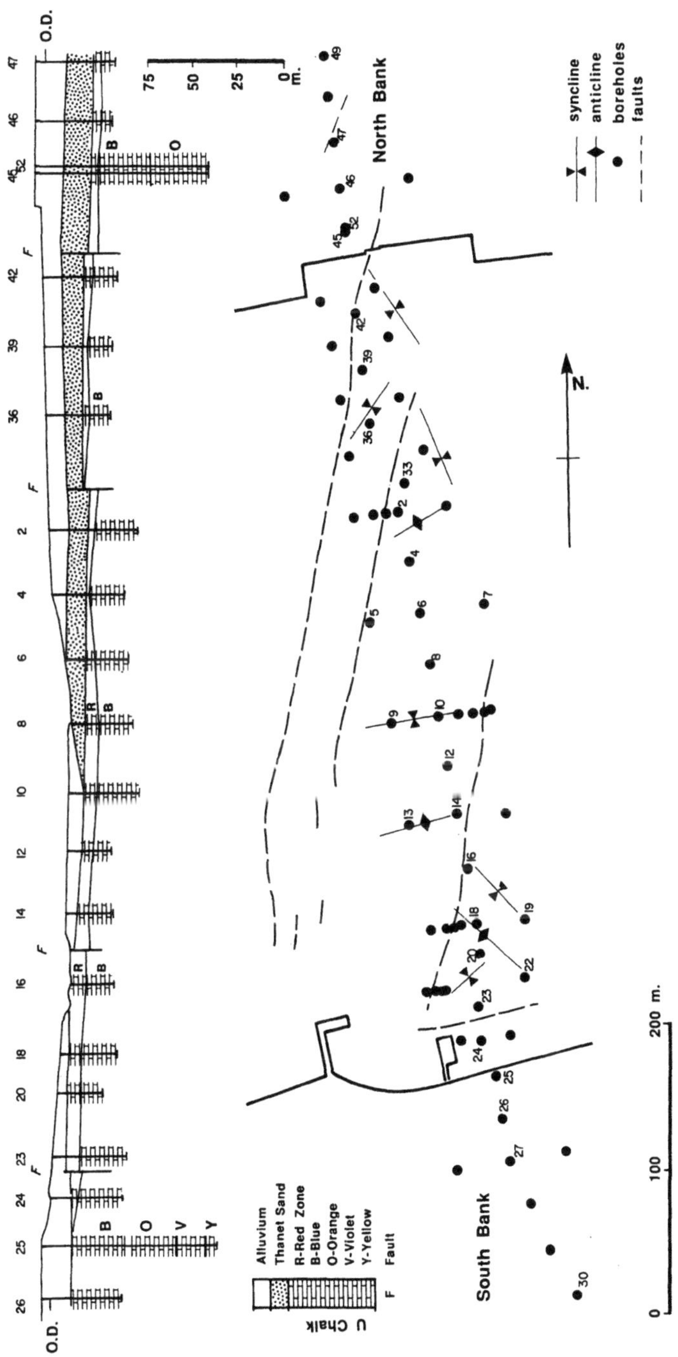


Figure 7.12: Location of borehole samples across the Thames Barrier site investigation. On the cross-section of the River Thames the south bank can be seen between Boreholes 24/25 and the north bank is located between Boreholes 42/45. It must be stressed that the structures indicated on the map are very small features.

In the U.100s the recognition of frost-shattering is extremely difficult as secondary fracturing is often produced as the tubes are driven in. When the chalk is examined closely the underlying structure tends to be obscured by a network of secondary fractures, but from a distance of about two metres *in situ* material gives an impression of general continuity of bedding. The top metre of unmoved, frost-shattered chalk may show solution-rounding of fragments and contain foreign materials from above. U.100s of *in situ* chalk taken immediately below uncased and unconsolidated cover are usually contaminated by infall; such contamination was minimised by ensuring that all samples for micropalaeontological examination were taken from the centre.

No typical solifluction chalks were encountered on site and frost shattered chalk was encountered only south of Borehole 10 where the Thanet Sand is missing. This suggests removal of the Thanet Sand prior to the most severe Plio-Pleistocene climatic deterioration. It is not known if the late glacial, flood plain gravels, which overlie the broken chalk and Thanet Sand, were in place at that time. Presumably shattering could have occurred through thin, saturated cover. However, the occurrence of undisturbed, late glacial gravels overlying frost-affected chalk passing laterally into solifluction deposits at Swanscombe, near Northfleet, Kent, indicate that the gravels were deposited later.

The frost damage affects the Chalk from the "Red" zone and the upper part of the "Blue" zone. The level to which frost shattering extended down was plotted on the geological sections.

The micropalaeontological study confirmed the expected stratigraphical sequence across the site, differentiated *in situ* from frost-shattered chalk, and located cross-cutting faults at foundation level for evaluation by the engineering geologists. The faults were not considered a hazard to the construction or to the operation of the Barrier. The cost of the micropalaeontological study was 1% of the total cost of the site investigation, and a minute part of the total cost of the whole project through to construction.

### 7.4 Cambridge-bypass site investigation

The route of the M.10 Cambridge Western-Bypass was designed, investigated, and part-constructed in the mid-1970's. The geology of the anticipated route and the profiles of the expected cuttings were based on a series of boreholes taken along the route (Figure 7.13).

In the area immediately to the south-west of Grantchester (Figure 7.14) these profiles indicated an almost complete admixture of chalk, chalky boulder clay, clay, pebbles, sands and other superficial sediments. During construction water seepage into the road cuttings became a problem and it was necessary to re-assess the geology of a part of the route. Several trench sections and trial pits were dug to examine the geology. Nearly all of these exposed an *in situ* mid-Cretaceous succession spanning the boundary of the Lower Chalk-Cambridge Greensand-Gault Clay (Figure 7.14). This is the same succession (Figure 7.13) as that exposed in the nearby Barrington Cement Works (Hart, 1973b; Weaver, 1978, 1982). This succession (Figure 7.15), like that at Arlesey (Bedfordshire), is one of the few surviving exposures of the Cambridge Greensand. In 1968-72, when these sections were investigated by the author, an error had been introduced into the foraminiferal zonation of the uppermost part of the Gault Clay. This was the result of a macropalaeontological error in the dating of the Copt Point (Folkestone) succession, due to a failure to identify a landslipped part of the succession. This misidentified material was used by Hart (1973a) to establish the zonation used to date the Gault Clay immediately below the Cambridge Greensand (Hart, 1973b).

Citing this misdated succession as evidence, Hart claimed that the uppermost Gault Clay was missing below the Cambridge Greensand. This is not, in fact, the case and the strata immediately below the Cambridge Greensand belongs to the middle (and upper?) part of Zone 6 of the Gault Clay succession (Carter & Hart, 1977a). The uppermost Gault Clay along the site contains floods of large specimens of *Globigerinelloides bentonensis* (Morrow) and this appears to be directly correlatable with Zone 6M (as shown in Figure 7.5). The succession of foraminifera from the Barrington succession (Figure 7.15) can be correlated very accurately with that from the Albian-Cenomanian succession (Figure 7.4) in the Folkestone-Dover area. This same sequence can be recognised in almost all the profiles recorded along the site. Part of this profile, together with the line of the road, is shown in Figure 7.14. This shows how the road intersects the Lower Chalk/Gault Clay boundary and how the regional dip provides the source of the water from the chalk succession. Remedial action was needed and extra drainage required. The contractor was able to obtain compensation for this additional cost as the problem should have been predicted in the initial geological survey. This error is all the more serious when one considers the availability of a nearby succession (Barrington) in which the full story is quite clearly visible.

Figure 7.13: Sketch map of the area south-west of Cambridge showing the location of the completed M10. The area considered in the site investigation is indicated immediately south-west of Grantchester. The quarry used as a reference section is located immediately to the north of the village of Barrington.

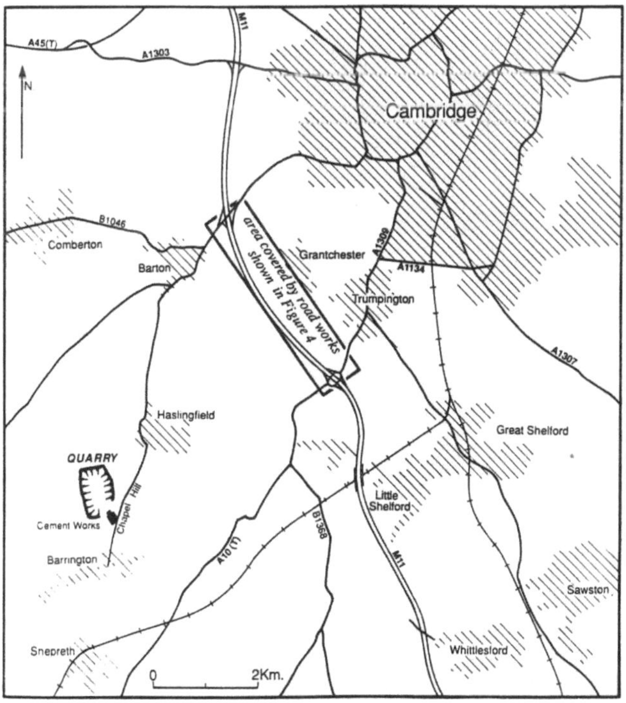

Figure 7.14: Sketch map of the M10 Cambridge-bypass site, located south-west of Grantchester. Two chainage points (13710 and 15310) are indicated and these allow the location of sample sites on the lower profile. This shows how the level of the road cuts across the regional dip of the chalk-Cambridge Greensand-Gault Clay succession, which is south-easterly in this area. Samples were taken at 1.0m intervals, with a closer spacing in the Cambridge Greensand. This was done for research and *not* engineering reasons. This was, when under construction, the best exposure of an interesting geological horizon (the Cambridge Greensand) which is now almost un-exposed in this area.

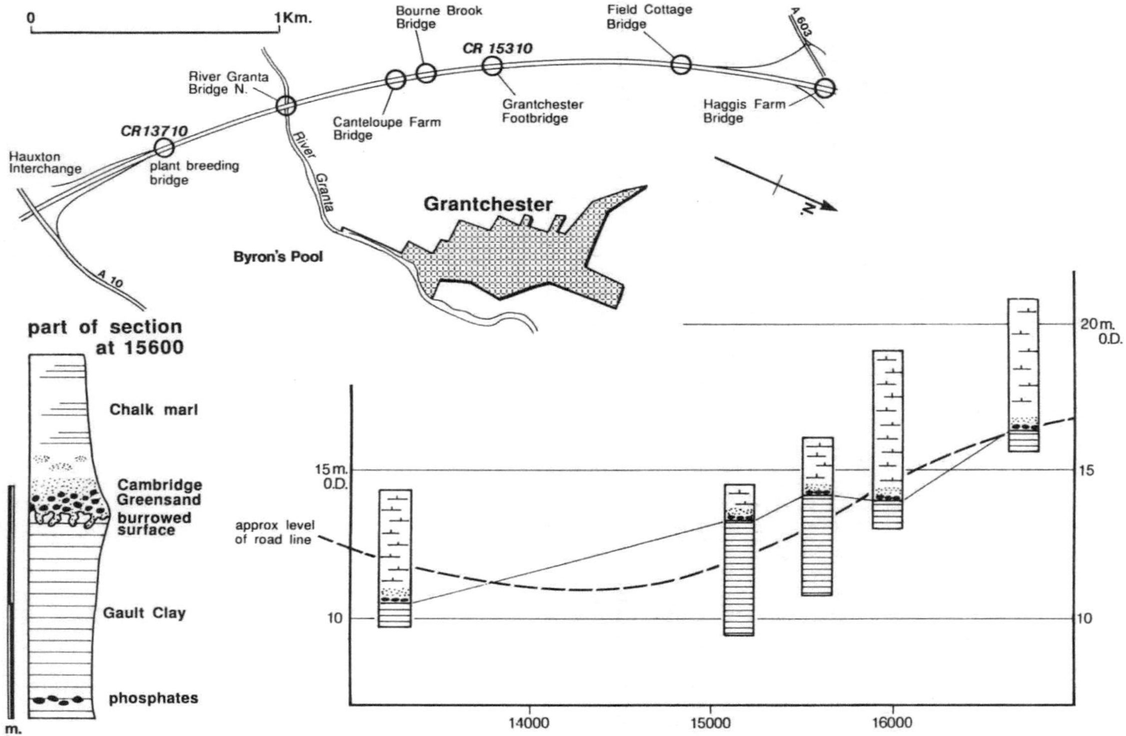

## 7.5 Isle of Wight landslips

At the southern end of the Isle of Wight (Figure 7.16) there is a series of classic rotational landslips, many of which give character to the Undercliff. These landslips are located on the southern limb of the Brixham/Sandown monocline which forms the E-W backbone of the island. The succession is dipping gently southwards and this has stimulated the development of major rotational landslips, particularly along the coastal area between Blackgang, St Catherines, Ventnor and Bonchurch. The old coast road (Figure 7.17) from Niton to Chale was breached in the 1920's and the Blackgang end of the same road (and several houses) disappeared in the early 1970's. The succession (Figure 7.18) involved in this slippage is the Lower Chalk-Upper Greensand-Gault Clay. This is not a particularly fossiliferous succession, with the very silty Gault Clay lacking many of the normal foraminiferal marker species. The Upper Greensand also contains an impoverished fauna, largely composed of agglutinated taxa. The basement beds of the Lower Chalk (Kennedy, 1969; Carter & Hart, 1977a) contain typical foraminiferal assemblages, together with a rich and varied macrofauna. In the landslip itself there is a complete mixture (Figure 7.17) of rotated blocks and isolated parts of the succession. The chalk basement beds and the top of the Upper Greensand succession (Chert Beds) usually remain as discrete units but the lower part of the Upper Greensand succession, the Passage Beds and the Gault Clay can be difficult to identify in the field. Samples collected from profiles from selected slips were used to try and identify the slippage horizons and allow restoration of the original profiles. The mechanisms involved in the slipage within the Gault Clay were investigated by Denness (1968) who used foraminiferal data to locate his clay samples, many of which were collected near the Blackgang end of the landslipped area.

## 7.6 Geological Resources

Exploitable resources of geological materials are invariably facies controlled, especially those that involve sedimentary rocks (ball clays, chalks, limestones, sands, etc.). Many come under the heading of low-cost industrial minerals, where the transportation costs are almost more significant than the cost of the product. The majority of such resources are worked in open pits where refined geological control might be absent. Micropalaeontology has a role in the identification of such reserves, the confirmation of extensions to present workings and the correlation of borehole data. The examples used to illustrate this type of application all involve the chalk succession, although the end uses do vary from site to site (cement, paper whitening, etc.).

*(i) Pitstone Cement Quarries*

At Pitstone (Hertfordshire) the Lower and Middle Chalk are quarried for the production of cement. There were two main quarries (Figure 7.19) working in the early 1970's when this investigation was undertaken. The quarry (No.2) adjacent to Pitstone Hill, south of the Upper Icknield Way, exposes the uppermost Lower Chalk, the Plenus Marls and the lower part of the Middle Chalk. These chalks, apart from the Plenus Marls, have a very high calcimetry value (Figure 7.20). Material from this quarry is "balanced" by the material from the larger quarry (No.3), located between the Upper and Lower Icknield Way, which exposes a largely lower Lower Chalk succession characterised by much lower calcimetry values.

Figure 7.15: Geological and foraminiferal succession at Barrington Cement Works, south-west of Cambridge. This is one of the two remaining exposures of the Cambridge Greensand but it is often below water level in the base of the quarry. The zonation is that described by Carter & Hart (1977a). It is important to note the flood abundance of *Globigerinelloides bentonensis* in the upper part of the Gault Clay which appears to allow a correlation with the middle of Zone 6m (Figure 7.5: despite statements to the contrary in Hart 1973b, GC = Gault Clay; CG = Cambridge Greensand; BR = Burwell Rock).

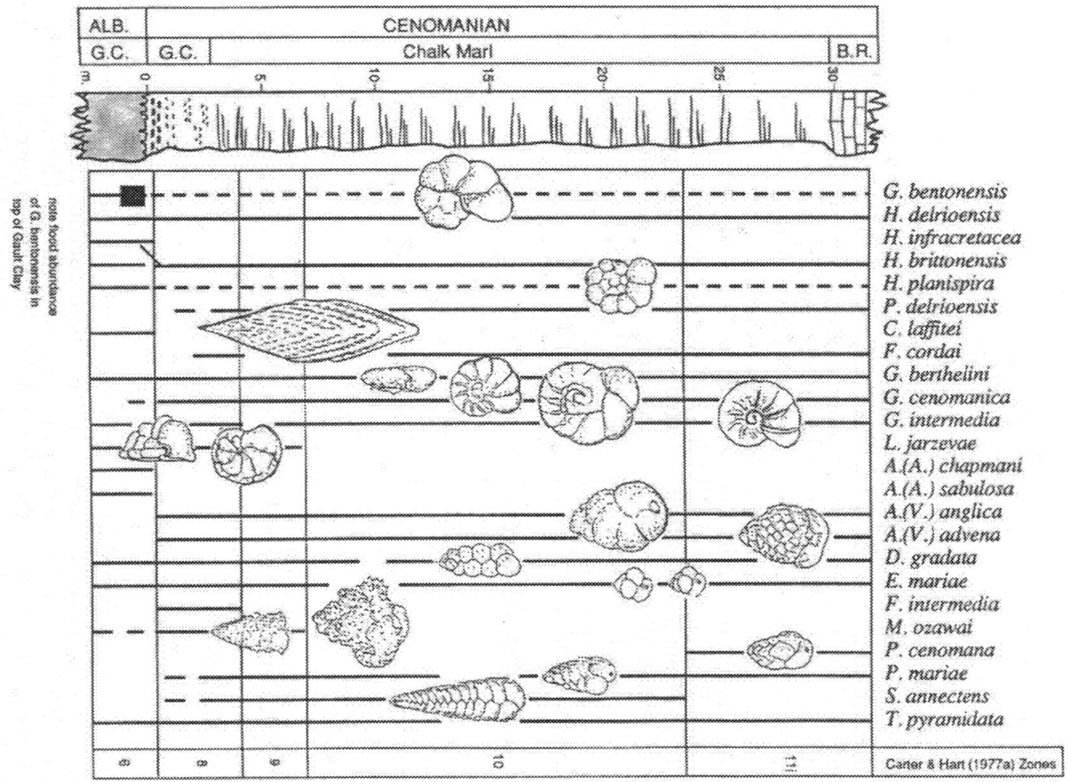

Figure 7.16: Sketch geological map of the Isle of Wight and two representative cross-sections. On Section B-B′, just south of St Boniface Down, the Blackgang-Undercliff landslip is indicated on the southward-dipping surface of the Gault Clay.

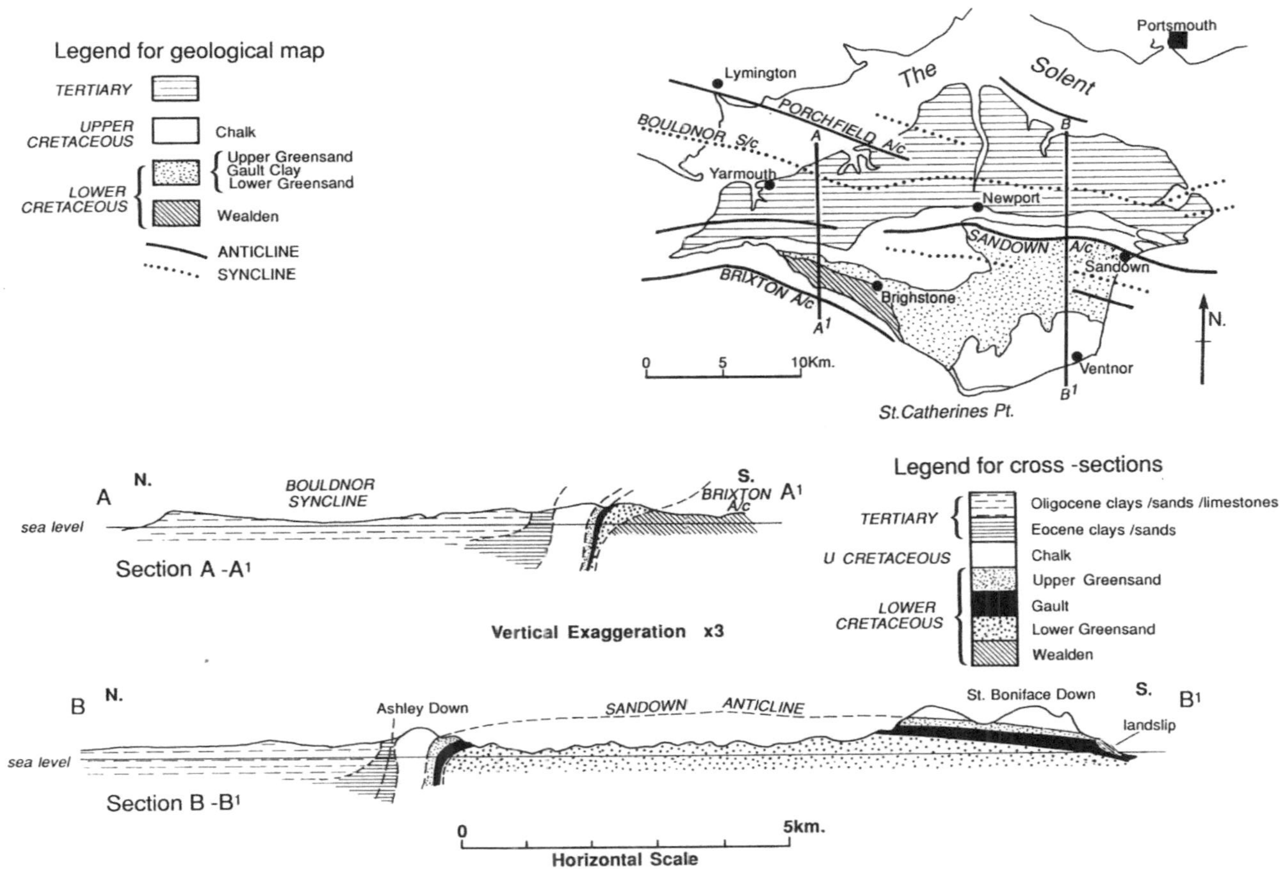

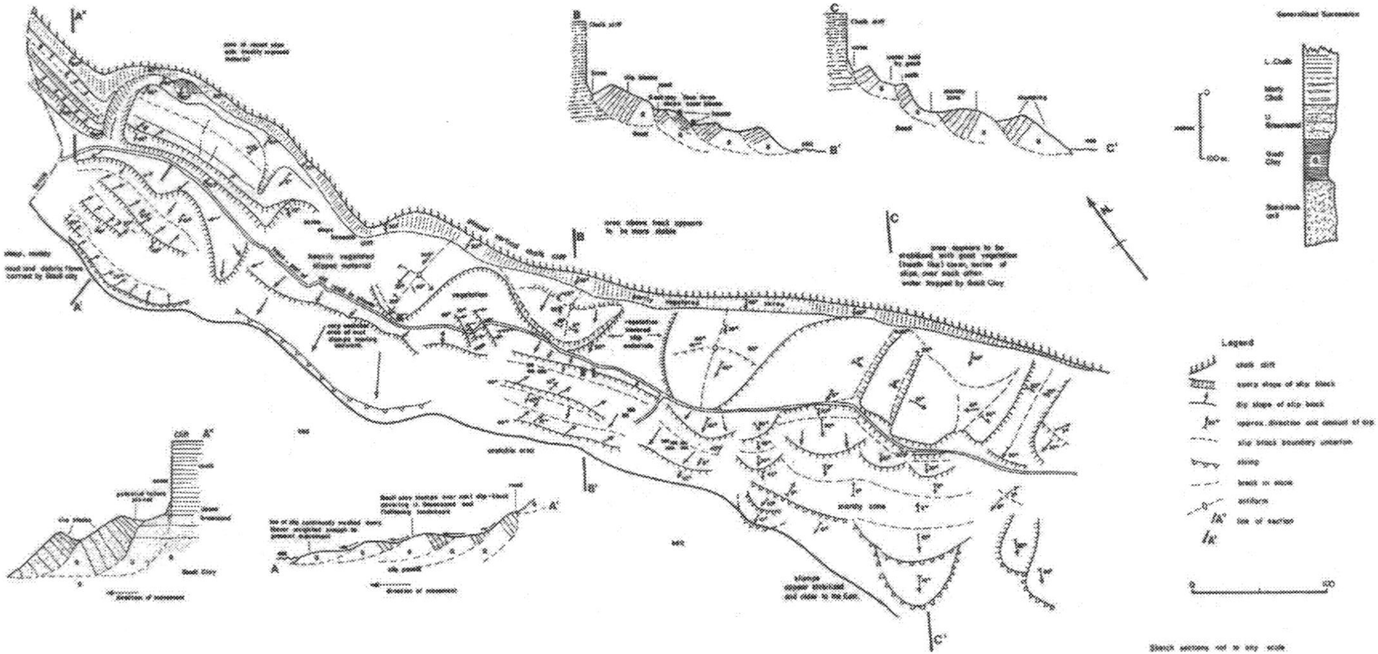

Figure 7.17: Sketch map of the Blackgang-Undercliff landslip showing principal rotated blocks, scarp slopes and fonds. Representative sketch cross-sections are also indicated. This map is not the result of an accurate surveying exercise but was developed by the author and undergraduate students Janet Pook and Tom Davies.

Figure 7.18: Geological correlation of the Chalk, Glauconitic Marl, "Chert Beds", Upper Greensand, Passage Beds and Gault Clay of the Isle of Wight. The landslipped area is represented by the Gore Cliff/Rocken End succession. The Lower Cenomanian succession at Dover is also shown (with the foraminiferal zones of Carter & Hart, 1977a) to indicate how it can be used to calibrate the landslipped material. Note that the Dover succession is drawn at a different scale.

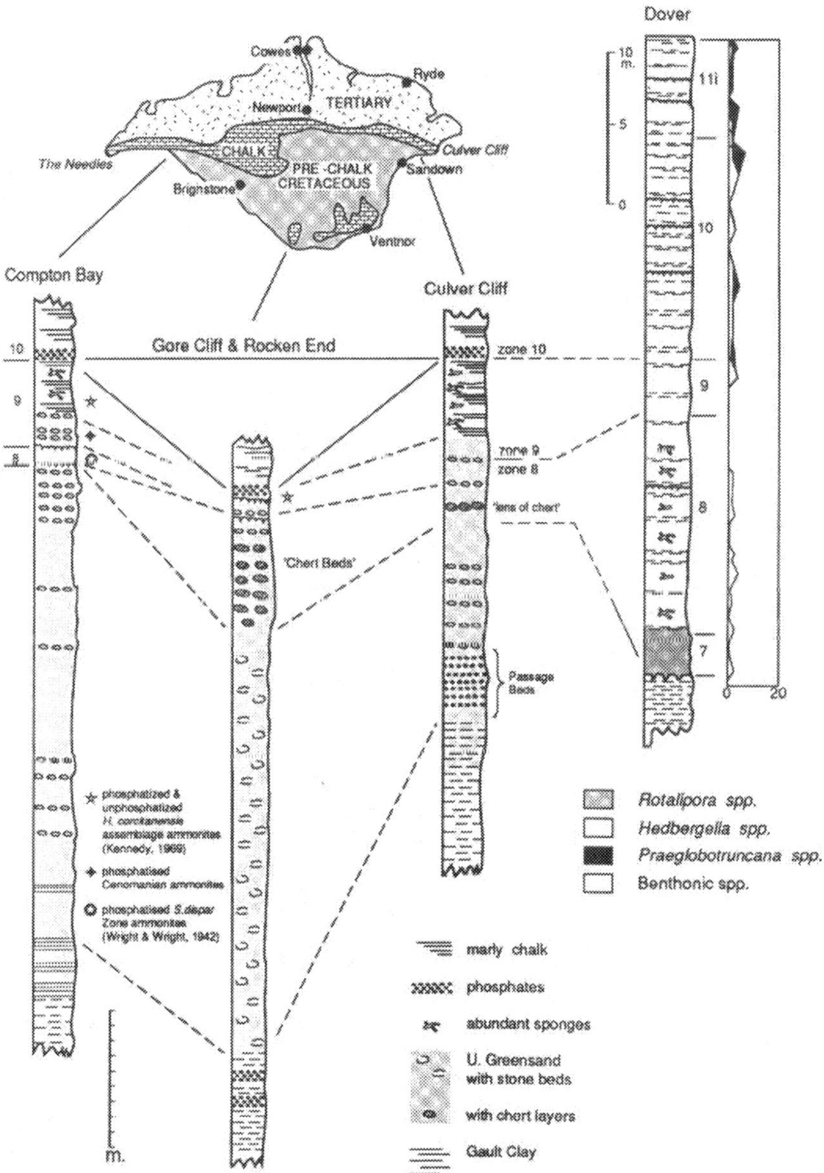

The chalk succession of both quarries was sampled in detail by the author with geologists from Engineering Geology Ltd, and together with borehole data the distribution of key foraminiferal taxa was documented (Figure 7.20). This stratigraphic succession and the foraminiferal zonation can be correlated directly with that recorded from the Folkestone-Dover area (Figure 7.7).

The upper part of the Cenomanian succession is comparable to that in the Folkestone-Dover succession but the Lower Cenomanian appears to be much thicker and clay-rich. Zone 11(i) is reduced to only one bed of hard chalk. There is no record of the base of the chalk and the Glauconitic Marl although the lowest samples investigated from the borehole material did record traces of glauconite and it is suggested that this must have been very close to the base of the local chalk succession. The foraminiferal zonation, once established in the quarries, was then used to allow the detailed correlation of the various faces and shallow boreholes throughout the site, especially in the area for the extension to quarry No.3.

During the course of the micropalaeontological investigation there was a parallel study of the calcimetry of the chalk (conducted by the cement company) and a full geotechnical assessment of the pit design (conducted by Engineering Geology Ltd). The eventual outcome was a series of quarry profiles (Figures 7.21, 7.22). These were subsequently used by the cement company in the formulation of an extraction plan appropriate to the capacity of their bucket-wheel excavators, extensive network of conveyor belts and the throughput of their kilns. Clearly all these factors control the maximisation of their resources.

Figure 7.19: Sketch map of the Pitstone quarries which are located near the Grand Union Canal between Tring and Dunstable at the foot of the chalk escarpment. Quarry 3 is located in the lower part of the Lower Chalk while Quarry 2 straddles the Lower Chalk/Middle Chalk boundary.

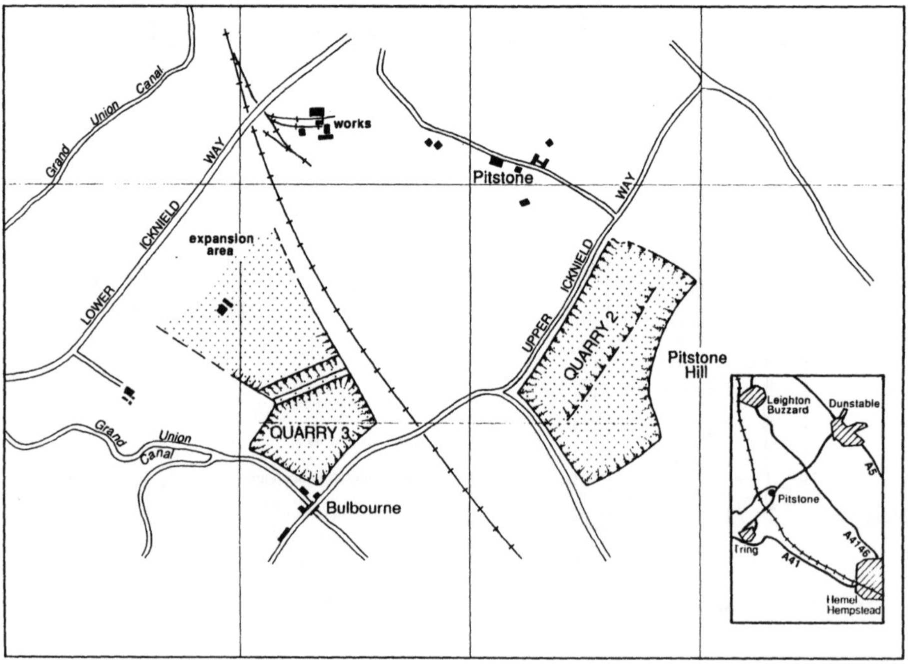

Figure 7.20: A composite succession of the Pitstone quarries and a few boreholes. The foraminiferal zones are based on Carter & Hart (1977a). The planktonic:benthonic ratio can be compared to the succession at Dover (Figure 7.7), which is reproduced here. The mid-Cenomanian non-sequence is quite marked in this correlation. The calcimetry graph (%$CaCO_3$) was provided by Tunnel Cement and Engineering Geology Ltd.

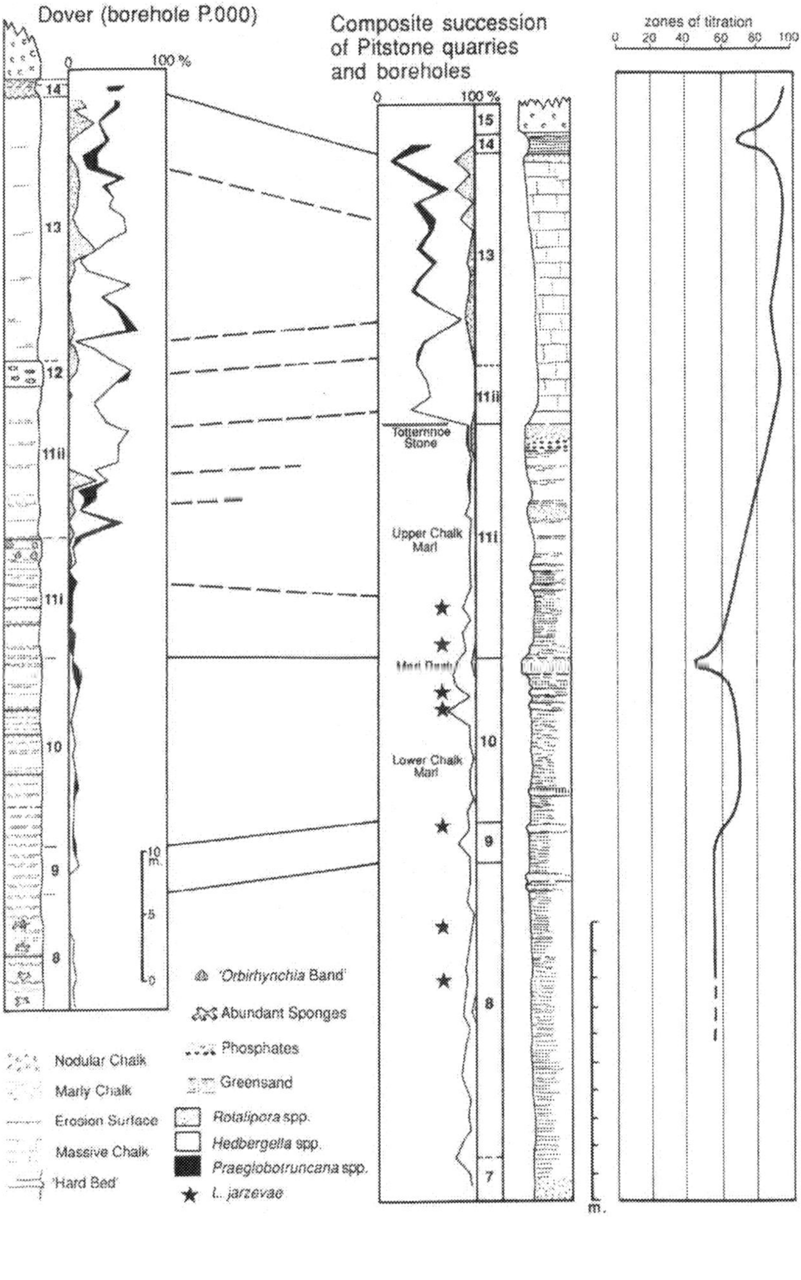

Figure 7.21: Sketch profile of Pitstone Quarry 2, showing how the foraminiferal zones (shown in Figure 7.20) and the calcimetry data can be used to define an extraction programme for the quarry. Outline data provided by Engineering Geology Ltd.

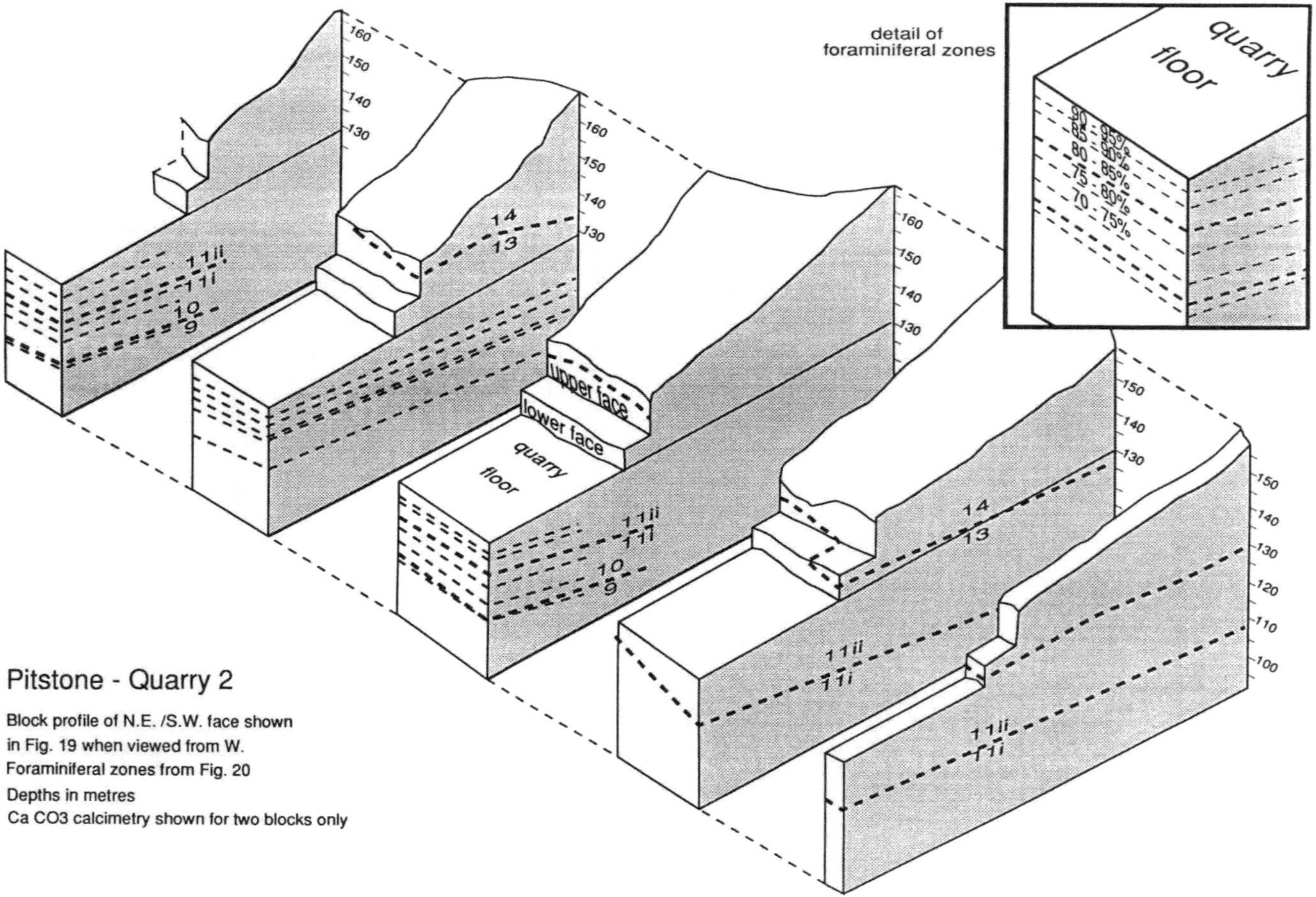

Figure 7.22: Sketch profile of Pitstone Quarry 3, showing similar data to that in Figure 7.21. This quarry provides clay-rich, zone 9/10 chalk which is then mixed with carbonate-rich chalk from Quarry 2 to provide the correct carbonate/silica ratio for the cement. The vertical line in the right-foreground represents a borehole that allowed calibration of the sub-surface succession.

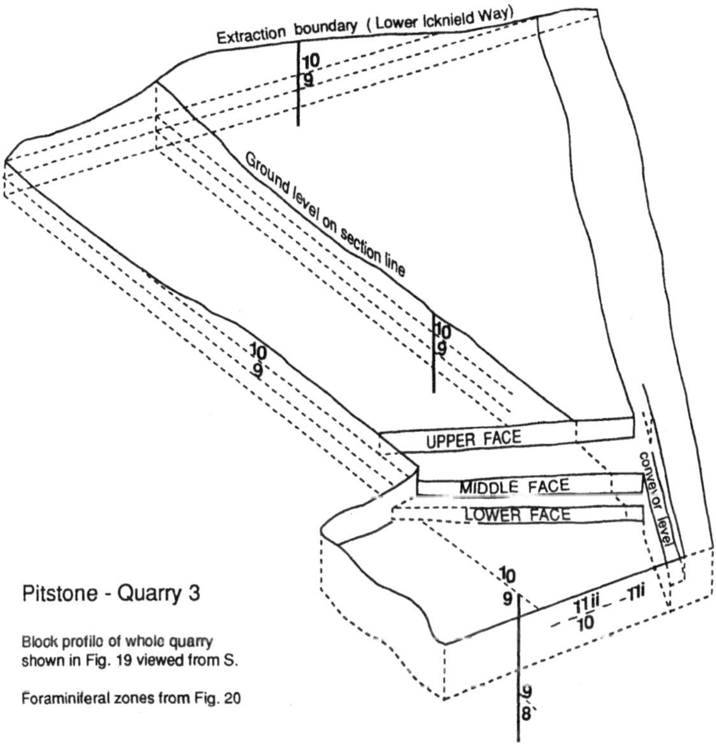


*(ii) Quidhampton (Wiltshire)*

This is a large, working quarry on the north-western edge of Salisbury (Wiltshire). It was described by Jukes-Browne & Hill (1904) as a small chalk pit in the *Micraster coranguinum* Zone. More recently, the succession was described by Bailey (1978) in a review of Coniacian-Santonian foraminifera. In 1976, the working face of the quarry exposed 28m of chalk with flints (Figure 7.23). In the lower half of the succession are numerous bands of irregular black flints. Six metres above the base, large "paramoudra-like" flints are common while a further 2m up-section there are a series of thin, impersistent tabular flints. In the upper half of the section only two flint bands are present. The uppermost flint band (21.5m above the base of the succession) is quite prominent, being formed of large, flattened, black nodules which merge in places to form an almost continuous layer.

Figure 7.23: Lithological and foraminiferal succession of Quidhampton Quarry, north-west of Salisbury. The zonation (D, E) is based on Bailey (1978) and can be compared to the succession described for the Thames Barrier (Figure 7.10).

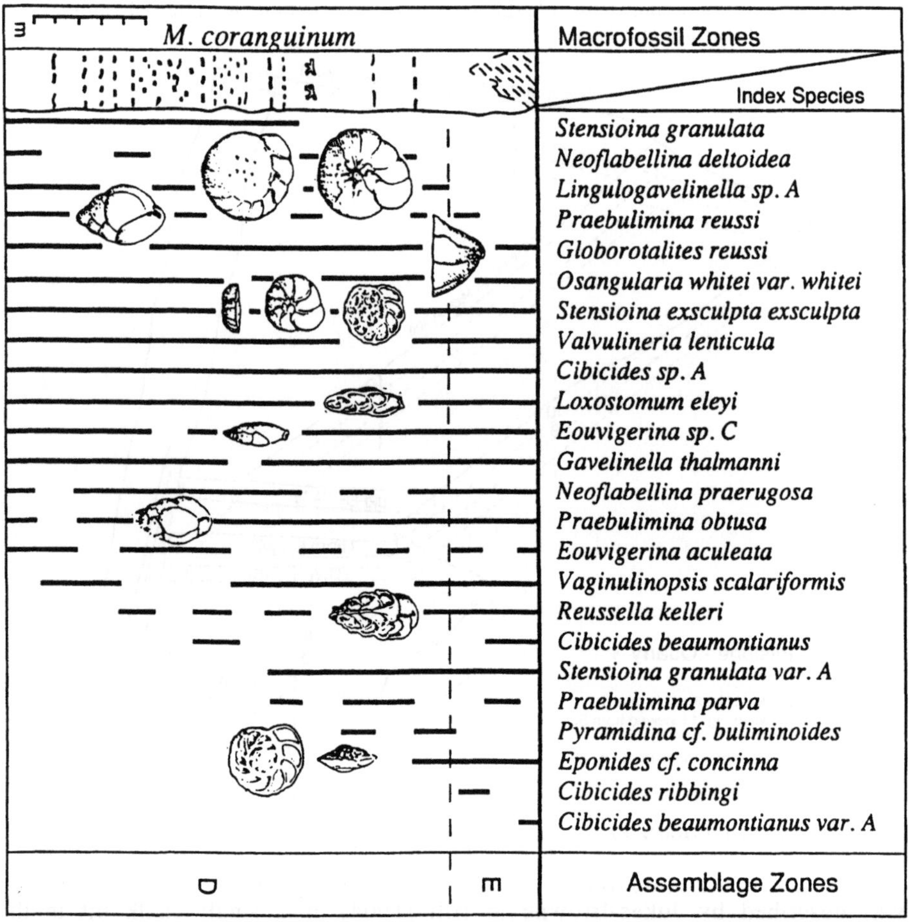


The chalk throughout the section is uniform, being an almost pure, fine-grained, biomicrite, with a few iron-stained horizons which appear to be associated with sponges. Few macrofossils have been recorded from the succession by either Bailey (1978) or the author. The uppermost 5-8m. of the succession has been strongly affected by periglacial activity, being penetrated by numerous solution pipes which contain reddish-brown clay-with-flints. Some of these pipes penetrate the chalk to a depth of more than 15m.

Figure 7.24: Map and profile of Quidhampton Quarry. The present working face (shown in Figure 7.23) can be correlated to the two boreholes within the limits of the samples available. By projecting the boreholes onto the ENE/WSW section line it can be shown that the 'bright' chalk being quarried at the present time extends into the area of the proposed quarry extension.

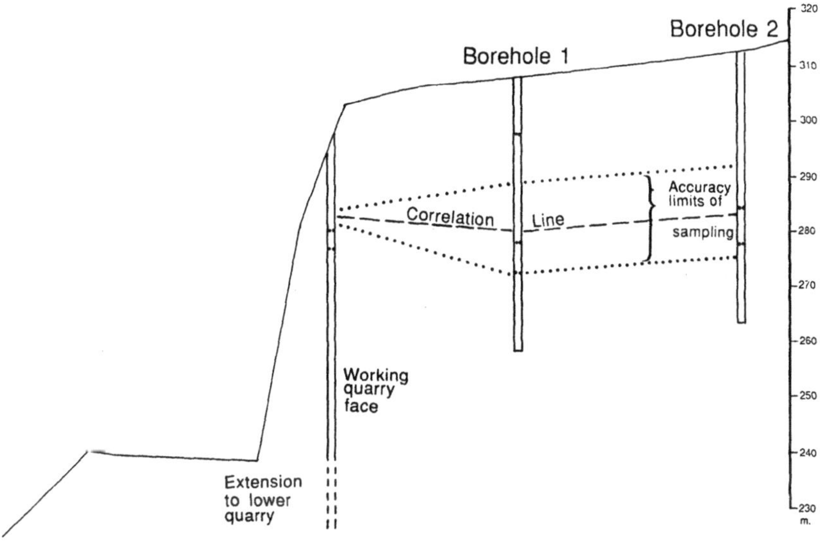


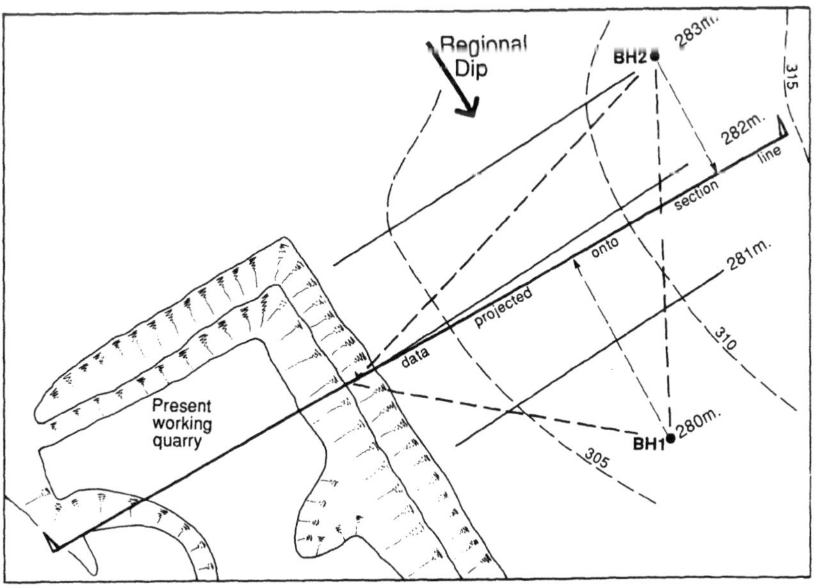

The problem at Quidhampton was to confirm the extension of the "bright" chalk seen in the quarry to the east of the present face. Two boreholes were drilled, but the samples were unfortunately pre-bagged into intervals ranging from one to several metres in thickness. These samples were broken up and a random set of fragments processed. To have taken a single lump from the initial sample would have yielded an unrepresentative fauna for the recorded sample interval. The distribution of the foraminifera in the two boreholes was compared to the known faunal succession from the quarry face (Figure 7.23). This succession shows an important microfaunal boundary near the top of the face, marked by the disappearance of *Lingulogavelinella* sp.cf. *L. vombensis* ("Red"/"Blue" boundary of the Thames Barrier site investigation) and an increase in the numbers of *Gyroidinoides nitida*, *Globorotalites micheliniana* and *Osangularia* spp. The planktonic fauna from this locality is relatively sparse but quite diverse. Planktonic species include *Dicarinella concavata, Globotruncana linneiana, Globigerinelloides rowei, Globigerinelloides ehrenbergi, Marginotruncana renzi, M. pseudolinneiana, M. marginata, Whiteinella brittonensis, W. baltica* and *Heterohelix globulosa*. The fauna compares well with that of SE Kent, Bailey's (1978) Zone D/E boundary being located near the top of the section. This is also described in some detail in the stratigraphical revision of Hart et al. (1989).

Using this boundary it is possible to correlate the two boreholes (Figure 7.24) with the quarry face, despite the problem of the sampling interval. Flint lines in the chalk are now known to be laterally persistent (Mortimore, 1986) and it is almost certain that the two flints in the borehole logs are those recorded in the quarry face. If this view is accepted a very fine correlation can be produced which confirms that the "bright" chalk, currently being extracted, continues eastward as far as currently tested by drilling. Using the data from the face and the boreholes it is possible (Figure 7.24) to calculate the strike and dip of the chalk across the projected quarry extension and if necessary design faces that would minimize the risks of face instability.

*(iii) East Grimstead (Wiltshire)*

Immediately south of the Salibsury-Romsey railway line, near the hamlet of East Grimstead is a large quarry which is currently exploiting "bright" chalk from the *Offaster pilula* Zone of the Upper Chalk. The quarry lies on the northern flank of the Dean Hill Anticline (Figure 7.25), the stratigraphy of which has been described by Williams-Mitchell (1956). The quarry section was logged in detail and a microfaunal assessment of its stratigraphic position completed (Figure 7.26). When first investigated, the nature of the faunal change at the *planoconvexa* bed was thought to be of prime importance to the success of the investigation. For a number of reasons, the change at this level is quite dramatic, while faunal turnover throughout the remainder of the succession is less obvious, although counts of the various taxa did indicate subtle variations. As the investigation proceeded it became clear that none of this succession was represented in the boreholes, drilled in a N-S transect to the south of the quarry. Again the problems created by 2-3m 'thick' samples randomly taken from the broken cores meant that it was impossible to predict the angle of dip on this N-S section with any real accuracy. In Figure 7.25 a N-S section from the quarry across the minerals prospect is shown, and this must be regarded as extremely tenuous as a result of the sample spacing. In Borehole 1 there is an important change (in numbers) roughly at half-depth, and this may be close to the top of the *M. testudinarius* Zone. This would give thicknesses very much in line with those of Williams-Mitchell (1956).

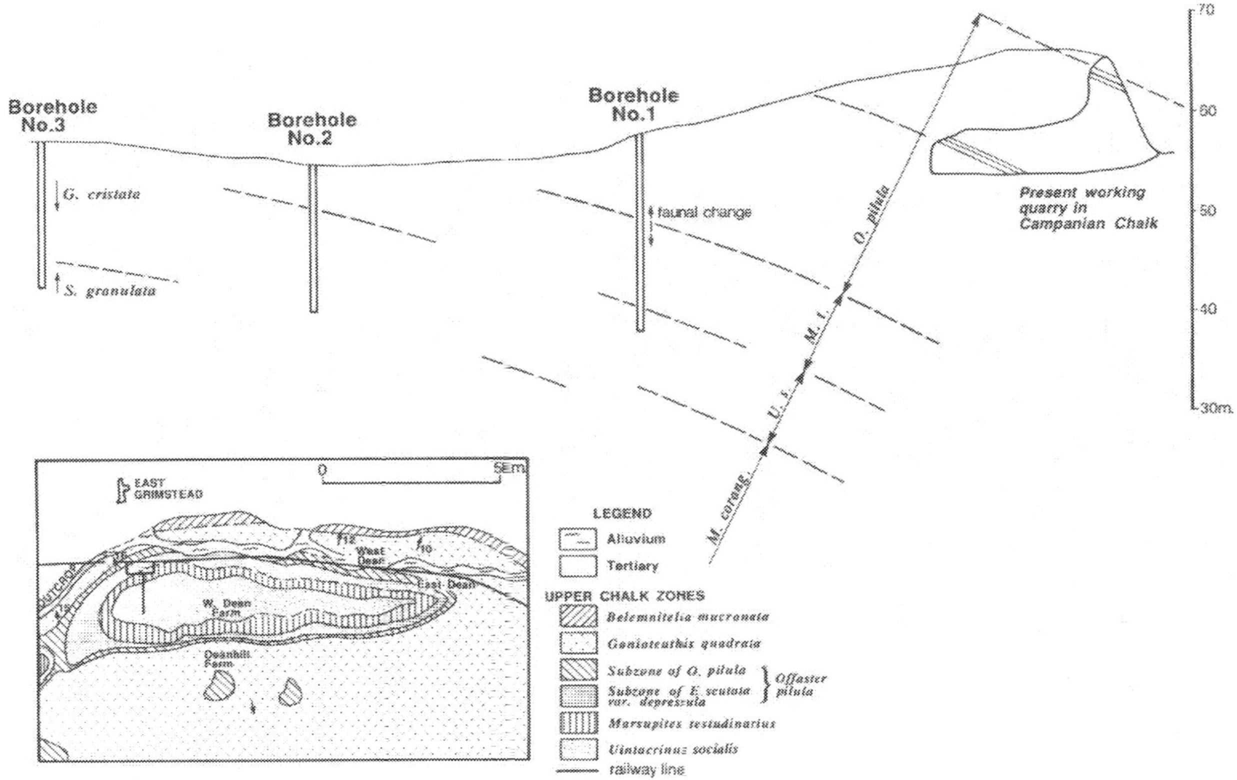


Figure 7.25: East Grimstead Quarry, which is located South-east of Salisbury, is represented by a rectangle on the inset map. The line running south of this rectangle represents the line of boreholes shown above. The sampling interval used in the boreholes was very crude and this created problems for the micropalaeontological investigation. The 'bright' chalk, currently worked in the quarry, can be seen not to extend into the area of possible quarry extension. The geological map of the Dean Hill Anticline is based on data in Williams-Mitchell (1956).

Figure 7.26: Lithological and foraminiferal succession of the present East Grimstead quarry. The species used in this zonation are described in Hart et al. (1989). This stratigraphic level has some of the brightest chalk in the UK and this is used in paper whitening.

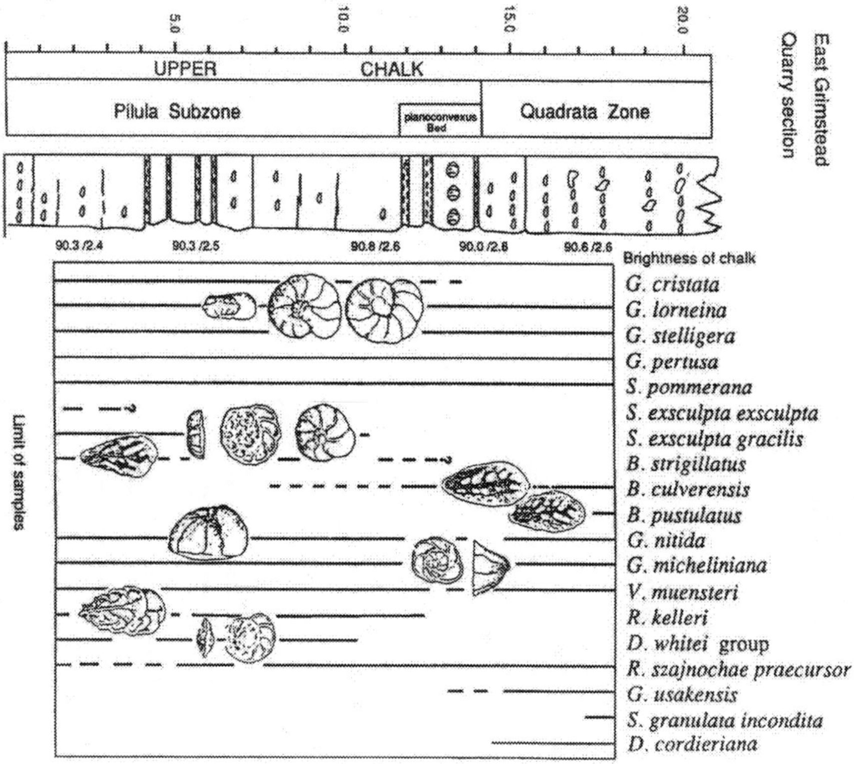


The appearance of *Bolivinoides strigillatus* (Chapman) in reasonable numbers at the bottom of Borehole 1 is significant as this species normally appears just above the base of the *M. testudinarius* Zone. Borehole 2 contains very few distinctive marker horizons with the whole of the fauna resembling that of the *Uintacrinus socialis* Zone. Borehole 3 records the lowest appearance of *Gavelinella cristata* s.l. and the uppermost occurrences of *Stensioina granulata granulata* and *Reussella kelleri* s.l. This would indicate that the lowest part of the borehole must be close to the top of the *Micraster coranguinum* Zone.

The Dean Hill anticline was mapped in detail by Williams-Mitchell (1956) and the thicknesses of the zones he recorded are shown in Figure 7.25. It is also clear from the available data that the "bright" chalk currently being exploited in the quarry does not continue over the area represented by the cross-section.

### 7.7 Conclusions

While several of the examples used in this account relied heavily on the micropalaeontological data, in others the role of the subject can only be described as marginal. In all the cases, however, the data from the micropalaeontological work was helpful and should be taken as a part of the general geological tools available to the client. This is the case in very few operations. As indicated at the outset, the Chalk is a rather special case as its lithology (to the un-enlightened geologist) appears rather uniform and featureless.

### Acknowledgements

Many of the projects described in this account represent team efforts and as such it is particularly difficult to identify everyone who has had an involvement. Particular mention must be made of the following, all of whom are thanked for their help, patience and encouragement; Haydon Bailey, Colin Bristow, Alistair Burnett, John Charman, Bruce Denness, Peter Fookes, Christopher Foster, Colin Harris, Robert Hiscock, Mike Saunders, Christopher Wood and Peter Wright. Particular thanks go to Dave Carter, whose training in the way of applied micropalaeontology was invaluable. The author also acknowledges the help given by Transmanche Link, the Department of the Environment, Foundation Engineering and Engineering Geology Ltd.
The author wishes to thank Marylyn Luscott-Evans for preparing the final manuscript and John Abraham who patiently prepared all the figures.

#### References

Amedro, F., Damotte, R. Manivit, H. Robaszynski, F. and Sornay, J. 1978. Echelles biostratigraphiques dans le Cénomanien du Boulonnais (Macro-, Micro-, Nannofossiles). Géologie Mediterranèe, 5, 5-18.

Amedro, F., Manivit, H. and Robaszynski, F. 1979. Echelles biostratigraphiques du Turonien au Santonien dans les Craies du Boulonnais (Macro- Micro- Nannofossiles). Annales de la Socëté Géologique du Nord, 98, 287-305.

Bailey, H.W. 1978. A foraminiferal biostratigraphy of the Lower Senonian of Southern England. Unpublished PhD Thesis, CNAA/Plymouth Polytechnic.

Bailey, H.W. and Hart, M.B. 1979. The correlation of the early Senonian in Western Europe using Foraminiferida. Aspekte der kreide Europas. IUGS. Series A. No. 6, 159-169.

Barnard, T. & Banner, F.T. 1953. Arenaceous Foraminifera from the Upper Cretaceous of England. Quarterly Journal of the Geological Society of London, 109, 173-216.

Barnard, T. & Banner, F.T. 1980. The Ataxophragmiidae of England: Part 1, Albian-Cenomanian Arenobulimina and Crenavereuilina. Revista Espanola de Micropalaeontologia, 12, no.3, 383-430.

Barr, F.T. 1962. Upper Cretaceous planktonic foraminifera from the Isle of Wight, England. Palaeontology, 552-580.

Barr, F.T. 1966. The foraminiferal genus Bolivinoides from the Upper Cretaceous of the British Isles. Palaeontology, 9, 220-243.

Berthelin, G. 1880. Memoire sur les Foraminiferes de l'Etage Albien de Montcley (Doubs). Mémoire de la Société Géologique de France, Paris, 1, 1-84.

Carter, D.J. 1966. Channel Study Group Site Investigation: Appendix 3 (Micropalaeontology Report), pp. MI-1 to MA-14; Appendix 4 (Figures).

Carter, D.J. & Hart, M.B. 1977a. Aspects of Mid-Cretaceous stratigraphical micropalaeontology. Bulletin of the British Museum, Natural History (Geology), 29, 1-135.

Carter, D.J. & Hart, M.B. 1977b. Micropalaeontological investigations for the site of the Thames Barrier, London. Quarterly Journal of Engineering Geology, 10, 321-338.

Carter, P.G. & Mallard D.J. 1974. A study of the strength, compressibility and density trends within the Chalk of South East England. Quarterly Journal of Engineering Geology, 7, 43-55.
Chapman, F. 1891-1898. The Foraminifera of the Gault of Folkestone. Journal of the Royal Microscopical Society, London, (Part 1) 1891: 565-575; (Part 2) 1892: 319-330; (Part 3) 1892: 749-758; (Part 4) 1893: 579-595; (Part 5) 1893: 153-163; (Part 6) 1894: 421-427; (Part 7) 1894: 645-654; (Part 8) 1895: 1-14; (Part 9) 1898: 1-49.
Denness, B. 1968. Fissures and related studies in selected Cretaceous rocks of SE England. Unpublished PhD Thesis, University of London.
Harris, C.S. 1982. Albian microbiostratigraphy (Foraminifera and Ostracoda) of SE England and adjacent areas. Unpublished PhD Thesis, Plymouth Polytechnic/CNAA.
Hart, M.B. 1970. The distribution of the Foraminifera in the Albian and Cenomanian of SW England. Unpublished PhD Thesis, University of London.
Hart, M.B. 1973a. A correlation of the macrofaunal and microfaunal zonations of the Gault Clay in South-east England. In: The Boreal Lower Cretaceous, Casey, R. & Rawson, P.F. (eds), Geological Journal Special Issue, No.5, 267-288.
Hart, M.B. 1973b. Foraminiferal evidence for the age of the Cambridge Greensand. Proceedings of the Geologists' Association, 84, 65-82.
Hart, M.B., Bailey, H.W., Crittenden, S., Fletcher, B.N., Price, R.J. & Swiecicki, A. 1989. Cretaceous. In: Stratigraphical Atlas of Fossil Foraminifera (2nd Edition), Jenkins, D.G. & Murray, J.W. (eds), Ellis Horwood, Chichester, 273-371.
Jukes-Browne, A.J. & Hill, W. 1904. The Cretaceous Rocks of Britain: The Upper Chalk of England. Memoir of the Geological Survey of Great Britain, London, 566pp.
Kennedy, W.J. 1969. The correlation of the Lower Chalk of south-east England. Proceedings of the Geologists' Association, 80, 459-560.
Kennedy, W.J. 1970. A correlation of the Uppermost Albian and the Cenomanian of south-west England. Proceedings of the Geologists' Association, 81, 613-677.
Loeblich, A.R. & Tappan, H. 1964. Protista 2. Sarcodina, chiefly 'Theocamoebians' and Foraminiferida. In: Moore, R.C. (ed.), Treatise on Invertebrate Paleontology, Part C, 1:1-510, 2: 511-590, Kansas.
Magniez-Jannin, F. 1975. Les Foraminiferes de l'Albien de l'Aube: paleontologie, stratigraphie, ecologie. Cahiers de Paleontologie, CNRS, 351pp.
Mortimore, R.N. 1986. Stratigraphy of the Upper Cretaceous White Chalk of Sussex. Proceedings of the Geologists' Association, 91, 97-139.
Owen, M. 1970. Turonian foraminifera from Southern England. Unpublished PhD Thesis, University of London.
Price, F.G.H. 1874. On the Gault of Folkestone. Quarterly Journal of the Geological Society of London, 30, 342-366.
Price, R.J. 1977. The stratigraphical zonation of the Albian sediments of north-west Europe, as based on foraminifera. Proceedings of the Geologists' Association, 88, 65-91.
Robaszynski, F., Amedro, F., Foucher, J.C., Gaspard, D., Magniez-Jannin, F., Manivit, H. & Sornay, J. 1980. Synthese Biostratigraphique de l'Aptien au Santonien du Boulonnais a partir de sept groupes paleontologiques: Foraminiferes, Nannoplancton, Dinoflagelles et Macrofaunas. Révue de Micropaleontologie, 22, 195-321.
Weaver, P.P.E. 1978. Cenomanian Ostracoda from Southern England their taxonomy, stratigraphy and palaeoecology. Unpublished PhD Thesis, CNAA/Plymouth Polytechnic.
Weaver, P.P.E. 1982. Ostracoda from the British Lower Chalk and Plenus Marls. Palaeontographical Society Monograph, Pulication No. 562, part of Vol. 135 for 1981, pp. 1-127, pls. 1-20.
Williams-Mitchell, E. 1948. The Zonal value of Foraminifera in the Chalk of England. Proceedings of the Geologists' Association, 59, 91-112.
Williams-Mitchell, E. 1956. The stratigraphy and structure of the Chalk of the Dean Hill Anticline, Wiltshire. Proceedings of the Geolologists' Association, 67, 221-227.

Malcolm B. Hart
Department of Geological Sciences
University of Plymouth
Drake Circus
Plymouth PL4 8AA
Devon, UK

# General index


Africa 24-26, 32, 35, 63, 64, 67, 185, 186, 188, 203, 223
Aggradation 197
Albian 32, 60, 138, 179, 203, 223, 229-232, 235, 245, 261, 262
Amazon 161, 185, 189
Amorphous organic matter 155, 158, 162
Andrew 34, 135, 138
AOM 162, 163, 165, 169, 170, 174, 175, 177, 178, 183
Arabian Sea 24, 28, 35
Atlantic Ocean 1, 2, 5, 8, 9, 12, 14-17, 19-27, 29, 30, 32, 33, 35-38, 75, 78-81, 168, 169, 171, 173, 187, 188, 190, 191
Australia 23, 28, 185, 186, 193, 213, 214, 218, 220, 221, 223
Azores 30

*B. reticulatum* zone 208, 210, 212
Balsfjord 16
Baltic 2, 15
Barremian 61-64, 94, 138, 145, 179, 188
Barton Clay 56
Barton Sands 48, 56
Bavaria 31, 34
Bay of Bengal 24, 28
Bay of Fundy 30, 35
Baylei ammonite zone 107, 108
Beatrice 104-106, 138, 148, 150
Beaufort-Mackenzie Basin 202, 222
Bembridge Limestone 43
Bembridge Marls 43, 56
Benthonic foraminifera 69-71, 73, 81, 94, 108, 114, 209, 211, 212, 215-217, 219, 229
Beryl 101
Bisaccate 167
Bonaparte 193, 203, 208, 210, 220
Boulogne 31
Brae 117, 128, 129, 132, 133, 135, 138, 144, 150
Brazil 64, 65, 67, 220
Brent 101, 110-114, 116, 117, 128, 129, 135, 138, 145, 148, 150, 151, 178, 188, 215
Britain 9, 30, 31, 65, 146, 147, 152, 191, 262
Brixham 247
Brockenhurst Beds 42
Brora Coal 105
Bruce 101, 261
Buchan 138
Bull Head Bed 242

*C. laticosta* zone 56
*C. delicata* 207
Callovian 104-106, 147-149, 169, 171, 177-179, 190
Cambridge 34, 36, 37, 145, 147, 185, 189, 226, 244-246, 248, 262
Canada 184, 186-188, 202, 222
Capricorn Basin 23, 28, 35
Caribbean Sea 24, 25, 38
Carnian 30
Caspian Sea 63
Caving 140
Celtic Sea 16, 17, 20-23, 31, 36, 37, 62, 65, 94
Cenozoic 1, 16, 29, 33, 34, 37, 70, 81, 96, 101, 139, 146-152, 159, 169, 184, 187, 197, 202, 221, 222
Central Graben 33, 100, 120, 138, 150
Chalk Sea 32, 33
Channel Tunnel 226-228, 230, 235, 236
Chanter 138, 150
Chesapeake Bay 162, 187
China 19-21, 23, 24, 30, 34, 38, 63, 64, 66, 67
China Sea 19-21, 23, 24, 30, 34, 38
Clastic 64, 99, 104, 107, 115, 116, 122, 123, 143, 147, 149, 155, 162, 163, 168, 173, 184, 187, 197, 214, 215, 217, 222
Claystone 100, 105, 107, 109, 132
Coastal barrier 49, 99, 115
Cocobeach Formation 63
Colwell Bay 46, 54, 58
Coniacian-Santonian 255
Cook Formation 111-113
Copt Point 244
Coralline Crag 19, 36
Core 24, 35, 64, 73-75, 79, 81, 83-86, 120, 128, 141, 183, 193, 207-213, 236
Cuticle 158, 164, 165, 177, 205, 207, 209, 211, 216, 217
Cyanobacteria 158, 162, 165
*Cyrena pulchra* Bed 50-52
Cyrus 135, 149

*D. lobispinosum* zone 207, 208
Dakota 63
Dampier 193, 208, 210, 212-214, 216, 218, 223
Danang Basin 79
Deep Sea Drilling Project 32, 187, 188, 190
Delta 21, 22, 26, 81, 149, 161, 164-166, 169, 173, 185, 188, 194, 198, 200, 202, 210, 211, 213-215, 217, 218, 220
Denmark 2, 11, 188
Depositional sequence 93, 96, 98, 103, 121, 149, 155, 198
Depositional episodes 99, 100
Dinocysts 94, 102, 114, 132, 139, 140, 143, 144, 163, 166, 169-172, 177, 183, 205
Dogger Bank 12, 14, 18
Don Field 111-113, 149
Dover 2, 9, 13, 15, 16, 19, 229, 231, 233, 234, 245, 251-253
  Harbour 233, 234
  Strait 9
Drake 114, 262
Dunlin 110-114, 138, 215
Dysoxia 217

*E. torynum* zone 212
East China Sea 20, 21, 23, 24, 30, 38
East Grimstead 226, 258-260
East Wear Bay 229
Ekofisk 10, 36, 70, 81, 117, 138, 139, 143, 145, 151
England 4, 9, 10, 16, 19, 36, 37, 41, 55, 60-67, 80, 81, 184-189, 223, 225, 227, 229, 230, 234, 237, 261, 262
English channel 2, 10, 13, 15, 16, 19, 21, 36, 37, 67, 229, 232, 234
Estuarine 3, 12, 16, 27, 35, 36, 70, 155, 161, 171, 172, 185
Europe 18, 21, 23, 24, 30, 32, 33, 36, 55, 61-63, 65, 66, 100, 101, 129, 139, 143, 145-152, 187, 188, 223, 261, 262
European shelf 18, 19, 36, 37
Eustasy 67, 197
Exmouth 193, 198, 200, 203, 208, 210, 211, 213, 214, 218-221
Exxon 98-100, 104, 106, 107, 148, 151, 197, 200, 201

Faeroe Isles 11
Fair Isle Channel 10, 11, 15
Falkland 32
Fastnet Basins 62, 65
Fault 54, 110-113, 116, 117, 122, 128, 175, 200
Faunicycles 61
Fladen 69, 70, 81
Flamborough Head 12
Flinders Fault Zone 200
Fluorescence 158, 159, 165, 183
Fluvio-deltaic 162-167, 169
Folkestone 229, 231, 244, 245, 252, 262
Forties 6, 10, 36, 70, 81, 114, 117, 129, 132, 138, 145, 147, 151, 184, 201
France 10, 62, 65, 101, 185, 187, 189, 221, 261
Frigg 129-132, 135, 138, 145, 147
Fulmar 100, 116, 128, 138, 148, 150

Galapagos Islands 26
Gamma 99, 100, 102, 106-109, 111, 201, 208, 210-212, 214, 215, 218
Gas-prone material 158
Gault Clay 228, 229, 244-249, 251, 262
Genetic stratigraphic sequences 99, 146, 149, 197, 221
Geochemistry 116, 153, 186, 187, 190, 191
Geochronological 96, 97
Germany 31, 34, 36, 60, 63, 188
Glyndebourne Borehole 229, 231
Gondwana 67
Gonyaulacoid dinocysts 171, 183
Grantchester 244-246
Great Barrier Reef 20, 24, 28, 30, 35
Great Fish River 25
Greensand 214, 244-248, 251, 262
Grimstead 226, 258-260
Guillemot 138
Gulf of Aden 24
Gulf of Aqaba 24, 38
Gulf Coast 44, 55, 56, 66, 99, 102, 189, 200, 202, 220-223
Gulf of Elat 24, 29
Gulf of Mexico 20, 24, 35, 71, 73, 80, 81, 179, 187, 202, 221, 222
Gulf of Panama 25, 37

Hampshire Basin 41, 42, 49, 50, 54-56, 66, 67
Hatherwood Limestone 52-54
Headon 43, 45, 46, 49-52, 54-56, 58, 66
Hebrides Basin 105
High supply 161
High relative preservation 161
Hills 69, 167, 189
HST 198, 202, 205-214, 217, 219
Hydrocarbon fields 93
Hydrodynamic equivalence 162, 163, 167, 173

Iberian Peninsula 32
Indian 1, 24, 25, 27-29, 32, 33, 35, 36
Indo-Pacific 24-28, 30, 33
Indonesia 202, 215, 217, 220, 223
Inert material 158
Irish Sea 20, 21, 36
Isle of Wight 42, 43, 46, 49, 51, 54-58, 65-67, 188, 226, 247, 249, 251, 261

Isotopic dating 71
Ivanhoe 108, 138, 149

Java Sea 19, 23
Jura 35, 62, 184, 185

K-Ar 96
Kerogen 109, 155, 156, 158, 159, 161-164, 170, 173, 177, 183, 185, 186, 188, 189, 190
Kimmeridge 31, 33, 35, 38, 108, 109, 116, 120, 179, 188, 190

Linstone Chine Member 53, 54
Lithostratigraphy 93, 96, 101, 104, 105, 109, 111, 114, 118, 120, 125, 138, 146, 151
Log character 99, 112, 113, 125, 132, 134
Lower Aptian 94
Lower Barremian 94
Lower Cenomanian 32, 251, 252
Lower Chalk 228, 229, 234, 236, 244, 245, 247, 252, 262
Lower Headon Beds 49, 51, 56
Lower Purbeck Beds 61, 63
Lower Warm Shale 108
LST 198, 210, 211, 213-216

Maceral 155, 156, 188
Maureen 116, 135, 139, 146
Maximum flooding surface 99, 100, 102, 104, 106-109, 116, 129, 132, 198, 205, 206, 208, 212
Mediterranean 1, 20, 189, 190
Mexico 20, 24, 30, 35, 71, 73, 80, 81, 179, 107-109, 202, 221, 222
*Micraster coranguinum* zone 238, 255, 260
Microplankton 105, 165, 166, 170, 177, 181-185, 191
Mid-Cenomanian 233, 234, 253
Middle Chalk 236, 238, 247, 252
Middle Headon Beds 54-56, 58
Minches 2, 3, 5, 8, 9, 12, 15
Mongolia 63
Moray Firth 4, 5, 8, 9, 12, 101, 102, 104-109, 117, 132, 138, 145, 147-151

Nematura Bed 55
New Forest 49
Ninian 129, 135, 145, 151
North Viking Graben 101, 106, 110, 116
North Sea 1-21, 23, 32-36, 39, 69-71, 73, 75, 78-81, 93-96, 98-102, 104-107, 109-111, 114-118, 120, 122-124, 126, 128, 129, 132, 135-152, 178, 186-188, 201, 215, 220, 222, 223, 234
North Celtic Sea 62, 65
North Atlantic 2, 12, 14, 17, 20-22, 24-27, 29, 32, 35, 37, 75, 78, 79, 81, 187, 188, 190

Northfleet, Kent 244
Norway 1, 2, 11, 12, 15, 37, 114, 117, 135, 145, 146, 150, 151, 185, 191
Norwegian Sea 1, 2, 11

*Offaster pilula* zone 258
Oil-prone material 158
Orinoco 164, 173, 188
Orkney 4, 5, 8, 9, 12, 14, 15
Osborne 43, 66
Oxygenation 108, 155

Pacific 1, 19, 20, 22-30, 32-37, 184, 186, 187
Palaeoenvironment 41, 67, 93, 101, 114, 115, 135, 142-144, 151, 159-161, 182, 183, 205
Palaeoprodelta 200, 211
Palaeozoic acme 169
Palynofacies 114, 135, 151, 153-155, 159-161, 168, 173-176, 183, 184, 190, 191
Palynomorphs 143, 153, 155, 158, 159, 162, 163, 165, 166, 170, 172-174, 183, 185, 186-188, 190, 203, 205, 207, 209, 211, 213, 216-218
Papuan Basin 214
Paris Basin 62, 150, 222
Particulate organic matter 153, 155, 162, 163, 186, 188, 189
Passage Beds 247, 251
Peridinioid 171, 183
Petronella 138, 151
Phytoclast particle size 165
Phytoclasts 155, 158, 159, 161-165, 174, 183
Piper 100, 108, 109, 128, 138, 148, 150, 190
Pitstone 226, 247, 232-233
Pockmark 73, 74
Polarity 155
Polyploidy 62
POM 153, 155, 161
Portlandian 60-62, 150
Priabonian 55
Progradation 197, 217
Psammobia Beds 49
Pteridophyte 166
Purbeck 41, 60-63, 65, 66, 187
Purbeck Beds 41, 60-63, 65
Pyrolysis 153, 158

Quaternary 15, 36, 38, 41, 69, 70, 81, 139, 148, 183, 186, 188, 190, 220, 222
Quidhampton 226, 255-258

Radiolaria 94, 97, 99, 102, 108, 109, 114, 116, 121, 132, 138-140, 143-146, 148
Rannoch Formation 111
Red Sea 1, 20, 24, 30, 38
Reef 20, 24, 28, 30, 35, 99, 115

Reservoir 93, 94, 96, 98-101, 104, 109-117, 120-122, 125, 128-132, 134, 135, 138, 143-147, 150, 151, 162, 186, 191, 193, 213, 218
Retrogradation 197
Reworking 49, 140-143, 164
Rhaetian 30
Rob Roy 108, 138, 149
Rogn 135
*Rosenkrantzi ammonite* zone 108

*S. areolata* zone 213
Salinity 2, 3, 11-15, 19, 24, 30, 32, 44, 48-52, 54, 55, 57, 59, 61, 62, 67, 70, 114, 155, 161, 171, 172, 180
Sargasso Sea 20, 25, 30, 36
Scapa 132, 138, 150
Scheldt 5, 13
Scotland 2, 5, 10, 12, 15, 36, 81, 105
Scott 108, 138, 153, 171, 189, 190, 220, 222
Sea level 24, 32, 37, 55, 66, 98, 99, 105, 108, 148-151, 161, 166, 168, 171, 173, 197, 198, 202, 221-223
Seal 98, 99, 110, 115, 116, 129, 144
Sedco Explorer 200
Sediment compaction 197
Sedimentology 67, 116, 144, 145, 147, 148, 153, 184, 185, 187, 221
Sendai Bay 30, 38
Serratum ammonite zone 108
Sgiath 100, 109
Shetland 1, 2, 4, 5, 8, 9, 11, 14, 15, 94, 100, 101, 117, 118, 145, 148
Siberia 63
Similarity index matrix 74, 75
Skagerrak 2, 12, 14-16
Sleipner 101, 135, 148
SMW 198
Snorre 135, 146
Solent 10, 16, 37, 43
Songliao Basin 63, 67
Sonic log 100, 104, 108, 111
Source 16, 63, 93, 96, 99, 109, 115-117, 141, 144, 153, 155, 158, 159, 164-167, 169, 173, 177, 183, 185, 187, 189-191, 245
South Africa 25, 185, 203, 223
South China Sea 20, 23, 34, 38
Soviet Union 30, 63
Spain 62
Sporomorph 147, 165-168, 177, 181
Statfjord 110-113, 116, 128, 129, 135, 138, 148, 149, 215, 222
Station Hotel 22, 26
Stockbridge 138, 150
Stratigraphic trap 98, 122, 123, 150
Stratigraphic sequence 149, 197
Structureless white chalk 227
Subarctic Pacific 22, 27
submarine fan 99, 115, 117, 134, 147, 150, 164, 211
Swatchway 69, 70
System tracts 208

T-Block 135
Tarbert Formation 106
Tartan 138
Tasman Sea 20, 23, 24, 28, 34
Tectonism 145, 197
Ternary 177, 181-183
Tertiary 36, 41, 56, 66, 81, 100-102, 104, 115, 135, 139, 142-148, 151, 171, 185, 223, 236, 242
Thames 226-228, 234, 236, 237, 239, 240, 243, 256, 258, 261
  Barrier 226-228, 234, 236, 237, 239, 240, 243, 256, 258, 261
Thanet Sand 236, 242, 244
Thecamoebian 217
Thistle 111, 116, 129, 135, 147, 152
Tithonian 31, 32, 34, 61, 62
Total Organic Carbon 158, 161, 162, 164, 183
Totland Bay 54
Transgressive surface 100, 106, 107, 198, 210
Translucent 158, 163, 164
Transmanche Link 228, 261
Trap evaluation 93, 98, 122
Triangular 49, 50, 177
Turbidite 202, 220, 222, 223

Ultra violet 159, 165
Undercliff landslip 226, 249, 250
Upper Headon Beds 51, 52
Upper Purbeck 61, 62
Upper Chalk 236, 238, 258, 262
Upper Icknield Way 247
USA 24, 30, 36, 99, 102

Valanginian 31, 35, 62, 193, 196, 200, 203, 205, 212, 213, 218, 221

Wadi Dahab 24
Washout 119, 140
Water column stability 155
Wealden 60-63, 65-67, 184, 187
Western Interior Basin 99
White Band 55
Whitecliff Bay 46, 48, 49
Wireline log 98, 102, 121, 125, 129, 193, 200, 201, 205-208, 210-213, 220
Wyoming 63

Yellow Sea 23

Zonal identification 142

# Micropalaeontological index


*Acanthoica quatrospinosa* 11
acritarchs 65, 97, 152, 169, 170, 177, 183, 184, 205
*agglutinated foraminifera* 81, 102, 108, 139, 145, 146, 220, 222
algae 18, 28, 36, 38, 97, 116, 152, 158, 162, 165, 169, 170, 183-185, 191
*Ammobaculites* 106, 217
*Ammobaculites coprolithiformis* 106
*Ammonia* 21, 35, 69, 70, 82-87, 90, 202
*Ammonia batava* 69, 70, 82-87, 90
*Ammonia beccarii* 70
ammonites 60, 96, 106, 107
*Amphistegina* 202
*Anthosphaera robusta* 11, 15
arenaceous forams 202
*Archaeozygodiscus keossensis* 30
*Archangelskiella cymbiformis* 32
*Asterigerina* 202
*Asterionella japonica* 12
Astrorhizida 76, 80, 84, 87, 88

bacteria 162
*Bairdia* 48
*Bathysiphon sp.* 209, 216, 217
*Beggiotoa* 73
belemnites 96
benthonic foraminifera 69-71, 73, 81, 94, 108, 114, 209, 211, 212, 215-217, 219, 229
*Biddulphia* species 14
*Biscutum constans* 32
*Biscutum novum* 31
*Bisulcocyprus postangularis postangularis* 64
bivalves 48, 96, 105, 106
*Botryococcus* 169, 185, 191
*Braarudosphaera bigelowii* 5, 7, 9, 11, 13, 30, 38, 39
*Braarudosphaera rosa* 30
*Braarudosphaeracea* 29
*Broinsonia* 32
*Buccella frigida* 69
*Bulimina gibba* 69, 86-88
*Bulimina marginata* 69, 70, 76, 80, 83-88
*Buliminacea* 238, 242
Buliminida 76, 80, 82-88

calcareous nannofossils 30, 33-37, 102, 114, 132, 140, 143, 149
*Calcidiscus leptoporus* 5, 7, 9, 10, 12, 23, 39
*Calciopappus caudatus* 15
*Calyptrosphaera hyalina* 11
*Caneosphaera mollichii* 11
*Cassidulina carinata* 70
*Cassidulina crassa* 69
*Cassidulina laevigata* 70, 76, 80, 82-88
*Cassidulina norcrossi* 69
*Cassidulina obtusa* 69, 70, 82-88
Cassidulinacea 238, 241
*Cellanthus* 202
*Cenosphaera* sp. 209, 214, 216, 217
*Ceratium* species 14
*Chaetoceros danicus* 12
*Chaetoceros* spp. 12
*Chara* 44, 52
*Chiasmolithus* spp. 10
chlorococcale algae 158, 165, 169, 183
*Cibicides* 70, 82, 84-87, 90, 202
*Cibicidoides* 85-87, 90, 202
*Coccolithus pelagicus* 5, 6, 8, 10, 11, 13-17, 29, 39
*Conusphaera mexicana* 32
*Coronosphaera mediterranea* 5, 6, 9
*Cribrononion excavatum forma clavatum* 69
*Crucicirhadus minutis* 30
*Cyclagelosphaera margerelii* 31
*Cyclogyra* 202
*Cypridea* 60-64, 66, 67
*Cypridea (Pseudocypridina) faveolata* 64
*Cypridea (Sebasianites) devexa* 64
*Cyprideis* 44, 67
*Cystammina* 202
*Cytheretta laticosta* 56
*Cytheromorpha* 44, 46, 49-51
*Cytheropteron* 61
*C. excavatum* 69

*Darwinula* 62
*Dentalina-Nodosaria* Assemblage 215
diatoms 12, 14, 25, 94, 97, 132, 139-141, 144, 145, 201
dinocysts 94, 102, 114, 132, 139, 140, 143, 144, 163, 166, 169-172, 177, 183, 205
dinoflagellate cysts 65,94, 101, 102, 104, 105, 109, 114, 116, 132, 135, 139, 184, 186, 187, 188, 190, 191, 222
*Dinophysis acuta* 6, 14
*Discorbis* 202
*Discosphaera tubifera* 11

*Egmontodinium torynum* 203
*Ehrenbergina* 202
*Eiffellithus* 33
*Ellipsagelosphaera* 1, 31-34
*Ellipsagelosphaera britannica* 31
*Ellipsagelosphaera lucasi* 31
*Elphidiella* 87, 202
*Elphidium* 69, 70, 76, 80, 82-87, 90, 202
*Elphidium bartletti* 69
*Elphidium clavatum* 69
*Elphidium ex gr. clavatum* 76, 80, 82-86, 90
*Elphidium excavatum* 70, 90
*Elphidium incertum* 70
*Emiliania huxleyi* 1, 5, 6, 10-29, 33, 34, 39
*Epistomina parastelligera* 106, 107
*Eucyrtis* 211
*Exuviaella baltica* 14

*Fabenella* 61
foraminifera 19, 25, 36, 52, 69-71, 73, 79-81, 94, 96, 97, 99, 102, 104, 105, 107-109, 114, 116, 132, 138-141, 143-147, 151, 172, 173, 201, 202, 207, 209, 211, 212, 215-217, 219-222, 229, 238, 241, 245, 255, 258, 261, 262
*Fursenkoina fusiformis* 70

*Galba* 44
*Galliaecytheridea* 61
*Gaudryina sp. B* 238
*Gavelinella beccariiformis* 114
*Gavelinella cristata* 260
*Gephyrocapsa* spp. 1, 5, 7, 8, 11, 16, 17, 21, 24, 29, 34
*Gephyrocapsa caribbeanica* 5, 7, 8, 11, 23, 24, 26, 27, 39
*Gephyrocapsa oceanica* 1, 11, 17, 19-21, 23-29, 33
*Globigerinelloides bentonensis* 245, 248
*Globigerinelloides ehrenbergi* 258
*Globigerinelloides rowei* 258
*Globobulimina* 202
*Globocassidulina* 202
*Globotruncana linneiana* 258
*Gonyaulax polyhedra* 19
*Gravellina* 202
*Gyroidina* 78, 85-87, 90, 202
*Gyroidinoides nitida* 258

*Hantkenina* 55
*Haplocytheridea* 46, 47, 49, 50, 56, 58, 61
*Haplocytheridea debilis* 56
*Haplophragmoides* spp. 108, 201-203, 205-207, 209, 211, 212, 216, 217
*Haplophragmoides canui* 108
*Haplophragmoides cf. infracalloviensis* 203, 205-207, 209, 211, 216, 217
*Haynesina orbicularis* 82-84, 87, 90
*Hazelina* 46, 48
*Hedbergella/Whiteinella* spp. 229
*Helicosphaera carteri* 5, 7, 8, 11, 39
*Hemicyprideis* 56, 58
*Heterohelix globulosa* 258
holoplankton 170
hormosinids 201
*Hyalinea balthica* 70, 76
*H.* sp. 217

*Idiocythere* 46, 48
*Islandiella teretis* 70

*Kamptneriums magnificus* 32

*Lagenammina* 201
*Lenticulina* ex gr. *muensteri* 205-207, 209, 211, 215, 216
*Lingulogavelinella vombensis* 240, 241, 258
*Lithastrinus floralis* 32
*Lithraphidites bollii* 32
*Lituolida* 76, 80, 83-88
*Lohmannosphaera michaelsari* 11
*Lotharingius imprimus* 31
*Lucianorhabdus* 32, 33

*Mantelliana* 61
*Marginotruncana marginata* 258
*Marginotruncana pseudolinneiana* 258
*Marginotruncana renzi* 258
megaspores 167
*Melonis* 70, 82-87, 90, 202
*Melonis barleeanus* 70
*Micranolithus* 29, 32
*Micranolithus obtusus* 32
microspores 166, 167
*Miliolinella* 202
*M. groenlandicum* 106

*Nannoconus* 31, 32
*Navicula* species 14
*Neochiastozygus dubius* 10
*Neocyprideis* 44, 46, 47, 49-51, 58, 66
*Nitzchia delicatissima* 12, 14
Nodosariida 76, 80, 82-88
*Nonion depressulum* 69
*Nonion labradorica* 70
*Nonionella* sp. 70, 84, 86, 87, 90
*Nuttallides* 202

*Oolithotus fragilis* 11
*Operculina* 202
*Ophiaster hydroideus* 15
*Ophiomorpha* 56
*Orastrum* 33
*Osangularia* spp. 258
ostracoda 41, 44, 46, 47, 49, 51, 52, 54-67, 105, 114, 132, 143, 149, 151, 262

*Paranotacythere* 61
*Pediastrum* 169, 186
*Pemma* 29
*Pentaster* 29
peridinioid 171, 183
*Petrobasia diversicostata* 64
*Phanolithus* 33
phycomata 170
planktonic foraminifera 19, 25, 96, 99, 102, 116, 139, 145, 146, 238, 241, 261
*Planorbina* 44
*Podorhabdus cylindratus* 31
*Pontosphaera* 10, 29
  *Pontosphaera discopora* 10
*Praeglobotruncana/Dicarinella* spp. 229
prasinophyte algae 158, 170, 183
*Prediscosphaera* 33
*Prinsiosphaera triassica* 30
*Protelphidium orbiculare* 69
*Protoperidinium species* 14
*Pseudorotalia* 202
*Pullenia* 83, 84, 87, 90, 202

*Quinqueloculina* 83-88, 202

radiolaria 94, 97, 99, 102, 108, 109, 114, 116, 121, 132, 138-140, 143-146, 148
*Recurvoides* 201, 202
*Reophax scottii* 70
*Retacapsa* 33
*Reticulofenestra* 10, 16, 29, 34
  *Reticulofenestra umbilica* 10
*Reticulophragmium* 201, 202
*Rhabdammina* 201, 202
*Rhabdosphaera clavigera* 5, 7, 9, 10, 11, 21
*Rhadocera suessi* 30
*Rhagodiscus asper* 32
  *Rhagodiscus splendens* 32
*Rhaxella* 106, 107
Robertinida 76
*Rotaliatina* 202
*Rotaliida* 16
*Rotalipora cushmani* 234
*Rotalipora* spp. 229
*Russellia* 33

*Saccammina sphaerica* 70, 84, 87, 88
saccate pollen 166, 167
*Schizosphaerella puncticulata* 30
*Schuleridea* 61
*Scriniodinium crystallinum* 107, 108
*Scyphosphaera* 10, 17, 29
*Seribiscutum primitivum* 32
silicoflagellates 97, 139, 147, 149
*Sphaeroidina* 202
sphaeromorph pollen 167
*Spirillina minima* 207
*Spiroplectammina spectabilis* 114
*Spirosigmoilinella* 202
spores 94, 97, 115, 140, 143, 144, 147, 165-168, 185-187, 190, 191, 202, 205, 207, 209, 211, 216, 217
sporomorphs 163, 165-168, 171, 173
*Stainforthia fusiformis* 83-88
*Staurorhabdus* 31
*Stensioeina exsculpt exsculpta* 240, 241
*Stradnerlithsi* 31
*Syracosphaera caudata* 11
  *Syracosphaera pulchra* 5, 8, 11, 17, 39

tetrads 168
*Tetralithus obscurus* 32
*Textularia wiesneri* 71
*Theriosynoecum* 61
*Trifarina angulosa* 69, 70, 76, 80, 82-88
*Trochammina* 71, 202, 217
  *Trochammina glabra* 71
*Turrilina* 202

*Umbellosphaera irregularis* 21
  *Umbilicosphaera hulburtiana* 17
  *Umbilicosphaera sibogae* 17
*Unio* 44
*Uvigerina peregrina* 69, 70, 76, 80, 82-88

*Virgulina loeblichi* 69

*Wanea fimbriata* 106
*Watznaueria* 1, 31, 32, 34
  *Watznaueria barnesae* 31-33
*Whiteinella brittonensis* 258
  *W. baltica* 258

*Zeugrhabdotus* 31
*Zygodiscus* 32, 33
  *Zygodiscus erectus* 32

MIX
Papier aus verantwortungsvollen Quellen
Paper from responsible sources
FSC® C105338

If you have any concerns about our products,
you can contact us on
**ProductSafety@springernature.com**

In case Publisher is established outside the EU,
the EU authorized representative is:
**Springer Nature Customer Service Center GmbH**
**Europaplatz 3, 69115 Heidelberg, Germany**

Printed by Libri Plureos GmbH
in Hamburg, Germany

